# VOLCANISM AND THE UPPER MANTLE

## Investigations in the Kurile Island Arc

# Monographs in Geoscience

General Editor: Rhodes W. Fairbridge
*Department of Geology, Columbia University, New York City*

B. B. Zvyagin
   *Electron-Diffraction Analysis of Clay Mineral Structures—1967*

E. I. Parkhomenko
   *Electrical Properties of Rocks—1967*

L. M. Lebedev
   *Metacolloids in Endogenic Deposits—1967*

A. I. Perel'man
   *The Geochemistry of Epigenesis—1967*

S. J. Lefond
   *Handbook of World Salt Resources—1969*

A. D. Danilov
   *Chemistry of the Ionosphere—1970*

G. S. Gorshkov
   *Volcanism and the Upper Mantle: Investigations in the
   Kurile Island Arc—1970*

*In preparation:*

A. S. Povarennykh
   *Crystal Chemical Classification of Minerals*

D. Carroll
   *Rock Weathering*

E. I. Parkhomenko
   *Electrification Phenomena in Rocks*

E. L. Krinitzsky
   *Radiography in the Earth Sciences and Soil Mechanics*

B. Persons
   *Laterite—Genesis, Location, Use*

# VOLCANISM AND THE UPPER MANTLE

## Investigations in the Kurile Island Arc

**Georgii S. Gorshkov**
*Director, Institute of Volcanology*
*Siberian Branch, Academy of Sciences of the USSR*
*Kamchatka*

Translated from Russian by
**Charles P. Thornton**
*Department of Geochemistry and Mineralogy*
*College of Earth and Mineral Sciences*
*The Pennsylvania State University*

PLENUM PRESS • NEW YORK - LONDON • 1970

*Georgii Stepanovich Gorshkov* was born in Irkutsk in 1921. A student of geology, he graduated from Moscow State University in 1943 and since 1946 has specialized in the study of volcanoes, on the subject of which he has written more than 120 studies, including *Catalogue of the Active Volcanoes of the World—Kurile Islands* (Naples, 1958; in English) and *Bezymyannyi Volcano and the Characteristics of Its Last Eruption in 1955-1963* (Moscow, 1965). Recognized as one of the leading volcanologists of the world, G. S. Gorshkov has made important contributions to the description of volcanic processes in many regions, including Japan, Indonesia, and the United States.

The original Russian text, published by Nauka Press in Moscow in 1967, has been corrected by the author for this edition. The present translation is published under an agreement with Mezhdunarodnaya Kniga, the Soviet book export agency.

VOLCANISM AND THE UPPER MANTLE:
INVESTIGATIONS IN THE KURILE ISLAND ARC

VULKANIZM KURIL'SKOI OSTROVNOI DUGI

ВУЛКАНИЗМ КУРИЛЬСКОЙ ОСТРОВНОЙ ДУГИ

*Георгий Степанович Горшков*


Library of Congress Catalog Card Number 69-12530
SBN 306-30407-4

© 1970 Plenum Press, New York
A Division of Plenum Publishing Corporation
227 West 17th Street, New York, N. Y. 10011

United Kingdom edition published by Plenum Press, London
A Division of Plenum Publishing Company, Ltd.
Donington House, 30 Norfolk Street, London W.C. 2, England

# Translator's Preface

The present volume seems to me to be a particularly important one for several reasons. Not least among these is the fact that it summarizes the work of two decades by G. S. Gorshkov, one of the world's leading volcanologists. In addition, it is the first general work of this length on the volcanism of what might be called a "narrow" island arc, a relatively simple megastructure as compared with the "wide" arcs such as Japan and Indonesia. Finally, in this volume Gorshkov has summarized and cited extensive evidence for his general ideas on the relation between volcanism and the earth's crust and mantle.

A few potentially troublesome items should be noted here. In the translation the Russian terms "suite" and "series" have been retained, though for American readers these might better have been translated as "formation" and "group." In almost all cases Russian place names have simply been transliterated rather than translated (e.g., "Yuzhnyi Isthmus" rather than "South Isthmus"); in a few cases the English equivalent has been given in brackets where this is essential to the understanding of the author's comments. The adjectives have retained their Russian case endings in the process (masculine -yi or -ii, feminine -aya or -'ya, neuter -oe) and this may occasionally lead to some slight confusion, for example, when the author calls a given feature Severnyi Volcano at one point and Severnaya Mountain at another.

I would like to take this opportunity to acknowledge the generous assistance of Mr. Sergius Theokritoff in translating a number of the older Russian quotations in the chapter dealing with the history of investigation of the volcanoes.

Gorshkov, like all Russian petrologists in the past 25 years, recalculates his chemical analyses to produce a set of petrochemical coefficients or numerical characteristics "according to the method of A. N. Zavaritskii." The method is presented in full in the latter's Introduction to the Petrochemistry of the Igneous Rocks (Zavaritskii, 1950); it is summarized rather briefly in the following paragraphs.

The weight percentages of the major oxides are converted to cation proportions, and those of ferric and ferrous iron are combined. These proportions (or, loosely speaking, "numbers of atoms") are then divided among the four principal characteristics $a$, $c$, $b$, and $s$, with $a + c + b + s$ set equal to 100:

$a =$ relative number of alkali atoms tied to aluminum, in $1:1$ ratio, in the feldspars and feldspathoids.

$c =$ relative number of calcium atoms tied to the remaining aluminum atoms (if any), in a $1:2$ ratio, in anorthite; if no aluminum is left from $a$ (i.e., if $Na + K > Al$), $\bar{c} =$ relative number of sodium atoms tied to iron atoms, in a $1:1$ ratio, in acmite.

$b =$ relative number of remaining cations other than silicon and titanium.

$s =$ relative number of silicon and titanium atoms.

In addition, a series of secondary characteristics are calculated. These are of two types, one expressing the relative proportions of the cations included in $b$, the other expressing the relative proportions of the alkalies making up $a$.

What cations are included in $b$ depends on the chemistry of the rock. Four cases are possible:

1.  In most rocks all the alkalies and aluminum are used in $a$ and $c$, so that $b$ is composed of all the iron and magnesium plus the calcium remaining from $c$. The secondary characteristics of $b$ then consist of $c'$, $f'$, and $m'$ (calcium, iron, and magnesium, respectively) where $c' + f' + m' = 100$.

2.  In the aluminum-oversaturated rocks, all the alkalies and calcium are used in $a$ and $c$; $b$ is then composed of all the iron and magnesium plus the aluminum remaining from $c$. The secondary characteristics of $b$ then consist of $a'$, $f'$, and $m'$ (aluminum, iron, and magnesium, respectively), where $a' + f' + m' = 100$.

3.  In moderately alkali-oversaturated rocks, all the aluminum is used in $a$ and the remaining alkalies (as-

sumed to be sodium) are used in $\bar{c}$; $b$ is thus composed of all the calcium and magnesium plus the iron remaining from $\bar{c}$. The secondary characteristics of $b$ then consist of $c'$, $f'$, and $m'$ (calcium, iron, and magnesium, respectively), where $c' + f' + m' = 100$.

4.    In strongly alkali-oversaturated rocks, all the aluminum is used in $a$ and all the iron is used in $\bar{c}$; $b$ thus is composed of all the calcium and magnesium plus the sodium remaining from $\bar{c}$. The secondary characteristics of $b$ then consist of $n$ , $c'$, and $m'$ (sodium, calcium, and magnesium, respectively), where $n + c' + m' = 100$.

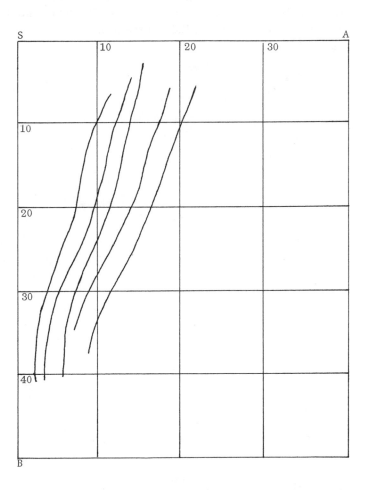

The secondary characteristic of $a$ is expressed by the ratio 100 $Na/(Na+K)$, which is called $n'$.

Gorshkov uses two of the primary characteristics, $a$ and $b$, to construct variation diagrams for the Kurile volcanics and for the volcanics in areas to which he wishes to compare the Kurile volcanics. The two perpendicular axes of these diagrams are labelled SA (horizontal) and SB (vertical); the distance along SA is equal to $a$ and the distance along SB, to $b$. The more acid (silica-rich) a rock or series is, the closer it will plot to the origin S. In his original presentation of this method, Zavaritskii plots a number of "type curves" (following Burri, 1926) for use as standards of reference; in order of increasing alkalinity these are Pelee, Lassen Peak, Yellowstone, Etna, and (Maros-)Highwood Mountains. These curves are illustrated on the diagram on page vii.

# Foreword

In recent decades the problem of the origin and development of island arcs has attracted constant attention from scientists in a variety of specializations. The highest energy contemporary geologic processes are restricted to just these regions; it is here that the conversion of one type of terrestrial crust to another takes place. Only after we completely understand present-day processes will it be completely safe for us to conjecture about the processes of the past.

Work on the problem of island arcs in our area (i.e., the Kurile-Kamchatka arc) was begun by Academician A. N. Zavaritskii, who beginning in 1946, repeatedly raised this question in the pages of scientific periodicals. He noted the "convenience" of the Kurile-Kamchatka arc for investigation: the Kamchatka part of the arc is accessible to investigation by geologic methods; the Kurile part is in part a single arc and in part a double one, yet it is not complicated by such phenomena as, e.g., the intersection of arcs that are found in Japan.

The processes of volcanism play a large role in the island arcs; for this reason Zavaritskii regarded the study of the volcanoes and volcanic phenomena of the Kurile Islands as one of his first tasks. To a large degree, completion of this mission was assigned to the present author, who, beginning in 1946, though with long interruptions, carried on the volcanological investigations here.

At first the tasks were rather limited, he was simply supposed to describe the volcanoes and the rocks composing them.

In 1957, as a result of work on the program of the International Geophysical Year in the region of the Kurile island arc, the basic outlines of the structure of the earth's crust and upper mantle were made clear; subsequently it began to be possible to com-

pare the distinctive features of volcanism with those of the deep
structure of the arc. Gradually there developed the concept of the
"through-the-crust" character of the volcanism, of its direct connection with great depths, with the upper mantle of the earth.

The first work of the author in this direction was published
in 1956 in "Upper Mantle Project." In this work, attention was
called to the phenomenon of absorption of transverse seismic waves
in the magma chamber of the Klyuchevskoi group of volcanoes. On
this basis, the depth to the volcanic hearth was estimated to be 60-
80 km. This work attracted considerable interest and was printed
in several publications abroad.

Subsequently the idea of a direct connection between volcanoes and the upper mantle was developed along independent lines
based on a petrochemical study of the volcanic rocks in the Kurile
Islands. The conclusions reached required additional petrochemical analysis of the volcanic lavas from many regions of the world.

At the same time similar ideas were advocated and developed
here by Yu. M. Sheinman and abroad by G. A. Macdonald. In recent years many scientists have begun to express support for this
point of view. Among the foreign investigators we may distinguish
Dr. Zen (Indonesia), who applied our ideas directly to the material
of the Indonesian volcanoes, and Dr. Ringwood (Australia), who
carried out a series of interesting physicochemical experiments.
Interesting investigations in petrochemistry, geophysics, and geochemistry have also been carried out in the USA and Japan. Apparently the idea of an exclusively mantle source of supply for
volcanoes and of only a small role for the processes of assimilation and contamination by crustal material is very popular.

The author's ideas along these lines were published in some
papers and reports in which the original factual material could
not be presented. Thus it may be that the author's ideas have not
always sounded sufficiently convincing. In this book abundant factual material will be presented on the petrochemistry of the whole
Pacific volcanic region. In the course of this work it developed
that most of the older chemical analyses suffer from errors; thus
we have tried to use only chemical analyses made mostly after 1940.

The present work actually consists of two parts. The first
part is a description of the Kurile island arc; the basic information

on the history of the study of the volcanoes, the submarine relief, and the deep structure of the arc are presented here, but the main portion of this part is allotted to a description of the volcanoes. A considerable amount of this material is presented for the first time.

The second (and final) part is devoted to an account of the factual material on the petrochemistry and deep structure of the volcanic zones of the Pacific Ocean and of the areas bordering it, and also to general questions of the relation between volcanism and the earth's upper mantle. While considering this work as the sum of the preceding investigations of the volcanoes of the Kurile island arc, the author regards it also as the beginning of a more detailed study.

What is primarily necessary is a quantitative study of the distribution of the different rock types, a detailed petrographic description of them, a comparative geochemical study of the lavas developed on the various types of crust, and a geophysical investigation of the deep structure of the volcanoes.

In conclusion, the author acknowledges his debt to his colleagues and assistants in the expeditions to the volcanoes of the Kurile Islands: to V. I. Lymareva, a participant in the first (1946), most extended, and probably most difficult expedition; to his coworkers in the former Laboratory of Volcanology of the Academy of Sciences of the USSR, N. K. Klassov, I. I. Tovarova, V. A. Bernshtein, I. I. Gushchenko, and K. I. Shmulovich; and to G. E. Bogoyavlenskaya, who was a division head in all the later expeditions.

# Contents

# Chapter I

# Structure of the Kurile Arc

The Kurile Islands chain extends from the south of Kamchatka to the island of Hokkaido, as though tying these two regions together and, at the same time, separating the more epeiric Sea of Okhotsk from the Pacific Ocean. On the whole, the chain of islands forms an arc that is slightly convex to the east and that belongs to the system of island arcs that borders the continent of Asia on the east. Like other island arcs, the Kurile Islands are separated from the ocean basin by a deep trench and from the continent by the basin of an epeiric sea.

The islands of the Kurile arc are divided into two chains, the Lesser Kuriles and the Greater Kuriles.

The Lesser Kurile chain is an extension of the Nemuro Peninsula (of Hokkaido) and extends northeastward for 105 km. Here belong eight rather small islets (Tanfil'ev, Anuchin, Yurii, Demin, Zelenyi, Polonskii, Lis'ya, and Shishka) that rise to heights of no more than 30-40 m, and the island of Shikotan, which has rather considerable dimensions (9 × 28 km) and a height of 413 m. To the northeast of Shikotan stretches a submarine ridge (Vityaz' Ridge).

The Greater Kurile chain extends for 1150 km from Cape Lopatka on Kamchatka to the Shiretoko Peninsula on Hokkaido; to it belong 16 more or less large islands and a series of rocks, of which the Lovushki Rocks are also counted as an island. The characteristic features of submarine topography and crustal structure (which will be described in greater detail below) allow us to divide the Greater Kurile chain into three parts: northern, central, and southern. The North Kurile Islands in turn are divided into two sections, northern and southern, by the wide Fourth Kurile Strait.

The islands of Shumshu and Paramushir belong to the northern section of the North Kurile Islands; to the southern section be-

1

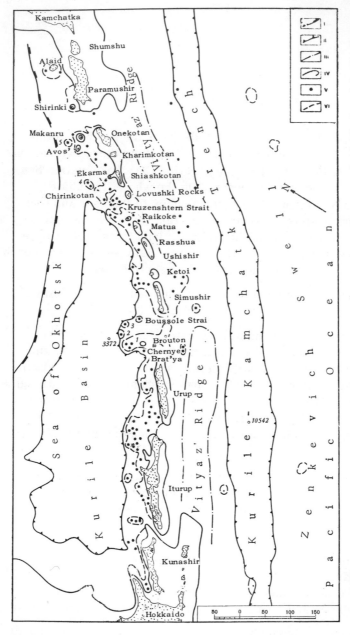

Fig. 1. Map of the structure of the Kurile arc and its environs. I) Edge of the continental shelf; II) edge of the Kurile Basin of the Sea of Okhotsk; III) outline of the Vityaz' Ridge; IV) edge of the island shelf (submarine terrace at −130 m); V) submarine volcanoes; VI) outer edge of the submarine volcanic system: 1) Vavilov Volcano; 2) Obruchev Volcano; 3) Mironov Volcano; 4) Edel'shtein Volcano; 5) Belyankin Volcano (after Zatonskii et al., 1961).

long Onekotan, Kharimkotan, Shiashkotan, and the Lovushki Rocks.
To the Central Kurile Islands are assigned Raikoke, Matua, Ras-
shua, Ushishir, Ketoi, and Simushir.  Finally, to the South Kurile
Islands belong the rather small islands of Chernye Brat'ya (Chirpoi
and Brat Chirpoev) and the large islands of Urup, Iturup, and
Kunashir.

The series of islands lying to the west of the main chain
(Alaid, Shirinki, Makanru, Avos' Rocks, Ekarma, Chirinkotan, and
Brouton) also usually are assigned to the Greater Kurile chain, but
details of the submarine topography and the petrochemical differences of their lavas, which will be described in more detail in the
appropriate parts of this monograph, compel us to consider this
group of islands as an independent zone:  the western zone of the
Greater Kurile chain.

Thus, the Kurile arc is a double one:  the chain of the Lesser
Kurile Islands and the Vityaz' Ridge form the outer, nonvolcanic
arc, the Greater Kurile chain forms the inner, volcanic arc.  The
latter, in turn, may be subdivided into a main zone and a western
zone.  Details of the structure of individual islands will be presented during the description of the volcanoes.

All of the islands of the Kurile chain are actually only the
summits of large mountains that are hidden beneath the surface
of the sea.  This is why it will be necessary to examine the submarine relief in more detail (Fig. 1).

Details of the structure of the submarine part of the Kurile
arc, the floor of the Sea of Okhotsk, and the adjacent parts of the
Pacific Ocean have been studied mainly by the expeditions of the
Institute of Oceanology of the Academy of Sciences of the USSR on
the R/V "Vityaz'" in 1949-1959 (Bezrukov and Udintsev, 1954;
Udintsev, 1954, 1955, 1957; Kanaev and Larina, 1959; Zatonskii
et al., 1961).

The submarine part of the Kurile arc has the character of a
rather complex mountain system, consisting of two parallel mountain ridges 75-100 km apart.  The inner ridge, whose peaks form
the chain of the Greater Kurile Islands, extends continuously from
Kamchatka to Hokkaido.  The outer ridge (Vityaz' Ridge) is almost completely concealed beneath the waters of the Pacific Ocean;
only at its southwestern end does a small chain of islands, the
Lesser Kurile chain, reach above sea level.

In the middle part of the Kurile chain, between Boussole and Kruzenshtern Straits, the outer ridge is broken, so that in fact there are two outer ridges, a northern and a southern one. At their ends these two ridges closely approach the inner one, forming a single mountain system.

A characteristic feature of both divisions of the outer ridge is their flat summit surface. This surface slopes toward the middle part of the ridge, being broken into a series of progressively lower steps. Between the Boussole and Nadezhda Straits, there are no traces of the outer ridge at all.

Near the ends of both sections of the outer ridge a clearly defined longitudinal trough occurs between them and the inner ridge, with its floor 2000 m below the crest of the outer ridge. A large submarine valley reaches to the floor of the northern longitudinal trough in the vicinity of Shiashkotan. Bending around the southern end of the northern section of the outer ridge, this valley reaches the side of the abyssal Kurile-Kamchatka Trench. The valley has steep walls and a narrow floor that is as much as 200 m deep. In the middle, gently sloping part well-formed levees are recognized, evidence of the activity of turbidity currents.

In contrast to the outer ridge, the basement of the islands of the Greater Kurile chain has the character of a single mountain system. The passages between the individual islands are rarely deeper than 500 m; only two passages are exceptions to this: Kruzenshtern Strait reaches a depth of 1920 m, Boussole Strait reaches 2318 m. These two straits cut deeply into the basement of the inner ridge, missing by a kilometer or so the floor of the Kurile Basin of the Sea of Okhotsk, and divide the ridge into three sections: a northern one from Paramushir to Shiashkotan, a middle one from Raikoke to Simushir, and a southern one from Chernye Brat'ya to Kunashir. These two deep straits are certainly tectonic formations of the graben type. The axis of Boussole Strait runs perpendicular to the trend of the arc and closer to the shore of Simushir. The axis of Kruzenshtern Strait graben has an almost north-south trend and cuts the island basement diagonally. The other rather deep passages (Fourth Kurile Strait, Vries Strait, Yekaterina Strait) most probably also belong to formations of the graben type.

In-shore shoals, the edges of which lie at an average depth

of 130-140 m, are clearly recognized near the coastlines of most
of the islands of the Greater Kurile chain.

On the Okhotsk Sea side, the inner ridge of the Kurile chain
is clearly bounded by the deep-sea Kurile Basin. The flanks of
the ridge have a rather steep slope (up to 10-20°). Locally the passage onto the floor of the basin is marked by a rather distinct break,
but in most cases the flanks of the ridge flatten out gradually into
the floor of the basin. The maximum depth of the Kurile Basin
(northwest of Boussole Strait) reaches 3372 m. Its floor is an
ideally flat surface 3350 m deep, with a slight slope to the southeast. The flat floor has the shape of a strongly elongated triangle
120 miles (about 220 km) wide and 600 miles (about 1100 km) long.
In the vicinity of Kruzenshtern Strait the flat floor narrows almost
to zero; it is prolonged farther to the northeast, as far as Alaid
island, by a large branching submarine valley. The upper part of
this valley (as far as the flanks of Avos') has steep walls and great
depth (up to 200 m) and is very probably tectonic. Farther down,
the valley expands and shows evidence of turbidity currents. The
limits of the Kurile Basin on the west is the submarine basement
of the Sakhalin-Hokkaido mountain system. On the northwest the
Kurile Basin ends against the scarp of the continental slope that
extends on to the southern end of Kamchatka, coming completely
to an end 30 km north of Alaid.

Southeast of the ridges of the Kurile chain lies the abyssal
Kurile-Kamchatka Trench, which belongs to the system of trenches
in the western part of the Pacific Ocean. This trench extends for
a distance of 2200 km from near Cape Kamchatka on the northeast,
where it joints the Aleutian Trench, to the southern end of Hokkaido, where it passes into the Japan Trench. In plan the trench
has a slightly arcuate form, and its axis is parallel to the trend
of the Kurile island arc. The distance between the axis of the
trench and the inner ridge varies from 180 km. in the middle part
of the chain to 200-220 km near the northern islands.

Inside the −6000 m isobath the length of the trench is 2000
km and its width is 20-60 km. The deepest part is located in the
south. The −9000 m isobath outlines the portion of the trench extending from the north end of Urup to the south end of Iturup, an
area about 550 km long and 5 km wide. The deepest point in the
trench, −10,542 m, is located 180 km southeast of Vries Strait.

The height of the volcanoes above the floor of the trench in this part of the arc amounts to about 11,700 m.

The Kurile-Kamchatka Trench has the typical V-shaped cross section: the northwest wall rises to a height of 6-10 km, the southeast wall to all of 2-5 km. The slope of the upper walls does not exceed 5-6°, but downward they steepen to 20-25°. Everywhere the walls are broken by tectonic scarps and terraces. The walls of the scarps reach slopes of 45° or more; the surfaces of the terraces are almost flat. Almost everywhere the floor of the trench is a narrow flat area produced by sedimentation. In the very deep southern part the width of the flat floor is up to 1 km, while in the rest of the trench it usually is not more than 5 km, though locally it widens to 8-10 km.

On the oceanic side of the Kurile-Kamchatka Trench, at the very edge of the basin of the Pacific Ocean, there is a gently sloping, wide structure called the Zenkevich Swell. This swell does not rise very high (200-300 m) above the floor of the northwest basin of the Pacific Ocean, but it is 300-400 km wide.

On the crest of the Zenkevich Swell are 16 submarine mountains up to 3400 m high; the tops of these mountains lie 1700-1900 m below sea level. In addition to these large mountains, lower broad hills 100-300 m high are also present.

On the whole, the transition zone from the Asiatic continent to the Pacific Ocean has a very complex structure: the outer and inner ridges of the Kurile arc form a double anticline separated by a medial syncline. This composite anticline is separated from the submerged continental platform of the Sea of Okhotsk by the deep synclinal Kurile Basin, and from the oceanic platform by the synclinal Kurile-Kamchatka Trench.

*Chapter II*

# History of Investigation
# of the Volcanoes

The Kurile Islands became known only at the beginning of the eighteenth century. True, Europeans caught sight of these islands 50 years earlier, but neither the voyage of the Polish explorers Vries and Skhep in 1643 (Witsen, 1946) near the southern Kuriles nor the expeditions of Dezhnev's companion Fedot Alekseev in 1649 (Krasheninnikov, 1755; Miller, 1758) and of the Cossack M. Stadukhin in 1656 (Sgibnev, 1869) through the First Kurile Strait in the northern part of the chain led to the geographical study of the island chain for various reasons.

The very first information on the existence of a chain of islands to the south of Kamchatka was received in Moscow in 1701 from the first explorer of Kamchatka, the 50-year-old Vladimir Atlasov. At the end of his first Kamchatka expedition, at the end of 1697, he emerged at the mouth of the present-day Golygina River and caught sight of the island of Alaid. In his second report, written on February 10, 1701, in the Siberian Command in Moscow, he stated, "And opposite the first Kurile River it looks as if there are islands in the sea, and the foreigners report that there are islands there..." (Ogloblin, 1891, p. 16).

Somewhat earlier, in December 1700, the famous Siberian cartographer S. U. Remezov met in Tobolsk with V. Atlasov, who was returning to Moscow, and obtained information from him on Kamchatka and the islands. On the basis of these data Remezov drew up the first map of Kamchatka and put the following note at the south end of the peninsula: "The Kurile land is on a lake and on islands." This was the first mention of the Kurile Islands.

In 1706 the Kamchatka agent Vasilii Kolesov guided a party of 50 Cossacks headed by Mikhail Nasedkin to the southern part of

the peninsula. This party reached to the southernmost tip of Kamchatka, Cape Lopatki. V. Kolesov wrote in Yakutsk to the governor,
D. A. Traurnikht, that "Vasilii sent military men into the Kurile
land; and on their return these men reported to him that they had ex-
plored the Kurile Island to the tip of the land, and that beyond the
headland there is no land but only the sea; however, in the sea on
the other side of the strait land is to be seen, but it was impossible
to explore this land because of the lack of ships and marine sup-
plies" (Memorials ... Eighteenth Century, pt. 1, p. 422).

Later M. Nasedkin in person, at the Yakutsk Command Post,
sent the same D. A. Traurnikht somewhat more detailed information (Memorials ... Eighteenth Century, pt. 2, pp. 502-503). Nasedkin's information greatly interested the Siberian agent, and the
Irkutsk governor, Prince V. I. Gagarin, wrote in instructions to the
Yakutsk governor, D. Traurnikht, on March 17, 1710 "to visit those
islands in the sea ... opposite the land of Kamchatka, and to explore
those islands very diligently" (Memorials ... Eighteenth Century,
pt. 2, p. 522).

In his turn Yakutsk Governor Traurnikht, under whose supervision Kamchatka lay, at the time of the departure of the next Kamchatka agent V. Savost'yanov, ordered him on September 9, 1710
"to build suitable ships, to explore as much as possible the lands
and peoples beyond the straits ... and to make a special map of
that land" (Memorials ... Eighteenth Century, pt. 1, pp. 422-423).

This order was carried out the following year. In August 1711
a party of Cossacks under the leadership of Ataman Danil Antsiferov
and Yesaul Ivan Kozyrevskii crossed the First Kurile Strait. The
first Russians landed on the Kuriles; the island of Shumshu was
annexed to Russia. In their petition to Peter I they reported that
"from that point we, your humble subjects, visited the islands in
the sea beyond the strait in shallow-draft boats and canoes" (Memorials ... Eighteenth Century, pt. 1, p. 462).

In 1713 the steward of the Kamchatka prison, V. Kolesov, sent
a party of Cossacks to the Kurile Islands under the leadership of
Ivan Kozyrevskii. Kozyrevskii was on the three northern islands,
Shumshu, Paramushir, and Makanru (Kukumiva), and formally annexed the island of Paramushir to Russia. By making inquiries he
collected information on all the islands as far as Matmai (Hokkaido). In answer to D. Traurnikht in 1713 Kozyrevskii wrote,

"... and he, Ivan, went from Kamchatka to the strait with the military men and submitted ... a report under his hand [signature] and a plan of these islands, even as far as Matmai Island" (Memorials ... Eighteenth Century, pt. 1, pp. 542-543; pt. 2, p. 46).

Information collected in 1713 by I. Kozyrevskii was one of the most important landmarks on the road to the study of the Kurile Islands, and for more than half a century all the information on the Kuriles were based almost exclusively on these data.

Kozyrevskii's description was sufficiently accurate with regard to the number and disposition of the islands, and many island names have been retained almost without change down to the present.

Kozyrevskii's data were used in the construction of many maps, among them the well-known Elchin map. The "Map of the Kamchatka Peninsula and the Oceanic Islands" made by I. Kozyrevskii in 1726 in Yakutsk for Captain Bering has survived down to the present day (Ogryzko, 1953). This map was later used by G. F. Miller who, according to the Yakutsk archival material, put together the first report on the Kurile Islands. This report to a large degree became the corresponding chapter in S. P. Krasheninnikov's (1755) "Description of the Land of Kamchatka," which was translated into many western European languages, and in the "Description of Ocean Voyages" by the same Miller (1758); in the original German it was published in 1774 as an appendix to a book by Steller (1774).

The first mention of the islands between Kamchatka and Japan in Russian is contained in a note concerning Ivan Kozyrevskii in the newspaper "Sankt-Peterburgskie Vedomosti" of March 26, 1730. In the same year the Swedish officer Tabbert (Strahlenberg), who was captured in 1709 below Poltava and who remained for 13 years in exile in Siberia, produced a description of northern and eastern Europe and Asia (Strahlenberg, 1730). In this work, along with much other information, he also used the data of Atlasov and Kozyrevskii on Kamchatka and the Kuriles.

In the "Journal" of I. Kozyrevskii was given the first information about a volcano on one of the Kurile Islands: "... the fourth island, called Araumakutan,* is uninhabited. On it there is a burn-

---

* The fourth island is now called Kharimkotan.

ing mountain." The fact that only one was noted among the dozens of Kurile volcanoes indicates that in the year of Kozyrevskii's expedition (1713) this volcano had an eruption. Rather vague information is also cited there that may indicate an eruption on Chirpoi: "... from Kitui cannon fire was audible on Chirpoi" (Krasheninnikov, 1755; Müller, 1774).

In 1721 the geodesists I. M. Evreinov and F. F. Luzhin were sent to the Kurile Islands on the personal orders of Peter I. They travelled as far as "Sixth Island" and carried out the first geodetic determinations of the position of the islands in the northern half of the chain. Evreinov and Luzhin established for the first time that the islands did not extend from Kamchatka directly to the south, but to the southwest. According to their data, they went as far as 49°18' N (Evteev, 1950), but judging from their location of Cape Lopatka a systematic error was included and they actually went as far as 48°06' N, i.e., as far as Matua.*

In 1724 I. K. Kirillov made a map of northeastern Asia; on the basis of the map of Evreinov and Luzhin and in part that of Kozyrevskii the Kurile Islands were plotted on it, but still without a name. On the orders of Peter I this map was turned over to I. Homan by Ya. V. Bryus; the former reproduced it in his atlas (Homan, 1725). This was the first publication of a map of the Kuriles.

The expedition of Vasilii Shestakov, who in 1730 visited the first five islands (Miller, 1758; Sgibnev, 1869), added nothing to Kozyrevskii's data.

In 1738 the head of the Kurile section of the Second Kamchatka Expedition, Captain Shpanberg, went from Kamchatka along the Kuriles as far as Urup on the hooker "Arhangel Mikhail." In 1739 Captain Shpanberg, on the same vessel, and Captain Val'ton (or Walton], on the double sloop "Nadezhda," passed through and mapped the whole chain of the Kurile Islands. In 1742 Captains Shpanberg and Shel'ting repeated the trip. Unfortunately, there is no information on the volcanoes in the journals of Shpanberg, Val'ton, and Shel'ting (Sokolov, 1851).

In 1738 S. P. Krasheninnikov sent the Cossack S. Plishkin to the North Kurile Islands with an interpreter, M. Lepekhin. They

---

*The data of various authors varies considerably in relation to how far Evreinov and Luzhin travelled.

made descriptions of four islands and supplied information on smoking volcanoes on Alaid and Paramushir (Krasheninnikov, 1755).

In the 1740's and 1750's tribute collectors reached the sixteenth island (Simushir), but their expeditions provided no new information.

In 1761 Soimonov, governor of Siberia, commissioned Sub-Colonel Plenisnev, commander-in-chief of the Anadyr, Okhotsk, and Kamchatka prisons, to obtain detailed information on all the Kurile Islands; detailed instructions were composed. Fulfillment of this order was begun only in 1766. In 1766-1769 Cossack Captain Ivan Chernyi visited the whole chain as far as northern Iturup and put together a very detailed and extremely sensible description of all the islands that he visited. Chernyi's "Journal" was the basis for a detailed review of the Kurile Islands that was put together by the head of the Irkutsk navigation School, Second Major Tatarinov (1785). The complete "Journal" was published a hundred years later in the review by A. Polonskii (Chernyi, 1871).

It can be said that the expedition of I. Chernyi was a very great contribution to the knowledge of the Kurile Islands. For more than a hundred years, down to the very end of the nineteenth century, Chernyi's data were the basis of our knowledge of the Kurile Islands. He presented descriptions of some eruptions, at times very colorfully. For example: "The twelfth island is Motuva (now Matua). On it is a peak which, according to the natives, has been burnt by fire in recent years; rocks were scattered over the whole island so that flying birds were killed by them in large numbers ... in some places the roots were burned out of the ground and were covered by rocks." Or: "The sixteenth island is Simusyr ... on it are four peaks ... the second peak is Itankioi (now Prevo Peak); its summit is flat; formerly it was on fire and as a result the stunted scrub vegetation and its roots were burnt out at its foot." Among the varied and very detailed geographical information he provides data on 26 volcanic peaks on 18 islands (not counting Alaid and Paramushir); information is included on recent eruptions or on continuing activity for 12 volcanoes.

On November 28, 1772, the Irkutsk governor, Lieutenant-General Bril', gave detailed instructions to the newly appointed commander-in-chief of Kamchatka, First Major Bem. One of the points in the instructions was "to visit and describe thoroughly all the Kurile Islands" (Polonskii, 1871).

In 1773 the "Siberian nobleman" Antipin (1775) went to the
Kuriles in a large dugout canoe, on orders from the Academy of
Sciences, collecting various "memorable items" for the Academy's
museum. In 1775 the Irkutsk merchant Lebedev-Lastochkin attempted to organize a trade in sea animals in the region of Urup;
at the same time a mission was commissioned by him, to make
citizens of the inhabitants of the South Kurile Islands and to establish a commercial relationship with Japan.

During 1775-1785 Antipin and Shabalin, persons trusted by
Lebedev-Lastochkin, sailed several times from Okhotsk and from
Kamchatka to Urup in sailboats and canoes. In the first year the
ship "Nikolai" was wrecked on the Urup coast. One other ship was
wrecked the following year on Kamchatka, and in 1780 a tsunami
grounded the brigantine "Natal'ya" on the Urup coast. With regard
to business, the whole enterprise resulted in some losses, but in
addition new data were obtained on the whole Kurile chain. In 1776
Shabalin and apprentice mate Ocheredin reached Atkis Harbor on
Matmai [Hokkaido] (42°20' N) where the first encounter of Russians
with the Japanese occurred. In 1779 Antipin and Shabalin again
reached Matmai.

From everything that has been written, it will be seen that
the whole Kurile chain, as far as Japan, became known in Russia
in 1713 following the journey of I. Kozyrevskii to Paramushir. In
1738 all the islands of the chain had been examined and placed on a
map by the members of the Second Kamchatka Expedition, and in
the 1760's and 1770's they were described in considerable detail.

At the beginning of the eighteenth century there was continuous Japanese settlement south of Matmai (Hokkaido). The northern part of the island and the Kurile Islands at this time were still unknown and were joined together in the rather undefined conception
of Ezo ("northern lands"). On the Japanese map dated 1730 (Teleki,
1909, Map XI) the Kuriles were not indicated. Apparently the Japanese pushed farther north along the coast of Matmai in the middle
of the eighteenth century; judging from the Japanese sources, provided in a book by A. Pozdneev (1909), the existence of the islands
of Kunashir and Iturup became known in Japan in 1750-1760. According to the description of I. Chernyi, the Japanese began to visit
Kunashir shortly before his voyage (i.e., in the middle 1770's).
Iturup apparently was visited by Chernyi and the Japanese at the

same time. The first Japanese official, Mogami Tokunai, visited
Urup in 1786, whereas the islands farther to the north were visited
only in the nineteenth century. In 1799 the Japanese began to colon-
ize Iturup; a *"de facto"* boundary was established between Japan
and Russian at the strait that separates Urup and Iturup. According to the Shimoda Treaty of January 26, 1855, this boundary was
recognized as the juridicial one.

But let us return to the results of the journeys of Antipin and
Shabalin. In 1781 Antipin went to Irkutsk and presented the new
data on the South Kuriles* and on the encounter with the Japanese
on Matmai. The information on the eruption of a volcano on northern Iturup and on an eruption on Raikoke in 1778 was newer than
Chernyi's data.†

The eruption on Raikoke became known even before Antipin's
report. The eruption broke out suddenly, and under the hail of volcanic bombs Cossack Captain Chernyi, brother of the author of the
"Journal," was killed while returning from Matua to Kamchatka
with 14 Kurile companions. In 1779 the information on this was
sent to Kamchatka and Kamchatka commander Reiniken sent Captain
Sekerin to Raikoke "to describe and show on a map the extent to
which the island consists of the eruption of a burning mountain"
(Pallas, 1781; Polonskii, 1871). As we said, this was the first
Russian, specifically volcanologic expedition. In the 1760's Sekerin
had come under the command of I. Chernyi and along with him had
visited Raikoke earlier. Thus he could describe in detail the appearance of the island before and after the eruption. Unfortunately,
the drawings to which there are references in the German edition
of the paper by Tatarinov have not been preserved.

The report by M. Tatarinov mentioned above joined together
the data of I. Chernyi and I. Antipin. The Irkutsk governor-general,
F. N. Klichka, submitted this description to the Academy of Sciences and it was published in 1785 (Tatarinov, 1785). But two years
earlier P. S. Pallas published it with insignificant changes and additions in German, without indicating the author, in the fourth volume
of his journal "Neue Nordische Beyträge," furnishing only the note

---

* Among these were samples of pyrite and chalcopyrite from Urup.
† As a result of various misprints and errors this eruption is also sometimes dated as
  1777 or 1780.

"according to the Russian original." Thus at first in western Europe and later in Russia, Pallas came to be cited as the author of this outstanding report. In 1792 this report was published anonymously in Ulm. Later the founder of the Russian-American Company, the merchant G. Shelekhov, included this description, with several abridgements, in one of his books (Shelekhov, 1812). Thus, one and the same work came to be attributed to three different authors. Comparison of the texts establishes their complete identity. Fairness requires us to point out that the actual author of the report was the least known of these, M. Tatarinov, and that at least three-fourths of this report was taken from the journal of I. Chernyi.

On the map of the Kurile Islands put together by Antipin and Ocheredin after 1780 (presumably in 1785) there is an additional indication of the existence of an active volcano also on Urup (Atlas, 1964, Map 159).

At the very end of the eighteenth century the Kurile Islands were visited by Cook's (Cook, 1784), La Pérouse's (1797), and Broughton's (1804) expeditions, but these added almost nothing that was new.

In 1790 Sergeant (of geodesy) Gilev, a member of the Billings and Sarychev Russian expedition, described the first eight islands; these data were included in the atlas by Sarychev (1826).

In 1805 I. F. Kruzenshtern, at the time of the first Russian round-the-world expedition, described the northern half of the Kurile chain, made a series of astronomical observations, and gave Russian names to many geographical objects, names that have lasted down to the present day (Kruzenshtern, 1812). In the atlas are reproduced drawings of many volcanoes (1809). In 1811 V. M. Golovnin extended the description and astronomical determinations in the Kuriles from Matua to Kunashir (Golovnin, 1819). As a result of the trips of Kruzenshtern and Golovnin, the map of the Kurile Islands took on its present form; in later years its precision was improved only to a rather small degree. The information on the volcanoes had also been supplemented, especially in the southern part of the chain, and the known number of active volcanoes increased fourfold.

In 1830 the Kurile Islands were transferred to the supervision of the Russian-American Company (Tikhmenev, 1861). On northern Simushir an office of the company was constructed on Brouton Bay, and in 1828 the settlement on Urup (which had existed from 1795

to 1805) was reconstructed. The islands belonged to the Russian-
American Company until its liquidation in 1869. Annual voyages
of company ships to the Kuriles allowed the map of the islands to
be made more accurate; new data was included particularly in the
atlas and "Hydrographic Notes on the Atlas" of Captain Teben'kov
(1852). Captain Teben'kov entered the service of the Russian-
American Company in 1825, and in 1845 he became manager of the
company. In his works he made use of the logs of the company
ships dating back to 1782. Teben'kov contributed much new information. A paper by P. Doroshin (1870), an employee of the
company beginning in 1848, was devoted specifically to the volcanoes and their eruptions.

In 1864 Perrey made the first attempt to put together a criti-
cal report on the eruptions of the Kurile Volcanoes. At the time
this was a very good work, one in which all the then existing literature in western European languages was used. Perrey attempted
to compare and evaluate critically the data of various authors and
to show up the accumulated inaccuracies and errors. To some degree he succeeded in this, but since he did not understand Russian,
he could not use the Russian original sources and he could not remove all the errors.

In 1875 the Kurile Islands were handed over to Japan, and the
first period of Russian investigation came to an end. At this time
25 active volcanoes had been described in the Russian literature.

In 1878 J. Milne, an English seismologist in Japanese employ,
toured the whole Kurile chain and, on the basis of shipboard observations, described the volcanoes (Milne, 1879). He counted 52 well-
formed peaks, of which nine were referred to as active volcanoes.
In 1885 Milne visited Kunashir and Iturup. In 1886 he published a
long report on the volcanoes of the Kurile Islands in which, besides
his own observations, he included data collected by the English fur
trader Captain Snow and all the literature data collected by Perrey.
According to Milne's report, there were 23 extinct and 16 active
volcanoes in the Kuriles. In 1897 Captain Snow published his "Notes
on the Kurile Islands" (there was a Russian translation in 1902), in
which all of the islands and some volcanic eruptions are described.
Subsequently, prior to 1946, no other volcanologic investigations
were carried out that included the whole chain, and all the later volcanologic reports were based mainly on Milne's work.

The Japanese literature on the volcanoes of the Kurile Islands is rather sparse and is limited mainly to short reports of eruptions (Tanakadate, 1925, 1931, 1934, 1935, 1939). A more detailed investigation was carried out only in the case of Alaid and its lateral crater Taketomi, which arose in 1934 (Kuno, 1935), of Kharimokotan relative to its violent eruption in 1933 (Nemoto, 1934), and of Trezubets (Jigoku) Volcano on Urup (Nemoto, 1937). In these works not only morphologic characteristics are given, but also the first information on the petrography and chemistry of the lavas is provided. Unfortunately, the latter two works, which were printed in Japanese and in limited editions, are still poorly known outside Japan. Brief reports on some volcanoes are found in geologic works and in brief volcanologic notes, mainly in Japanese. A geologic map of the Kurile Islands with data on the distribution of volcanoes and a composite work on the Kurile volcanoes appeared in the 1950's and 1960's (Nemoto, 1958). Our data were already used in these works.

In 1945, according to the conditions of the Potsdam Agreement, the whole chain of the Kurile Islands was returned to the Soviet Union. The data existing in 1945 on the volcanism of the Kurile Islands were correlated by Academician A. N. Zavaritskii (1946). He noted 23 active volcanoes, three points of submarine eruptions, and seven dormant volcanoes.

In 1946 the third stage in the study of the Kurile Islands began, this time in detail and comprehensively.

A hydrographic expedition of the Pacific fleet worked in this region first, beginning in April 1946. Following the suggestion of A. N. Zavaritskii, the Academy of Sciences of the USSR attached the author of the present work to this expedition to study the volcanoes. During the seven months of the expedition's work the following islands were visited: Alaid, Shumshu, Paramushir, Onekotan, Kharimokotan, Shiashkotan, Ekarma, Matua, Rasshua, Chernye Brat'ya, Urup, and Shikotan. In addition, all the rest of the islands of the chain were examined from shipboard.

At the same time a complex expedition of the maritime branch of the Geographical Society was working in the southern islands of the chain (Urup, Iturup, Kunashir, Shikotan). The geomorphologist on this expedition, G. V. Korsunskaya, obtained a quantity of new information on the volcanoes, the results of which were published in

a series of papers (Korsunskaya, 1948, and others). The geographer
Yu. K. Efremov also provided observations on some of the volcanoes
of Iturup.

The results of our investigations enlarged considerably existing ideas on the volcanoes of the Kurile Islands. The number of
these volcanoes was estimated to be about 80 (Gorshkov, 1948), and
36-37 of these (not counting two submarine ones) were referred to
the active category. The brief results of our work were prepared
for publication at the beginning of 1947, but the paper was not published until 16 years later (Zavaritskii and Gorshkov, 1963). In
1954 the author published an account of the eruptions of the volcanoes
from the beginning of the eighteenth century to 1952 (Gorshkov, 1954).
In this account a brief description of all the active volcanoes and
some of the dormant ones was included, a description put together
by the author mainly on the basis of his own observations in 1946.
Here for the first time photographs of many of the volcanoes were
included.

In 1947-1948, under the supervision of G. M. Vlasov (Far
Eastern Geologic Administration), the sulfur deposits on Kunashir
and Iturup were studied. Analogous works, beginning in 1953, were
published by the Kamchatka Geologic Administration on Paramushir
(Vlasov, 1958, 1960).

In 1950 the Institute of Oceanology of the Academy of Sciences
of the USSR, on the R/V "Vityaz'," began to study the waters adjoining the Kurile chain. The results of these investigations are the
numerous, very important works on the morphology of the southern
part of the Sea of Okhotsk and the portions of the Pacific Ocean adjacent to the Kuriles (Bezrukov and Udintsev, 1953; Udintsev, 1955,
1957; Zatonskii et al., 1961). One of the works was specifically devoted to the submarine volcanoes (Bezrukov et al., 1958).

In 1951 the author took part in the voyage of the R/V "Vityaz'."
The islands from Álaid to Urup were examined, and for the first
time petrographic information was obtained on the islands of Shirinki and Chirinkotan.

In 1952 the author studied the geologic structure of Shikotan
and the volcanoes of southern Kunashir. In 1953 the volcanoes on
Paramushir were studied. In 1954 work was carried out on Groznyi
and Teben'kov Volcanoes (on Iturup) and on the island of Matua, and

later all the islands of the Greater Kurile chain were examined from the air. As a result of this "tour" the "Catalogue of Active Volcanoes of the Kurile Islands" was published, a compilation dealing with the active volcanoes of the Kuriles, as were works on the volcanoes and the petrology of the lavas of Paramushir (Gorshkov, 1954, 1957, 1958; Gorshkov and Bogoyavlenskaya, 1962).

The volcanoes of Kunashir were studied in 1954-1955 by E. K. Markhinin (1959). The main results of this work were the first petrographic descriptions of the three volcanoes on this island.

In 1951 and in the following years the geology of the Kurile Islands was studied by Yu. S. Zhelubovskii and others; their work was included in Volume 31 of the "Geology of the USSR" (1964).

In 1957-1958 large-scale geophysical work included in the IGY program was carried out here by the Institute of the Physics of the Earth (Gal'perin, Kosminskaya) and other organizations; in subsequent years this same institute continued the seismic study of the South Kuriles (S. A. Fedotov, head of the expedition). As a result of this work a very important scientific contribution was made.

From 1958 to 1964 large-scale volcanologic investigations were carried on by the Laboratory of Volcanology of the Academy of Sciences and later by the Academy's Institute of Volcanology, particularly in the Central Kuriles, under the direction and to a considerable degree with the personal participation of the author:

1958 — Kunashir and Simushir (G. S. Gorshkov and I. I. Tovarova).

1959 — Simushir, Ketoi, Ushishir, Iturup, and Paramushir (G. S. Gorshkov, I. I. Gushchenko, G. E. Bogoyavlenskaya, and K. K. Zelenov).

1960 — Iturup (I. I. Gushchenko and G. E. Bogoyavlenskaya); Paramushir (G. S. Gorshkov and G. E. Bogoyavlenskaya).

1961 — Onekotan, Chirinkotan, Kharimkotan, Shiashkotan, and Simushir (G. E. Bogoyavlenskaya and K. I. Shmulovich).

1962 — Onekotan, Simushir, and Paramushir (G. S. Gorshkov, G. E. Bogoyavlenskaya, and K. I. Shmulovich); Chernye Brat'ya (G. E. Bogoyavlenskaya and K. I. Shmulovich).

1963 — Kharimkotan, Shiashkotan (G. S. Gorshkov and G. E.

Bogoyavlenskaya); Chirinkotan and Ekarma (G. E. Bogoyavlenskaya); Paramushir and Simushir (K. I. Shmulovich).

1964 — Onekotan, Kharimkotan, Shiashkotan, Ekarma, Chirinkotan, Matua, and Raikoke (G. E. Bogoyavlenskaya).

The results of this work have been only partly published (Gorshkov, 1960, 1962; Bogoyavlenskaya and Gorshkov, 1965).

From 1960 to 1963 the volcanologists of the Sakhalin Complex Institute of the Academy of Sciences, under the direction of V. N. Shilov, were working in northern Paramushir. Only the preliminary results of this work have as yet been published (Rodionova et al., 1963, 1964). The volcanologists of this institute also visited several other islands.

In 1962 E. K. Markhinin and D. S. Stratula, co-workers of the Institute of Volcanology, conducted routine investigations in the Central Kurile Islands. Rocks from the islands of Makanru, Raikoke, and Brouton were collected for the first time by them (Markhinin and Stratula, 1965).

The volume of work carried out after 1945 and especially after 1953-1957 is incomparably greater than what was done earlier. Now the basic characteristics of the geologic structure of the islands have become known, the structure of the earth's crust and the character of variations in the gravitational and magnetic fields have been studied, a picture has been given of the topography of the floor of the Sea of Okhotsk and the Pacific Ocean, and so on.

The already published information on the Kurile Islands gives a general idea of their volcanism; however, an important part of the volcanologic material obtained in recent years by our investigations still has not been published.

In the present work are presented the results of the volcanologic investigations carried out in the period 1946-1964 under the direction of the author in the Laboratory of Volcanology and later in the Institute of Volcanology of the Academy of Sciences. The data of other investigators published up to the end of 1965 are also used.

*Chapter III*

# Structure of the Earth's Crust and Geophysical Fields

In 1957–58 during the program of the IGY a complex geophysical investigation was carried out of the transition zone from the Asiatic continent to the Pacific Ocean, an important part of which is the Kurile Islands region. These projects included deep seismic sounding (DSS) and aeromagnetic and gravimetric surveys. In addition, in 1948 on Kamchatka and later on the Kurile Islands a network of regional seismic stations was developed that, along with the data of the seismic net of the USSR, made it possible to obtain many new characteristics of the seismic regime of this region. Beginning in 1958 the Pacific Ocean Seismic Expedition worked in the South Kuriles and produced detailed seismic investigations here.

As a result of all these investigations, the Kurile Island arc region is geophysically one of the most thoroughly investigated.

## SEISMICITY

The Kurile Islands region, along with Kamchatka, is the most seismic region of the USSR. The zone of intensity 8-9 earthquakes includes all the islands. Rather commonly the earthquakes are accompanied by the development of tsunamis that are even more destructive than the earthquakes themselves.

It is no wonder that the southern part of the Kurile Islands, along with Japan, was the first region in which the existence of earthquakes with focal depths greater than 60 km, the so-called deep-focus ones (Wadati, 1935), were recognized.

The first Russian scientist to concern himself with the problem of the deep-focus earthquakes, particularly in the Kurile Islands, was A. N. Zavaritskii (1946). He noted the existence of a zone of earthquake foci dipping at about 40° from the "Tuscarora Basin" (the present Kurile-Kamchatka Trench) toward the continent and extending to depths of up to 700 km, as well as the important fact that the active volcanoes are located at points where the focal plane lies at depths of 100-150 km. Zavaritskii (1946), comparing the scale of geologic processes in the crust with the scales of processes within the mantle, first came to the conclusion that "the deep movement of lithospheric tectonics may be the (primary) cause, and folding, fracturing, and alpine tectonics in the earth's crust may be the (secondary) effects" (pp. 7-8). Thus, Zavaritskii anticipated the ideas that later were considered to be the basis of the Upper Mantle Project.

In later years, with the development of a regional seismic net, the conclusions of Zavaritskii as to the existence of a zone of earthquake foci based on teleseismic data were confirmed and improved upon (Monakhov and Tarakanov, 1955, and others). Especially interesting are the results of the detailed observations made in the South Kuriles in 1958-1962 (Fedotov et al., 1961, 1963). According to these data, the foci of earthquakes are distributed even more compactly and regularly than was established earlier from the records of more distant seismic stations. Most foci are located on the continental side of the abyssal Kurile Trench at depths of 0-200 km.

The earthquake foci form a clearly defined zone dipping beneath the islands. In cross section this zone has a wedge-shaped form, 40-50 km "wide" in its upper part and decreasing to zero at a depth of about 200 km. Beneath the islands of the Greater Kurile chain the focal zone extends to a depth of 150-180 km (Fig. 2).

A second, much weaker seismic zone extends along the Okhotsk coast of the Greater Kurile Islands. Here weak, shallow earthquakes often occur at depths of not more than 20-30 km.

The tensions at the earthquake foci (Balakina, 1962; Aver'yanova, 1965; Udias and Staudes, 1964) act dominantly in a horizontal direction, at right angles to the trend of the arc. The fractures formed trend parallel to the arc, but dip at various angles (45-75°) toward the continent. Movement has the character of upthrusting, during which the continental part is shoved onto the oceanic part; the region of the abyssal trench is subjected to subsidence.

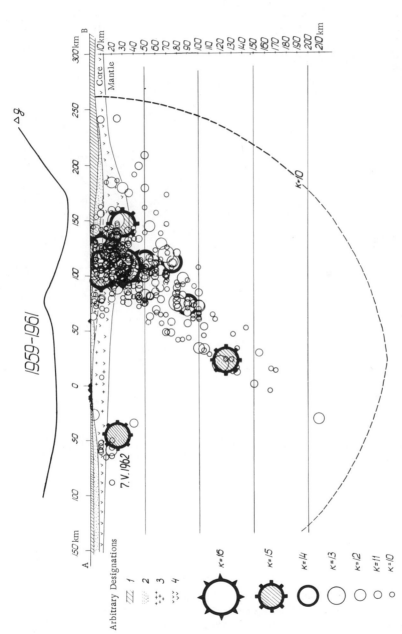

Fig. 2.  Distribution of earthquakes in the Kurile Islands area.  1) Water; 2) sediments; 3) "granitic layer;" 4) "basaltic layer;" 5) seismic stations (Fedctov et al., 1963).  Circles indicate earthquake epicenters and energy classes.

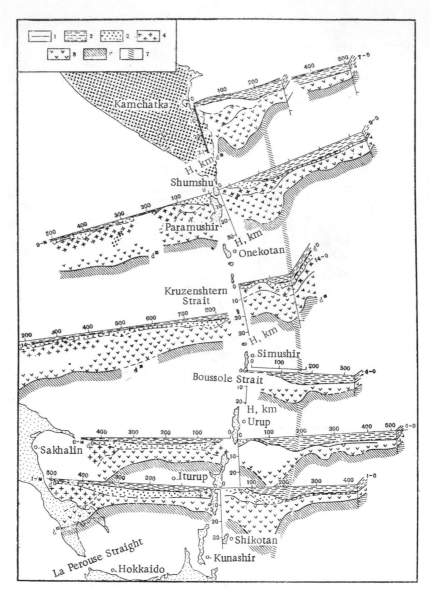

Fig. 3. Map showing the locations of DSS profiles. 1) Line of DSS profile; 2) water; 3) sediments; 4) "granitic layer;" 5) "basaltic layer;" 6) M discontinuity; 7) axis of abyssal trench (Veitsman, 1965).

## STRUCTURE OF THE CRUST FROM
## DEEP-SEISMIC SOUNDING DATA (DSS)

In the Kurile part of the transition zone 15 DSS profiles were made, with a total length of 5430 km (Fig. 3), of which five ran perpendicular to the arc into the Sea of Okhotsk (from north to south, profiles 9-M, 14-M, 6-M, 1-M, and 5-M), six profiles ran into the ocean (from north to south, profiles 9-0, 14-0, 4-0, 6-0, 1-0, and 5-0), four profiles cut Iturup island from north to south and from east to west (profiles 2-M, 2-0, 3-M, and 3-0), and one profile 15-0) ran along the Kurile-Kamchatka Trench between 9-0 and 14-0. In all, the work of DSS resulted in the rather complete mapping of the structure of the crust in this region (Gal'perin, 1958; Gal'-perin et al., 1958; Vasil'ev et al., 1960; Aver'yanov et al., 1961; Veitsman et al., 1961; Kosminskaya et al., 1963). Complete results were published in 1964 in the collection "Structure of the Earth's Crust in the Transition Region from the Asiatic Continent to the Pacific Ocean."

On the basis of the DSS data it was established that the crust has a mosaic structure here. The western limit of the region of typical oceanic crust coincides approximately with the floor of the Kurile–Kamchatka Trench. Typically continental crust, roughly speaking, extends from southern Sakhalin to the area of the Kruzen-shtern Passage; south of this line lies a broad area with a crust of the so-called "transition" type in which I. P. Kosminskaya subsequently distinguished "suboceanic" and "subcontinental" types. As is well known, at the end of the 1940's and the beginning of the 1950's the division of the earth's crust into continental and oceanic types had been established by various methods of DSS.

In oceanic regions the crust consists of one layer of consolidated rock 5-8 km thick with $V_p = 6.6-7.0$ km/sec. Above this lies a thin (1 km) layer of sediments and a 5 km layer of water. The usual thickness of the oceanic crust, inclusive of water, is 10-15 km.

Crust of the continental type has a much more complex structure and a much greater thickness. Above the M discontinuity, as in the oceans, lies a layer with $V_p = 6.6-7.0$ km/sec, and above this is a thick layer with $V_p = 5.5-6.0$ km/sec. The boundary between these two layers is called the Conrad discontinuity. Higher still are the unconsolidated sediments. The thickness of the two main layers is not constant, but the overall thickness of the continental crust averages 35 km.

The layer with $V_p = 5.5-6.0$ km/sec is often provisionally called "granitic" and the underlying layer, "basaltic." However, no petrographic significance whatsoever should be attached to these names; ignorance of this fact has led to many misunderstandings. We also will use the terms "granitic layer" and "basaltic layer" without attaching any petrographic significance to them, but using them only as synonyms for "layer with seismic velocity of 5.5-6.0 km/sec" and "layer with seismic velocity of 6.5-7.0 km/sec." Keeping all this in mind, we can represent the oceanic crust as a thin, unitary "basaltic layer." The continental crust is thick and two-layered, consisting of "granitic and basaltic layers."

As was stated earlier, Kosminskaya distinguished two more types of crust in the transition zone: suboceanic and subcontinental.

The suboceanic type of crust is distinguished from the oceanic type mainly by a significantly large thickness of sediments, 3-6 km as contrasted with the 0.5-1.0 km in the oceans. The thickness of the "basaltic layer" in the suboceanic crust is also somewhat greater.

The subcontinental crust is similar to the suboceanic; it also consists of sedimentary and "basaltic" layers. However, the velocity in the subcontinental crust, in the so-called "basaltic layer," has a value intermediate between that in the "granitic" and the "basaltic" layers, i.e., 6.3 km/sec. Locally the velocity increases to the normal "basaltic" value of 6.6 km/sec, and in this case a thin layer with a velocity close to the granitic value (5.0 km/sec) is usually present. These data allowed Kosminskaya to state that the crustal type in question was more similar to the continental than to the oceanic type.

Following these introductory remarks, let us return to the consideration of the structure of the crust in the Kurile Islands region. As noted previously, the structure of the crust is inhomogeneous to a significant degree.

The whole area to the east of the axis of the Kurile-Kamchatka Trench is underlain by typically oceanic crust. In the northern part of the chain, along profile 9-0, the thickness of the sediments in the oceanic crust is about 1 km, and the thickness of the "basaltic layer"

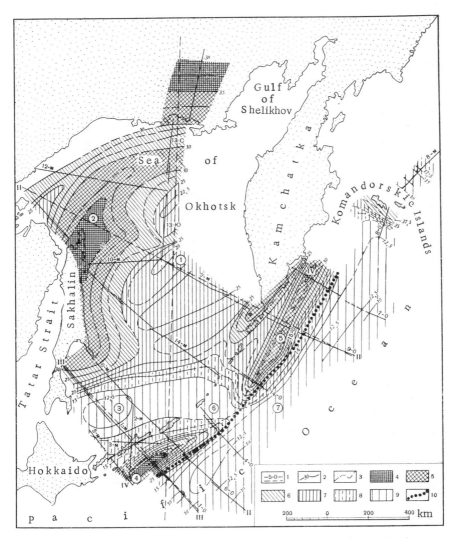

Fig. 4. Map showing depths to the M discontinuity. 1) Lines of profile; 2) depth contours based on accurate data; 3) depth contours based on less accurate data; 4) depths to M discontinuity greater than 35 km; 5) depth to M discontinuity 30-35 km; 6) depth to M discontinuity 25-30 km; 7) depth to M discontinuity 20-25 km; 8) depth to M discontinuity 15-20 km; 9) depth to M discontinuity less than 15 km; 10) axis of abyssal Kurile-Kamchatka Trench. On the map the largest structures are indicated by numbers in circles: 1) Okhotsk Rise (plateau); 2) West Okhotsk downwarp; 3) South Okhotsk upwarp; 4) South Kurile downwarp; 5) North Kurile downwarp; 6) Central Kurile upwarp; 7) Pacific Ocean upwarp.

with a limiting velocity of 6.4 km/sec is 6–8 km. The M discontinuity lies at a depth of 12–14 km. The M discontinuity dips gently toward the islands, and beneath the trench it reaches a depth of 18 km. The limiting velocity in the mantle beneath the ocean is 8.5 km/sec; beneath the Zenkevich Swell this drops to 7.7 km/sec. A similar picture is found in the central (profile 4–0) and southern (profiles 1–0 and 6–0) parts of the chain. Thus, for example, on profile 1–0 at a distance of 310 km from the coast of Iturup the depth to the M discontinuity is 11 km, and as the axis of the trench is approached it gradually increases. A very interesting peculiarity on profile 1–0 in the interval between the axis of the trench and a point 300–350 km eastward is the alternation of narrow blocks (20–50 km wide) of the mantle with distinctly differing limiting velocities: 7.8–8.0 km/sec and 8.9 km/sec. Farther out, from the 350-km point to the end of the profile (at 450 km), the velocity remains constant at 8.7 km/sec.

In the northern part of the chain along profiles 9–0 and 14–0 at the axis of the trench the oceanic-type crust is clearly replaced by continental crust. Above the "basaltic layer" a "granitic layer" suddenly appears, with a limiting velocity of 5.0–5.5 km/sec; its thickness here is not great: 2–6 km. At the trench axis this layer is overlapped by sediments up to 1.5 km thick. Beneath the west wall of the trench and beneath the island shelf the thickness of the sediments and of the "basaltic layer" increases markedly (the former up to 6 km, the latter up to 22 km) and the M discontinuity drops to a depth of 30–33 km. Closer to the islands the M discontinuity rises somewhat. On the other side of the islands, in the Sea of Okhotsk, on profile 9–M we also find continental crust up to 30 km thick, with a thin "granitic layer" and a thick "basaltic" one.

Thus, in the northern part of the Kurile-Kamchatka Trench a distinct replacement of oceanic-type crust by continental type occurs. The latter has a rather thin (up to 8 km) "granitic layer" and a markedly thicker (up to 20–22 km) "basaltic" one here. We note that in a continental area, on the Magadan-Kolyma profile, the correlation between the thickness of the layers is substantially different: the "granitic layer" has a thickness of 20 km, the "basaltic layer" a thickness of 12–13 km.

In the central part of the chain (profile 4–0) the thickness of the "basaltic layer" near the axis of the Kurile-Kamchatka Trench

amounts to 9-10 km. Beneath the west wall of the trench the thickness of the crust increases to 17-20 km; the increase takes place
in the sediments (up to 3.5 km) and in the "basaltic layer" (up to
14 km). Such a crustal type, "basaltic-sedimentary," with a somewhat increased thickness, has been called "suboceanic" (Kosminskaya et al., 1963). Somewhat farther north of profile 4-0, judging
from an examination of the gravity anomalies, crust of typically
oceanic type forms a deep "embayment" west of the axis of the
abyssal trench and reaches almost to the island of Simushir. At a
distance of 50-60 km from Simushir the usual thickness of the crust
is 13 km, of which 7 km belong to the "basaltic layer," 3 km to the
loose sediments, and 3 km to the water layer.

Closer to the islands the surface of the "basaltic layer" (and
apparently the M discontinuity) rises gently (at 4°) and the thickness
of the unconsolidated sediments correspondingly decreases. It must
be pointed out that the limiting velocity in the layer beneath the unconsolidated sediments here is 6.3 km/sec and is, so to speak, intermediate between the "granitic" (5.5-6.0 km/sec) and the "basaltic" (6.5-7.0 km/sec) velocities. The usual crustal thickness is
less than 15 km. As we have seen, the structure of the crust in the
central Kuriles region is very peculiar, although in publications of
recent years it has also been referred to the "suboceanic" type.

In the southern part of the Kurile zone (profiles 1-0 and 6-0)
west of the axis of the abyssal trough the M discontinuity dips
steeply, and 120 km from Iturup it reaches a depth of 36 km. For a
time this marked thickening alone formed the basis for the recog-
nition here of a continental-type crust, although its structure is
clearly different from the continental type: the thickness of the "basaltic layer" ($V_p = 6.6$ km/sec) is 26 km; above it lies a 7-km layer
of unconsolidated sediments ($V_p = 2.8$ km/sec) and 3 km of water.

Nearer the islands the M discontinuity rises rather steeply
(8°), a "super-basalt" layer is produced, and the thickness of the
sediments decreases. In the description of the first data from deep
seismic sounding (Tulina and Mironova, 1964) the "super-basalt"
layer was tentatively called "granitic." However, even then the
anomalously small seismic velocity (5.0 rather than 5.5 km/sec)
was pointed out. In all probability the "super-basalt sequence" is
composed of volcanic rocks. This sequence is not present everywhere, but is found only in spots. One such "spot" is related to the

submarine Vityaz' Ridge. Along the axis of the ridge the thickness
of the "super-basalt sequence" is 7 km, and sediments are practically absent. The thickness of the "basaltic layer" here is 17-18
km, and the depth to the M discontinuity reaches 25-26 km. Still
closer to the islands the thickness of the "super-basalt sequence"
again decreases, and at a distance of 40-45 km from the islands it
disappears. Thus the block of crust in the southern Kuriles region
that was called continental in the description of the results of the
deep seismic sounding (Tulina and Mironova, 1964; Kosminskaya
et al., 1964) is comparable to the continental type only in its general
thickness, which is up to 36 km. In fact it may be thought of as a
"suboceanic" layer (a "basaltic" sequence 20-26 km thick) supporting a thick (up to 7 km) layer of volcanic rock.

Finally, in the island part of the South Kuriles the structure
of the crust is also peculiar. Here the crust is again two-layered.
The upper, sedimentary layer has a thickness of not more than 3-4
km; the second layer, up to 16 km thick, has an "intermediate"
seismic velocity of 6.3 km/sec. Locally in the area of the southern
islands thin sheets are found with velocities similar to "granitic"
ones (4.3-5.5 km/sec). The depth of the M discontinuity in the vicinity of Urup is 16 km, and near Iturup it is 18 km. Kosminskaya
called this type of crust "subcontinental."

West of the southern Kuriles the M discontinuity rises
smoothly from the 15-18 km level to a depth of about 13 km. The crust
in the Kurile Basin of the Sea of Okhotsk has a two-layered basalt-
sediment structure with a normal velocity for the "basaltic layer"
of 6.6-6.8 km/sec. On profile 1-M the thickness of the sedimentary
and "basaltic" layers are about equal and total 10 km. On profile
6-M the thickness of the sediments decreases to 2.5-3 km and the
thickness of the "basaltic layer" increases to 7-10 km. This type
of crust has been called "suboceanic," although it is substantially
different from the "suboceanic" crust in the central part of the
islands or in the wall of the trench. Essentially we have a thin
oceanic crust supporting a thick layer of unconsolidated sediments
in the southern part of the Sea of Okhotsk. The thickness of the
consolidated crust is very clearly seen on the map (Fig. 5).

Speaking of velocities at the M discontinuity, we have already
pointed out that according to the deep seismic sounding data these
velocities have different values in different parts of the seismic

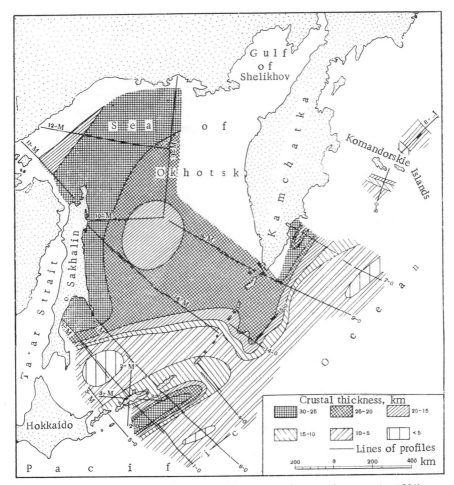

Fig. 5. Map of the thickness of the consolidated crust (Kosminskaya et al., 1964).

profiles. In this relation the results of the detailed study of the
seismicity of the South Kuriles mentioned earlier (Fedotov, 1963;
Fedotov et al., 1961, 1963, 1964) are extremely interesting and important. According to these data, in the region of the Kurile island
arc the velocity of the longitudinal seismic waves at the M discontinuity is lower than usual and equals 7.7 km/sec (instead of the
"normal" 8.1 km/sec). Such a velocity figure is maintained to a
depth of 80 km, where it increases slightly. Only at a depth of 125
km does the velocity reach 8.1 km/sec. The zone of decreased

seismic velocities within the upper part of the mantle (Gutenberg zone), which is observed everywhere in the continental and oceanic regions of the earth, is either absent here in the Kuriles or is very poorly expressed. Against this background of low velocities at depths of 60-110 km and especially at 80-90 km we can set the strong increase in absorption of transverse seismic waves. The zone of such anomalous structure in the upper mantle very clearly coincides with the zone of modern volcanism. In both directions from the belt of contemporary volcanoes, i.e., in the Sea of Okhotsk and in the Pacific Ocean, the mantle has its usual, normal structure and composition with seismic velocities of 8.0-8.1 km/sec at the M discontinuity.

Thus the structure of the upper mantle in the volcanic zone of the Kurile arc is defined by very distinct characteristics: a layer of decreased velocities within the mantle (Gutenberg or wave-guide zone) is absent; in addition, seismic velocities at the M discontinuity are substantially lower than in the adjacent, nonvolcanic zones. If we judge by analogy with Kamchatka, Japan, and other volcanic regions which will be described in their proper places, such a mantle structure is inherent in the whole Kurile arc, from north to south.

In addition, as we said, the crustal structure along the arc is not homogeneous. In the northern Kuriles the crust has a more or less typically continental structure. The central Kuriles are underlain by crust of the "suboceanic" type, actually very similar to the usual oceanic crust, but with an increased thickness of volcanic rocks. Here typically oceanic crust pushes into the arc as a deep "gulf," reaching almost as far as the islands of Simushir and Ketoi.

The crustal structure of the southern Kuriles is also inhomogeneous. The consolidated crust here is single-layered; seismic velocities within it have an intermediate value (6.3 km/sec). Above it lies a 3-4 km layer of sediments, and the total thickness scarcely reaches 18 km. This type of crust has been called "subcontinental."

## GRAVITATIONAL AND MAGNETIC FIELD

Gravimetric investigations were carried on in the Kuriles in 1951-1954 (Gainanov, 1955) and during the IGY (Veselov et al., 1961;

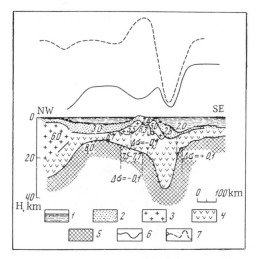

Fig. 6. Composite section of the crust and gravitational anomalies for a profile from Sakhalin through the South Kurile Islands into the Pacific Ocean. 1) Water; 2) sediments; 3) "granitic layer;" 4) "basaltic layer;" 5) mantle; 6) Bouguer anomaly curve; 7) free-air anomaly curve.

Gainanov and Smirnov, 1962; Gainanov, 1963). On the basis of the Bouguer anomalies the whole region can be divided into four zones with the following characteristics: (1) the northern part of the Sea of Okhotsk, including Sakhalin, has anomalies close to zero, passing into weakly positive or weakly negative ones such as are usually characteristic of the continental platforms; (2) the Kurile Basin of the Sea of Okhotsk, in which strong positive anomalies are found such as are typical of oceanic areas; (3) the Kurile Islands and the abyssal Kurile Trench, the most interesting region for us, which are characterized by a double minimum in the anomalies, separated by a secondary maximum, with the very deep minimum corresponding to the west wall of the Kurile Trench and the maximum gradient of $\Delta g$ corresponding to the axis of the trench; (4) the Pacific Ocean region beyond the trench, where strong positive anomalies again are observed.

The picture is somewhat simpler in the free-air reduction. The free-air anomalies are close to isostatic and depart sharply from this only in the Kurile Islands zones. Beneath the Sea of Okhotsk the anomalies are weakly positive; under the east flank of the inner ridge there is a small maximum that changes to a deep minimum beneath the west edge of the trench. Then comes a small maximum and the flattening out of the anomalies. Figure 6 gives the characteristics of the structure of the crust and the variations in the Bouguer and free-air anomalies in a section from Sakhalin through Iturup to the Pacific Ocean.

It has long been known that the Bouguer anomalies depend on the depth at which the M discontinuity lies. The pertinent correlation curves were constructed, curves that are used to determine the thickness of the crust in the absence of direct data. In the Kurile Islands area the deep seismic sounding profiles coincide with the gravimetric profiles; this makes it possible to estimate the pertinent Bouguer anomalies from these two profiles. In the region of the Kurile Islands and the Zenkevich Swell a very marked difference was found between the observed and calculated anomalies, which can be explained only by a change in the density of the subcrustal material. The residual anomalies disappear if it is assumed that beneath the Kurile Islands the density of the mantle is decreased by $0.1$ g/cm$^3$ down to a depth of 50-60 km and that passing into the oceanic region its density is increased by the same amount to a depth of 30-40 km (Gainanov, 1964; Gainanov et al., 1965). According to another interpretation (Livshits, 1965) a decrease in density, as in the first variant, affects the mantle beneath the islands and amounts to $0.11$ g/cm$^3$, whereas the increase in mantle density coincides with the earthquake focal zone and amounts to $0.06$ g/cm$^3$.

An aeromagnetic survey was carried out during the IGY; it included the whole area of the Sea of Okhotsk, the Kurile Islands, and the adjacent parts of the Pacific Ocean. The survey was flown at a constant height of 2000 m (Solov'ev, 1962; Solov'ev and Gainanov, 1963). According to the data, two types of anomalous fields are clearly manifested: (1) isometric anomalies with rather small intensities (100-200 $\gamma$) in the northern part of the Sea of Okhotsk and in the Kurile Basin, and (2) intense (up to 700 $\gamma$), narrow, belt-like anomalies along the Kurile chain and in adjacent parts of the Pacific Ocean.

The Kurile Basin of the Sea of Okhotsk corresponds to a negative area that is enclosed by positive anomalies toward the northwest. The narrow zone of the clearly disturbed field of the Greater Kurile chain, with its intense positive and negative anomalies, stands out against the background of this deep magnetic depression. No relationships can be established between the locations of the islands and the sites of the local anomalies. However, there is a very definite relationship between the low values of $\Delta T$ and the Greater Kuriles and between the high values and the Lesser Kuriles.

To the east the Kurile chain is bordered by a linear zone of intense positive anomalies that locally are separated by narrow magnetic minima. On the basis of its position, this zone includes the west (continental) edge of the abyssal trench, and it can definitely be traced from Hokkaido to the Komandorskie Islands. This band apparently represents a zone of vertical fractures and faults that are reflected in the submarine topography (see Chapter I).

The axis of the abyssal trench is not very clearly reflected in the magnetic field; it occurs in the weakly negative zone. Near the South Kuriles the trough is bordered on its oceanic side by a new system of linear magnetic anomalies reminiscent of the Kurile anomaly zone. However, the character of the field here is not typical of the oceans as a whole, and it dies out opposite the Boussole Strait.

In a large part of the Pacific Ocean a low-relief field is found, with anomalies close to zero. Locally, linear positive anomalies of up to 400 $\gamma$ are encountered. These may represent fissure eruptions of lava onto the floor of the ocean.

There is no general, direct relation between the character of the magnetic field and the structure of the crust or the sea-floor topography. The local anomalies in the area of the Greater Kurile chain probably are related to eruptions of basic lava. Inasmuch as a reversal of the magnetic field took place at the Pliocene-Pleistocene boundary (Pospelova, 1960), the eruption of lavas of different ages might cause strong positive and negative anomalies.

The upper ($h_1$) and lower ($h_2$) limits of the magnetic masses causing the disturbances were calculated for different regions of the transition zone (Gainanov and Solov'ev, 1963; Gainanov, 1964). For the oceanic region adjoining the southern and central islands of the Kurile chain, the following values were determined: $h_1 = 4$–$16$ km, $h_2 = 18$–$34$ km; and near the abyssal Kurile Trench $h_2 = 40$ km. Thus, in this part of the Pacific Ocean the upper edge of the magnetic masses causing the disturbances lies in the "basaltic layer" or even in the upper mantle, while the lower edge lies in the upper mantle. The restriction of the band-like anomalies with negative values of $\Delta T$ to zones with higher than usual (up to 8.7-8.9 km/sec) velocities in the upper mantle may also serve to confirm the fact that the causes of the magnetic anomalies in the oceanic region are inhomogeneities in the upper mantle (Gainanov, 1964; Tulina, 1965).

In the area of the Greater Kurile chain it is found that $h_1 = 1$-4 km, $h_2 = 6$-9 km. These data indicate above all the influence of lava flows of different thicknesses and ages. In addition we must note that with a great amount of detailed magnetometric work it would probably be possible to discover the conduits feeding the volcanoes, "columns" rising directly from the mantle, as was done on Hawaii (Malahoff and Woollard, 1965).

Geophysical investigations in the Kuriles allow us to characterize rather completely the composition and structure of the crust and upper mantle. We have gone into the geophysical aspects in great detail because in the following chapters these data will serve as a basis for correlating the characteristic features of the volcanism with the crustal structure and with the upper mantle.

*Chapter IV*

# Geology of the Basement

## STRATIGRAPHY

The oldest rocks in the Kurile Islands are Upper Cretaceous deposits developed in the Lesser Kurile chain. They were first recognized only at the end of the last century (Jimbo, 1890), and later they were characterized faunally by Sasa (1934) who distinguished the "Shikotan formation" and the "Matakotan eruptive rocks."

At present Soviet geologists (Zhelubovskii and Pryalukhina, 1964) place all the Upper Cretaceous deposits in the Shikotansk series, which is divided into two suites, the lower Matakotansk suite and the upper Malokuril'sk (Lesser Kuriles) suite.

The Matakotansk suite is composed of volcanogenic rocks: lavas (including spherical and pillow varieties), tuff-conglomerates, and tuff-breccias of basaltic to andesitic composition. In the upper part there are tuffogenic sandstones. Based on their mineralogic and chemical characteristics all these rocks are referred to the same calc-alkali series as the present-day rocks. The thickness of the Matakotansk series is 400 m.

The Malokuril'sk (Lesser Kuriles) suite overlaps the Matakotansk suite conformably in a number of places and with slight angular nonconformity in others. This suite is composed predominantly of tuffogenic sandstones, siltstones, and claystones with layers and lenses of marl and calcareous sandstones. *Inoceramus shikotanensis*, remnants of other inoceramids, belemnites, and ammonoids are found in these deposits and give the age of the suite as Campanian-Maastrichtian or Santonian-Campanian (Sasa, 1934; Zhelubovskii and Pryalukhina, 1964). The thickness of this suite is up to 200-300 m.

A strongly deformed series of effusive-pyroclastic rocks rest with a distinct angular nonconformity (Kamchatkan phase of folding) on the deposits of the Shikotansk series; these make up the Zelenovsk suite, which is tentatively referred to the Paleogene. The lower part of the Zelenovsk suite consists of tuff-breccias and tuffs of basic composition with interlayers of basalt and pyroxene andesite; in the upper part pyroxene andesite and andesite-basalt are dominant. The thickness of the suite is 400-500 m.

Gabbroid intrusions are emplaced at the contact between the Upper Cretaceous and Paleogene (?) rocks on the island of Shikotan. Their age is Late Cretaceous or early Paleogene. The central parts of these massifs are composed of olivine gabbro with 49% $SiO_2$; they consist of plagioclase (60-65%), olivine (25-28%), and clinopyroxene (2-5%) with scattered grains of biotite and quartz and with apatite and magnetite (up to 5%) in accessory amounts. At their margins the olivine gabbros pass into gabbro-norites or leucocratic gabbros with 51.5-52.5% $SiO_2$. They are distinguished from the olivine gabbros by their greater content of plagioclase and by the dominance of hypersthene or sometimes of hornblende. In addition to the gabbroid intrusions, small massifs of gabbro-diorite and diorite are found on Shikotan.

The basement of the Greater Kurile chain is composed only of Tertiary rocks. Detailed stratigraphic analysis of these is made difficult by their small content of organic remains and the predominantly volcanogenic character of the succession. According to the verdict of the Okha Stratigraphic Conference (1959) the Tertiary formations of the Greater Kurile chain are divided into three series: Sredneparamushirsk [Middle Paramushir] (late Oligocene (?) − early Miocene), Iturupsk (middle Miocene), and Utesn (late Miocene− Pliocene).

The Sredneparamushirsk series was first recognized in 1955 by G. M. Vlasov.

Rocks of this series are known on all the large islands, where they have various local names: on Paramushir and Shumshu they are the Khamadinsk, Kaparinsk, and Shumnovsk suites; on Urup, the Urupsk suite; on Iturup, the Zhemchuzhnaya River suite; on Kunashir, the Kunashirsk suite. All these sequences are characterized by the presence in the lower part of effusive and tuffogenic rocks of basic and intermediate composition, which are almost everywhere deformed and propylitized; an early Miocene fauna is found in the tuffogenic sandstones in the upper part: *Palliolum (Delectopecten) kriljonensis*, L. Krisht., *Pododesmus schmidt* L. Krisht., *Mytilus miocenum* L. Krisht., *Modiolus solla* Slod., *Yoldia plivonensis* Slod., *Telina aragonia* Dall, and others. The thickness of this series reaches 2500 m.

The analog of this series in Japan is the "Green Tuff Formation," which is referred to the early Miocene.

The Iturupsk series (middle Miocene) includes the following: the Kuibyshevsk suite on Iturup; the Goryachii Plazh, Lovtsovsk, and Alekhinsk suites on Kunashir; the Tokotansk suite on Urup; and the Okhotsk suite on Paramushir. This series nonconformably overlaps the underyling Sredneparamushirsk series, which was deformed at the time of the Kurile phase of folding. The development of tuffs and tuffites of acid composition, of bleached tuffogenic rocks, and of tuffaceous diatomites is characteristic of the Iturupsk series. The thickness of the sequence averages about 2000 m. The rocks of this series have not undergone regional hydrothermal alteration, and propylitization has occurred only along individual tectonic zones. The remains of a marine fauna and flora often occur in the tuffitic sandstones and siltstones: *Acila (Truncacila)* sp., *Yoldia thraciaeformis* (Storer), *Palliolum (Delectopecten) peckhami* Gabb., *P. (D.) randolphi* Dall, *Lima (Limatula) kovatchensis* Ilyina, *Cuspidaria (Cardiomya) kova-tchensis* Ilyina, *Solemia tokunagai, Nuculana crassitelloides* Laut, and others.

These forms are often found in the Miocene of Japan, in the Vayampol'sk series on Kamchatka, and in the middle Miocene of Sakhalin. The rocks of the Iturupsk suite were deformed in the Aleutian phase of folding, but the overlying Pliocene is almost undeformed.

The Utesn series (late Miocene–Pliocene) was recognized in 1951 by Yu. S. Zhelubovskii. The rocks of this series are widely developed on Iturup, Paramushir, and probably Urup; they are also present on Kunashir, where the Pliocene still is not separated from the Quaternary. This series is dominantly an effusive-pyroclastic one; liparite and dacite pumice is often found. The series is characterized by marked variability in facies, associated with rather regular stratification. The thickness of the succession reaches 2800 m.

Various studies of the Utesn series disclosed the presence of some suites similar in age and in lithology, but territorially separate: the Okeansk and Cape Okruglyi suites on Paramushir; the Derbashevsk and Paramushirsk suites on Paramushir and Shumshu; the Bystrinsk suite on Urup; and the Rybakovsk, Cape Przheval'sk, Parusn, and Osennaya River suites on Iturup.

Fossil finds are numerous here; among these the following forms are represented: *Pecten* cf. *kurosavensis* Yok., *P. (Swifto-pecten)* cf. *nanakitaensis* Nakamura, *P. (Pallium)* ex. gr. *swiftii* Bern., *Modiolus Wajampolkensis* Slod., *Septifer* cf. *hirsutus* Lam., *Cardium (Cerastoderma)* cf. *tiglense, Natica janthostoma* Desh, and others.

The fossil assemblage in the Utesn series has much in common with the Alneisk series on Kamchatka and also with the Wakkanai series on Hokkaido. Sergeev (1963) recognizes one more series, the Kuril'sk, between the Sredneparamushirsk and Iturupsk series; to it he assigns the Goryachii Plazh (Kunashir) and Shumnovsk suites. According to him, the stratigraphy of the Tertiary deposits of the Greater Kurile chain is represented in the following way: 1) Sredneparamushirsk series (Oligocene—early Miocene), thickness not less than 2500-2800 m; 2) Kuril'sk series (early and middle Miocene), thickness 1300-1500 m; 3) Iturupsk series (middle and late Miocene), thickness 2200-2500 m; 4) Utesn series (late Miocene—Pliocene), thickness not less than 1600 m.

The usual thickness of the Tertiary deposits, according to Sergeev, amounts to 3000 m in the south, and reaches 7000 m in the north.

The Utesn series, in turn, according to Vergunov and Pryalukhina (1963), can be subdivided into two conformable suites: the Paramushirsk, which is exposed mainly on the flanks of the arc, and the Parusn, which forms the basement of the Quaternary volcanoes in the central part of the arc.

On many islands of the Greater Kurile chain outcrops of granitoid intrusions are known. They were discovered at the end of the last century on Kunashir, and later were studied on Urup (Nemoto, 1936) and Paramushir (Hirabayashi, 1941). Pebbles of granitoid rocks are found on Iturup and also on Onekotan, Shiashkotan, and other islands of the Central Kuriles that Hirabayashi noted earlier. However, the neo-intrusions were studied in detail only

in later years (Neverov et al., 1963; Sergeev, 1963; Sergeev and
Sergeeva, 1963; Vergunov and Vlasov, 1964).

On Kunashir the granitoid massifs are complex and multiphase: they are formed of quartz diorites (60% $SiO_2$) and plagio-
granites (up to 73.5% $SiO_2$) and also of granodiorites (65-68% $SiO_2$).
On Urup the intrusive massifs are composed of quartz diorites (60%
$SiO_2$), but at the contacts these grade into diorites (56.5-57.5% $SiO_2$),
gabbro-diorites (55% $SiO_2$), and finally leucocratic gabbros (47.5%
$SiO_2$). Analogous rocks are found in the form of xenoliths in the
caldera-stage pyroclastics in the Central Kuriles.

There is no consensus regarding the age of the granitoid intrusions. Zhelubovskii refers all the young intrusions to the middle Miocene, but Vergunov recognizes early Miocene and late Miocene intrusions, but convincing evidence of their early age is absent.

Data on absolute age determinations made in the Magadan geo-
chronological laboratory (Firsov, 1964) disagree with the field evidence. K−Ar determinations on six samples from Kunashir and
Urup give ages of $6 \pm 1.5$ million years, i.e., early to middle Pliocene.

## TECTONICS

Information on the tectonic structure of the Kurile Islands is
still more limited than data on the stratigraphy. There is not even
any consensus among the investigators on the very general question
as to whether or not the Kurile arc is a geosyncline.

Some geologists (Petrushevskii, 1964) deny the geosynclinal
character of the Kurile arc; others (Goryachev, 1960) assume an
essentially geosynclinal regime here, but state this with large reser-
vations. Belousov and Rudich (1960) refer the Kurile arc to the
"arcs of the second type," which are developed on the site of ancient
geosynclinal systems.

However, perhaps the majority of geologists consider the
Kurile arc to be a modern geosynclinal system. The present author belongs to this group.

It has already been noted above that one can see the reflec-
tion of a complex system of synclines and anticlines in the submarine topography. This system is distinguished even more clearly
if we remove the cover of unconsolidated sediments.

During the deep seismic sounding along the main profiles the thickness of the sedimentary succession was also determined (Zverev, 1964). According to these data, the thickness of the sediments is rather small on the tops of the outer and inner ridges, amounting to less than 2 km in general. Work on the more closely spaced set of profiles in the region of the South Kurile Islands indicated a very complex structure of the sediments here. Zones of decreased sediment thickness (1 km or less) apparently include Kunashir, part of Iturup, and a considerable portion of the near-Kurile waters of the Sea of Okhotsk northwest of Urup. Right around Urup an increase in the thickness of the unconsolidated sediments of up to 3-4 km was noted. Because of the numerous sheets of lava (with velocities of about 5 km/sec) that are interlayered with the pyroclastic and sedimentary rocks, the distribution picture of the sediments in the inner arc is very complex; it cannot be interpreted in detail on the basis of the existing net of profiles.

In the southern part of the outer ridge near Shikotan the thickness of the sediments amounts to about 2.5 km, opposite Iturup the thickness is less than 1 km, and opposite Urup the sedimentary succession on the surface of the outer ridge is practically absent. In the northern part, near Shiashkotan, the thickness of the sediments amounts to about 1 km.

In the longitudinal trench between the outer and inner ridges the thickness of the sedimentary sequence clearly increases (to up to 4 km in the axial part); two large downwarps are recognized, though more clearly in the upper surface of the consolidated crust than in the topography of the sea floor — the North Kurile and South Kurile downwarps. The depth of these downwarps reaches 4 km near Iturup and 8 km on the Shiashkotan traverse.

The thickness of the sediment on the floor of the Kurile Basin of the Sea of Okhotsk is very significant. In the southwestern part of the basin the sediments reach 5.5-6 km. A local, narrow zone with a sedimentary thickness of up to 4.5 km is noted opposite Matua. A 3-km isopach on the sediments outlines a broad depression in the upper surface of the consolidated crust, the South Kurile Basin, that extends far to the northeast beyond the limits of the Kurile Basin of the Sea of Okhotsk and that reaches as far as southern Kamchatka. The depth at which the top of the basement occurs in this basin is 8-9 km over a considerable area; the thickness of the consolidated crust in the sothern part of the basin barely reaches 5 km.

The sediments in the upper and middle parts of the west wall of the Kurile-Kamchatka Trench have a very considerable thickness (up to 7 km). As the axis of the trench is approached, the thickness of the sediments decreases to 2 km. The east wall of the trench and the ocean floor are underlain by a sedimentary succession 0.5-2 km thick. We must note the difference in the structure of the sediments in the eastern and western walls of the abyssal Kurile-Kamchatka Trench. On the ocean floor and in the adjoining east wall of the trench the bottom of the layered sediments practically coincides with the top of the consolidated crust, i.e., the whole sequence of sediments occurs here in its originally stratified condition. On the western, continental slope of the abyssal trench and on the island shelf only the upper 0.5-1 km of the sediments are layered. The lower, larger part of the sedimentary sequence is characterized by complex structure, and the basement has a velocity lower than usual for the "basaltic layer."

Such a difference in the relationship of the layered sediments with the total thickness of the laycred sequence on the eastern and western slopes of the trench may testify both to substantially different conditions of sedimentation and to differences in the tectonic conditions that led to the destruction of the normal bedding and the thorough reworking of the sediments on the west slope of the trench but which, for all practical purposes, did not affect its eastern slope.

In the general area of the abyssal Kurile-Kamchatka Trench in the ocean floor, then, there is a large linear downwarp in the top of the consolidated crust, the Kurile Trench; the east side of this latter trench practically coincides with the east wall of the trench in the ocean floor and with the edge of the ocean-marginal swell, but its axis is somewhat offset to the west relative to the axis of the deep-water trench. The depth to the top of the basement here is 6-11 km.

Thus the main anticlines and synclines of the Kurile geosynclinal system are reflected, but in stronger contrast, in the relief of the upper surface of the consolidated crust.

This system includes the South Okhotsk basin, the double anticline of the Kurile arch, and the Kurile trough.

The Kurile arch is broken up by numerous longitudinal and transverse fractures and apparently may be thought of as a horst-anticline.

With regard to the individual islands, those of the Greater Kurile chain are thought of (Goryachev, 1960) as large symmetrical anticlines complicated by fracturing and separated by synclinal straits. The axes of the anticlines (islands) diverge somewhat from the general trend of the arc; they have a more northerly trend and form an *en echelon* system, to which Tokuda (1934) called attention earlier. The *en echelon* form of the folds is noted best in the northern and southern islands of the Greater Kurile chain; the central group of islands is stretched out in chain-like fashion parallel to the general Kurile trend.

To the north the Kurile structure passes into Kamchatka; to the south the links of the volcanic chain pass through the Shiretoko Peninsula into Hokkaido, where they are distinctly offset to the west. At the west coast of this island they intersect the Nasu volcanic zone.

## SOME PROBLEMS OF QUATERNARY GEOLOGY

From the point of view of volcanology, the stratigraphy of the Ouaternary deposits is of exceptional interest. Unfortunately, this problem is practically untouched in the islands of the Kurile chain.

We will try briefly to deal with two generally closely related problems: the correlation of the marine terraces and the history of glaciation. Our conclusions as to the history of this or that volcanic center have been based on correlations with terraces and with furrowed topographic forms, i.e., geomorphologic methods have been used. No determinations of flora or fauna have been made.

The author apparently was the first one to call attention to the two-stage glaciation of Paramushir (Gorshkov, 1954), and he used this to distinguish between volcanoes and volcanogenic formations of different age. By analogy this same division was carried out on some other volcanoes (Gorshkov, 1958). In the first paper it was noted that both glaciations were of the mountain-valley type. The first glaciation was the more intensive, and its glaciers reached as far as the present coast. The glaciers of the second period were simply imposed on the topography produced during the first period.

Vlasov (1958, 1959) has called much attention to the glaciation of the Northern Kurile Islands. According to him, the first glaciation (supposedly early Quaternary) formed partial icecaps; the glaciers did not extend to the present coasts but ended on high (250–400 m) marine terraces. The second glaciation (supposedly late Quaternary) was a valley glaciation and the glaciers extended down to the coast, uniting with the 15-20 m terrace. Both glaciations, according to Vlasov (1959), were synchronous with large-scale transgressions.

On the basis of general considerations, some interesting ideas have been expressed by Chemekov (1959, 1961). In the northern Pacific Ocean, particularly in the Kurile-Kamchatka zone, four cycles of terraces are noted, according to him, that are related to eustatic changes in sea level: 1.5-3 m, 4-8 m, 15-20 m ("Kurile"), and 25-30 m ("Hawaiian"). In addition, a submarine terrace is noted at a depth of 70-100 m ("Japanese"). The two lower (first) terraces are assigned to Holocene times and correlate with the postglacial climatic optimum. The terrace at 15-20 m corresponds to the "second late interglacial" that separates early and late Würm or to the interstadial between the Iowan and Wisconsin glacial stages. The terrace at 25-30 m is referred to the Sangamon (Riss-Würm) interglacial. We note, by the way, that a terrace of analogous height on Chukchi is characterized faunally as Riss-Würm (Petrov, 1963). Finally, the 70-100 m submarine terrace Chemekov assigns to the "first interglacial," i.e., to the early Würm. We shall say something later regarding the position and age of the submarine terrace. Chemekov's thoughts on the above-water terraces seem completely convincing to us.

According to Chemekov's data, there were three glaciations in the Far East: the earliest, middle Quaternary one corresponds to the Illinoian in North America or to the Riss in Europe, and the two later ones are probably phases of a single late Quaternary (Wisconsin or Würm) glaciation.

Vlasov cites the 15-20 m terrace as the most prevalent one in the Kurile Islands; on that basis Chemekov has called it the "Kurile" terrace. Kanaev (1960), who determined the height of the terraces on all the Kurile Islands, notes that "among the great number of terraces of different heights ... there is no clearly expressed level that is followed on all the islands. Only the terrace at 20-30 m is widely developed on most of the islands of the chain" (p. 221). Thus the characteristic terrace of the Kurile Islands is apparently not a 15-20 m terrace, but a 20-30 m one, i.e., the one corresponding to the "Hawaiian" terrace of Chemekov.

In later years moraines have been found not only on Paramushir, but on the island of Makanru, Iturup, Simushir, Rasshua, and Urup. On the southern islands only one glaciation has been recognized (Kanaev, 1960, 1961).

According to our latest studies, including the massive analysis of aerial photographs, there are traces of two glaciations on the Kurile Islands. This evidence takes the form of both moraines and erosional landforms. The evidence of glaciation is seen most clearly on the northern islands (especially on Paramushir), but it also is found on the islands of the central part of the chain and on Urup, Iturup, and northern Kunashir.

The first glaciation was the more intensive; on Paramushir and possibly on Onekotan it produced a partial icecap. All the main river valleys are trough-like. On the upper ridges only isolated nunataks and remnants of the interfluve plateaus stuck out above the surface of the glaciers. The whole central part of the island, the area of Levinson-Lessing Ridge, was almost completely ice-covered. The glaciers overrode the flat interfluve areas, and along the troughs they ran down not only to the present coast but in many cases they stretched beyond the present shoreline. In particular the glaciers of the northern Vernadskii Ridge completely overlapped the present island of Shumshu.

Farther south the area covered by the glaciers grows smaller and they take on a purely mountain-valley character. However, even on Urup they covered large areas near the base of the ridges and on Iturup, for example, they completely filled the broad Medvezh'ya Caldera. Glacial landforms are found as far south as northern Kunashir.

The moraines of the first glaciation have been reworked and redeposited along with the material of the 20-30 m terrace. Only the moraines of the first glaciation are reworked; as Vlasov pointed out, the moraines of the second glaciation "pass into the deposits of the 20-30 m terrace."

The second glaciation in the northern chain was of the mountain-valley type. Most glaciers of the second glaciation used the U-shaped valleys of the first glaciation, but they were absent from the interfluve areas. The glaciers reached to the present sea shore; locally they appear to have reached even farther.

The moraines of the second glaciation, in contrast to those of the first, are enclosed within the 20-30 m terrace and have not been affected by that transgression.

In the Central Kuriles the second glaciation left almost no

traces, and beginning on Urup, we find only small moraines at a height of 1000-1200 m. Such small moraines of the second glaciation are also found on Iturup, but they have not been discovered on Kunashir.

Thus our studies indicate that the first glaciation included all the islands of the Greater Kurile chain; in the north it produced small icecaps, and in the south it formed mountain-valley and often "Malaspina-type" glaciers. The second glaciation also included almost all the islands; in the north it produced mountain-valley glaciers, but in the south they were more similar to cirque glaciers.

The age of the second glaciation is undoubtedly Würm (Wisconsin). The first glaciation preceded the transgression that produced the terrace at 20-30 m. According to present ideas, this terrace is Sangamon (Riss-Würm) in age, and in this case the first glaciation is identified as Illinoian (Riss). However, the possibility is not excluded that the terrace indicated has a younger age and that its formation is related to an intra-Würm interstadial. In this case the first glaciation would be considerably "rejuvenated" and should be referred to the early Würm.

Henceforth, on the basis of more detailed studies, we will refer the first glaciation to the Illinoian (Riss) and the second to the Wisconsin (Würm).

The volcanoes and volcanic formations may be divided into (1) late glacial — Holocene (this is the best established group, whose age does not exceed 10-11 thousand years); (2) interglacial, corresponding to Sangamon time, 80-170 thousand years B. P.; and (3) preglacial, older than 220 thousand years (age limits according to Romankevich et al., 1964). In some cases still other subdivisions may be recognized, for example, late glacial (11-30 thousand years B. P.). In other cases only two age groups are identified: Holocene and preglacial, the latter including in this case all formations older than 11 thousand years. As far as possible, the age relationships will be given for each volcanic edifice.

The correlation of this or that volcano with the submarine terrace may provide additional evidence as to age, especially for the smaller islands and the isolated volcanoes.

We mentioned above that Chemekov recognized a 70-100 m submarine terrace that is, according to him, of early Würm age

(50-80 thousand years B. P.). There is no submarine terrace at
that depth in the Kurile Islands, but a 130-m terrace is found almost everywhere (Zatonskii et al., 1961). A submarine terrace at
the same depth is known in Alaska and along the coast of North
America. The age of plant remains obtained from drill holes into
the surface of this terrace is indicated as 17-25 thousand years
B. P. (Emery, 1958), i.e., it is dated as late Wisconsin (late Würm).

Thus, absence of evidence of this terrace around individual
islands or coastal volcanoes fixes their age as Holocene. This additional criterion clearly coincides with the absence of any traces
of glacial action on the summit or flanks.

The form of this terrace indicates directly the late Wisconsin
paleography, and changes in its level indicate Holocene tectonic
movements.

The limits of the 130-m submarine terrace on Fig. 1 are
taken, with some revisions, from the map of Zatonskii, Kanaev, and
Udintsev (1961). On this diagram it is seen that in late Wisconsin
time Paramushir and Shumshu formed a single unit with Kamchatka;
Kunashir and the Lesser Kurile chain formed a single unit with
Hokkaido (and Sakhalin). A single large island extended from
Onekotan to Shiashkotan. Rasshua and Ushishir also were joined
together, as were Chernye Brat'ya and Urup. The other islands had
completely different outlines, and such large islands as Alaid did
not yet exist. More detailed observations, including echo-sounding
work on the R/V "Geolog," indicate that on the Okhotsk side of the
islands the 130-m terrace in many cases is cut off by faults of
great displacement. In particular, a large fault or system of faults
occurs along the western edge of the Central Kuriles from Matua
to Simushir.

The most detailed data on the depth to the edge of the island
shelf has been published for the Paramushir region (Kanaev and
Larina, 1959, Fig. 2). According to this, the western edge of the
shelf ends at a depth of 120-130 m, the eastern edge, corresponding
to the Vityaz' Ridge, at a depth of 140-150 to 260 m. Starting from
this, one can maintain that in the Holocene the area of the Greater
Kurile chain was not subjected to significant tectonic movement,
whereas the submarine Vityaz' Ridge area has subsided up to 130 m.
This subsidence apparently is related to the continued downwarping
of the Kurile-Kamchatka Trench.

*Chapter V*

# Volcanoes of the Kurile Islands

Our main task was the study of the volcanoes of the Kurile arc, and to their description we have devoted most of our attention. In all, 160 Quaternary terrestrial volcanoes and volcanic groups are described in the present work, i.e., twice as many as in our earlier summaries. Of these, 104 volcanic edifices have been active in the Holocene. In one of the earlier summaries (Gorshkov, 1958) we estimated this number at 51-52; thus, the results of our later investigations have allowed us to double this number. If we compare the number of now known volcanoes with that known prior to 1946 (cf., for example, Zavaritskii, 1946), the result of our work on the study of volcanism in the Kurile arc seems rather impressive; more than 100 new volcanoes, among them some active ones, have been discovered and described. In addition, dozens of other volcanoes have been described that earlier were known only by name. Perhaps mentioning our "services" is not very modest, but looking back at the road travelled, I cannot help but feel some satisfaction.

When speaking of the number of volcanoes discovered and described, one must keep in mind that individual cones of the so-called "linear-clustered" type are included in this number, but complex isolated volcanoes such as Ketoi Volcano or Medvezhii Caldera (Iturup), for example, with their many centers, are treated as single volcanoes. However, subsidiary cones were taken into account.

The new field of submarine volcanism in the Kuriles was discovered by the Institute of Volcanology of the Academy of Sciences of the USSR using the vessels "Vityaz'" and "Krylatka." Although earlier one could guess from the marine charts that dozens of submarine volcanoes were present here, in the 1958 report (Bezrukov

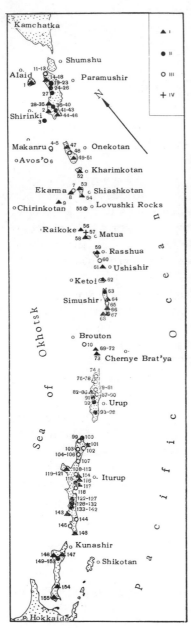

Fig. 7. Distribution map of the Holocene and principal Pleistocene volcanoes of the Kurile island arc.

I) Active volcanoes; II) Holocene volcanoes; III) Pleistocene volcanoes; IV) points of dated submarine eruptions. (Pleistocene structures in linear-clustered groups are not included individually in the list.) Volcanoes of the western zone: 1) Alaid; 2) Fuss Peak; 3) Shirinki; 4-5) volcanoes of Makanru island; 6) Avos' Volcano; 7-8) volcanoes of Ekarma (one active and one Pleistocene); 9) Chirinkotan; 10) Brouton. Volcanoes of the main zone — Paramushir: 11-13) volcanoes of the Vetrenyi group (at least three cones); 14-18) volcanoes of the Ebeko group (five Holocene cones, including one active one); 19-23) volcanoes of the Bogdanovich group; 24-26) volcanoes of the Vernadskii group; 27) Fersman Volcano; 28-35) volcanoes of the Chikurachki-Tatarinov group (eight Holocene cones, of which two are active); 36-40) volcanoes of the Lomonosov group; 41-43) volcanoes of the Arkhangel'skii-Belousov group; 44-46) volcanoes of the Karpinskii group (three Holocene volcanoes, of which two are active); Onekotan: 47) Nemo Peak; 48) Shestakov Volcano; 49-51) Tao-Rusyr group of volcanoes (one — Krenitsyn Peak — active; two Pleistocene); Kharimkotan: 52) Kharimkotan Volcano; Shiashkotan: 53) Sinarko Volcano group; 54) Kuntomintar Volcano; Lovushki Rocks: 55) Lovushki Volcano; Raikoke: 56) Raikoke Volcano; 57) 1924 submarine eruption; Matua: 58) Matua Volcano; Rasshua: 59) Rasshua Volcano; Srednii Rocks: 60) Srednii Volcano; Ushishir: 61) Ushishir Volcano; Ketoi: 62) Ketoi Volcano; Simushir: 63) Uratman Volcano; 64) Prevo Peak; 65) Ikanmikot Volcano; 66) Zavaritskii Caldera; 67) Milne Volcano; 68) Goryashchaya Sopka; Chernye Brat'ya: 69-72) volcanoes of Chirpoi (one active, two Pleistocene); 73) Brat Chirpoev Volcano; Urup: 74) Desantnyi Volcano; 75-78) volcanoes of the northern Shokal'-skii Ridge and Antipin Volcano; 79-81) volcanoes of the southern Shokal'skii Ridge; 82-86) volcanoes of the Kolokol group (five Holocene cones, of which two are active); 87-90) volcanoes of Petr Shmidt Ridge; 91) Tri Sestry Volcano; 92) Rudakov Volcano; 93-98) volcanoes of Krishtofovich Ridge; Iturup: 99) Kamui group of volcanoes; 100) Demon Volcano; 101) volcanoes of the Medvezh'ya Caldera group; 102) Tsirk Caldera; 103) Sibertovo massif; 104-106) Tornyi-Golets group of cones; 107) Vetrovoi Isthmus caldera; 108-113) Pleistocene cones of Groznyi Ridge; 114) Baranskii Volcano; 115) Teben'kov Volcano; 116) Machekh Crater; 117) Groznyi group of volcanoes; 118) Motonupuri Volcano; 119-121) Chirip Peninsula group (three Holocene structures, of which two are active); 122-127) Burevestnik group of cones; 128-132) Bogatyr group of cones; 133-142) Stokap group of cones; 143) Atsonupuri Volcano; 144) Urbich Caldera; 145) L'vinaya Past' Caldera; 146) Berutarube Volcano; Kunashir: 147) Tyatya Volcano; 148) Rurui Volcano; 149-153) Smirnov group of volcanoes (a preglacial volcano with four Holocene centers); 154) Mendeleev Volcano; 155) Golovnin Volcano.

et al., 1958) only 47 submarine cones were described, and on the later map (Zatonskii et al., 1961) 89 submarine cones were plotted in the region of the Kurile arc proper. In addition, more dozens of submarine cones were discovered in the area of the Vityaz' Ridge and on the sides of the Kurile Trench.

The total number of terrestrial and underwater volcanoes in the Kurile Islands region thus amounts to at least 250 (see the map of volcano distribution, Fig. 7).

## SUBMARINE VOLCANOES

A significant number of submarine volcanoes have been discovered within the Kurile arc proper and in the area surrounding it. The first work of the R/V "Vityaz'" and "Krylatka" in 1949-1955 gave us information on 47 submarine volcanoes which were described in some detail (Bezrukov et al., 1958). As a result of the work on the IGY program the number of submarine volcanoes more than doubled and on the final map of the submarine structure of the Kurile-Kamchatka arc 102 submarine volcanoes are plotted (Zatonskii et al., 1961). Unfortunately, no description of the submarine volcanoes discovered since 1955 has been given. The distribution of these volcanoes is indicated on Fig. 1. Undoubtedly further studies will bring to light still other submarine volcanoes, though the volcanic nature of others may not be obvious; however, the general regularity of their distribution already is rather clear.

To the east of the Greater Kurile chain, volcanic mountains are very sparse. On the 1960 map (Zatonskii et al., 1961) only four volcanoes were located on the northwestern flank of the Kurile Trench. Submarine volcanoes are absent on the Vityaz' Ridge, but in the break in the ridge east of the Central Kuriles there are eight volcanoes. In this region to the east of Simushir, on the submarine basement, there are four volcanic cones which, properly speaking, already belong to the Greater Kurile chain.

An overwhelming number of the submarine volcanoes (89) belong to the inner volcanic ridge and to its western flank. In the axial part of the inner volcanic ridge, in the passages between the islands of the Greater Kurile chain and in the immediate vicinity of the islands, there are 15-18 volcanoes.

The main mass of the volcanoes (70-75 cones, or three-fourths of all the known submarine volcanoes)* are located to the west of the main zone of the Greater Kurile Islands, on the northwestern flank of the inner ridge and partly even on the floor of the Kurile Basin of the Sea of Okhotsk. The greater part of these volcanoes are joined together into chains of volcanic massifs, though some rise as isolated cones.

Many of these volcanoes were described by Bezrukov and his co-authors (1958); thus, to avoid repetition, we will record only the main characteristics and the most interesting volcanoes and volcanic groups.

Being located along the west flank of the inner ridge (Greater Kurile chain), the submarine volcanoes are concentrated to a great degree in the zones of the very deep straits (Kruzenshtern, Boussole, and Vries). In addition, some of the submarine cones are situated even beyond the mountain structure, directly on the floor of the Kurile Basin of the Sea of Okhotsk. To the latter belongs the very high submarine volcanic massif named after S. I. Vavilov. This massif is located 30 km northwest of Brouton Island. The highest peak (of three) rises 2400 m above the floor of the Kurile Basin and reaches the −681 m mark. The slope of the upper part of the cone reaches 21-22°, but at the foot it flattens out to 12-14°.

To the north of the Vavilov massif two isolated submarine volcanoes, Obruchev and Mironov, rise from the floor of the Kurile Basin. Obruchev Volcano is located near the deepest part of the Sea of Okhotsk. It is a massif with a base 15 × 19 km and a height of up to 2000 m. The summit reaches a depth of −1227 m; its upper sides have slopes as high as 24.5°. Mironov Volcano is relatively small; its basal diameter is about 10 km, its height is 700 m above the basin floor (depth to summit, 2540 m).

A large group of volcanoes to the north of Kruzenshtern Strait also occurs on the floor of the Kurile Basin. We may mention among these the isolated massif of Edel'shtein Volcano. It is located 22 km north of Chirinkotan (which, by the way, is itself the top of a submarine volcano on the floor of the Kurile Basin). The basal diameter of Edel'shtein Volcano is 15 km; the slope of its sides exceeds 16°. The volcano's height above the basin floor is 1600 m, and its peak rises to a depth of −660 m.

---

*Not counting the submarine parasitic cones on Alaid Volcano.

Belyankin Volcano occupies an absolutely exceptional position among all the submarine volcanoes. It lies 24 km northwest
of Makanrushi and is located not only outside the mountain system
of the inner arc, but even on the far side of the Kurile Basin. The
basal diameter of Belyankin Volcano is 6 km, the slope of the flanks
is up to 21°; it rises to a height of 1500 m above the sea floor, the
depth to its summit being 552 m.

In the Kurile Islands only a single submarine eruption is recorded; it occurred on February 15, 1924 near Matua. Confusion
has arisen over fixing the position of this submarine eruption, so
we will give a literal translation of the original report (Tanakadate,
1925): "At the time of activity of Raikoke, a submarine eruption
was observed at two points in the vicinity of Banzio-iwa [now
Toporkovyi] located near Matua (155°E, 48°20'N), 12 nautical miles
from Raikoke." Here the point of the eruption is closely and definitely stated; however, the coordinates of Matua are given with a
large error or there is a misprint, because they indicate a point
in the Kurile Trench 70 nautical miles from Raikoke (instead of 12
nautical miles as in the text), and far from Matua and from the
island of Toporkovyi.

Different authors have arbitrarily "transferred" the point of
this eruption to different places — to the north of Raikoke, to the
Lovushki Rocks, and even to the First Kurile Strait. On the present
nautical charts a distinct small ridge of rather shallow depth extends northward from Matua and the islet of Toporkovyi. Apparently the two submarine volcanoes that erupted in 1924 are located
on this small ridge.

Information on the petrographic composition of the rocks of
the submarine volcanoes is lacking or at least has not been published.
In all probability the rocks of the submarine volcanoes on the ridge
crest match the rocks of the islands of the Greater Kurile chain, and
the submarine volcanoes on the west flank of the ridge, the volcanoes
of the western zone (see below).

## TERRESTRIAL VOLCANOES

The islands of the Greater Kurile chain, and along with them
the volcanoes, can be divided into two zones. The first of these,
which can be called the "Main Volcanic Zone," corresponds to the
axial part of the inner volcanic ridge. This zone, including all the
large islands and a large part of the smaller ones, extends continu-

ously from Shumshu and Paramushir on the north to Kunashir on the south and includes 16 islands.

To the west of the northern half of the main zone, six more islands (seven including Brouton) are scattered more or less randomly to make up the second or Western Volcanic Zone. The islands of this zone are located on the west flank of the inner ridge and on the flank of the Kurile Basin of the Sea of Okhotsk, though Chirinkotan and Avos' Rocks rise directly from the floor of this basin.

This group of islands is distinguished from the first one not only by its geographic position but by the characteristic petrographic and chemical features of the lavas as well, and these have led us to recognize a "Western Volcanic Zone." In it occur the islands of Alaid, Shirinki, Makanru, Ekarma, Chirinkotan, Brouton, and Avos' Rocks. On the basis of the characteristic petrographic and chemical features we will refer the isolated Fuss Peak, on the west coast of Paramushir, to this same group. The majority of the submarine volcanoes belong to the western zone.

We will begin our description with the volcanoes of the western zone, but to preserve geographic unity Fuss Peak will be described along with the volcanoes of Paramushir, and Brouton along with the volcanoes of the southern group of the Main Volcanic Zone.

The following description of the volcanoes was made in the winter of 1963/64 almost completely from the materials of my own investigations. From published work I have borrowed only descriptions of eruptions and some petrographic descriptions of rocks, particularly of Alaid, Trezubets Volcano on Urup, and Tyatya Volcano on Kunashir.

In the following two years (1964 and 1965) preliminary results of the detailed study of the volcanoes of Vernadskii Ridge on Paramushir appeared, a study that was carried out over a period of years by volcanologists of the Sakhalin Complex Institute of the Siberian Division of the Academy of Sciences of the USSR (V. N. Shilov, V. I. Fedorchenko, and others). In the main these results agreed with ours; some additions by the volcanologists of the Sakhalin Complex Institute have been recorded in the footnotes. In these same years material was published that had been gathered in the Central Kuriles in 1962 by Markhinin and Stratula; from this data were borrowed on the petrography of Brouton and Makanru, which were not visited by our expeditions.

The description of all the volcanoes was compiled anew, taking into consideration an analysis of aerial photographs. In many cases the present description disagrees with data published earlier by us or by other investigators. These differences in most cases are not specifically noted, and the new data presented here should be considered as the more accurate.

The diagrams illustrating the construction of all the volcanoes and volcanic groups were made by us. One must note that these diagrams in many cases omit some items of topography, but those details are emphasized that are important for the volcanologic relationships. Also, the scale of the diagrams is not always maintained: usually the near-crater parts are drawn on a larger scale than the areas at the base of the cone. Nevertheless, these diagrams ought to help in the description of this or that complex volcano.

## Volcanoes of the Western Zone

Seven islands are included in the western zone: Alaid, Shirinki, Makanru, Avos', Ekarma, Chirinkotan, and Brouton. Here also belongs Fuss Peak (Paramushir).* In all there are ten volcanoes in this group, of which five are Holocene (and four belong to the active category).

Alaid Island

Alaid is the northernmost and highest island of the Kurile archipelago. This unique volcano, forming an isolated island that rises from the waters of the Sea of Okhotsk 20 km northwest of the coast of Paramushir, has an oval outline with the long dimension running southeast to northwest at about 300-310°. The length of the island is 17 km, its width is 12-13 km, its area is 158 km$^2$.

Alaid Volcano is an almost perfect, strongly truncated cone whose flanks form the typical somewhat concave profile of a logarithmic curve (Fig. 8). The old summit crater, with a diameter of about 1.5 km, is badly destroyed and widely open to the south; its

---

* The two last-named volcanoes will be described in other sections.

Fig. 8. Alaid Volcano seen from Paramushir.

rim is well preserved as a semicircle only in the northern half. The
highest point on the volcano, Glavnyi Peak, 2339 m above sea level,
is situated on the northeastern part of the rim. Submarine terraces
are absent in the vicinity of the island and the cone actually rises
directly from the floor of the sea, whose depth here reaches approximately 600 m. Thus the total height of the cone is about 3000 m.

Beneath the steep cliff of the surviving part of the crater rim
there is a rather small inner scoria cone; one can regard this as a
central cone and the rim as the somma of the volcano. The basal
diameter of the central cone is about 1 km. At the peak is a closed
crater with a diameter of about 0.5 km and a height of about 100 m.
In the depression between the cliff of the somma and the sides of
the central cone is a small semicircular firn glacier. Another firn glacier descends from the inner cliff in the western part
of the somma, bending around the cone to the west. In addition, on
the floor of the crater of the central cone where, it would seem,
snow ought to accumulate, there is almost none, and on the crater's
rather gentle outer slopes snow is altogether absent. Very probably this is related to the still continuing thermal activity of the
central cone, although steam from the fumaroles is not important
there now (according to the summer studies of 1954).

The upper slopes of the volcano are covered by an unbroken
mantle of loose scoria and scoriaceous bombs. The general color
of the summit is slightly more pinkish than the mass of the red
bombs. The color of the bombs definitely indicates not only that
they were ejected in a molten state, but also that they remained so
hot for a long time that oxidation of their ferrous iron content to
the ferric state could go on. This phenomenon is also observed on
some of the other volcanoes of the Kuriles.

Alaid Volcano, gladdening the eye with the perfection of its
form and the ideal preservation of its slopes as seen from the north,
presents a different picture from the south. Here not only the rim
of the older crater, but also the upper slopes of the volcano are in
ruins. A wide, deep depression extends from the summit to the
south-southeast at an azimuth of about 170°. In the cliffs of this
depression a section of lavas and pyroclastic products typical of
stratovolcanoes is seen. In all probability this structure developed,
along with the old summit crater, as the result of a violent directed
explosion. The weakening of this part of the volcano by tectonic
stresses apparently also played a role. This is suggested by a scarp

at the head of Krivaya Creek which extends to the north-northwest the depression referred to and which resembles a tectonic scarp. The width of this depression amounts to 1.5-2 km, and its depth in the upper part is as much as 300 m. Products of the central cone's eruption fill the depression; downslope the height of its walls gradually decreases and the depression melts into the slope; the ejecta of the central cone (probably together with material from the directed explosion) here form a fan-shaped train that runs down to the coast.

The lower, more gentle parts of the flanks of the volcano were overgrown with high, thick brush, and are cut through by a system of radiating gullies. For the most part the coast of the island falls to the sea by steep cliffs (up to 250 m high in the south) in which the layered structure of the volcano, with a considerable dominance of ejecta over lava, is also well shown.

The base of the cone, like its sides, is covered with loose pyroclastic material. The lavas of the main structure are exposed only in the southeastern part of the island, in the area of Capes Lava and Siandrion. In addition, lavas "show through" the blanket of soil and pyroclastics in the eastern part of the island in the vicinity of Cape Serdityi.

Near the base of the cone and on its lower slopes are rather numerous, small parasitic craters with scoria cones (Fig. 9). One can count 32 such cones (actually, their number is probably larger), but the majority of them are badly destroyed. Traces of craters are more or less clearly preserved only on 11 cones. Most cones are entirely of scoria, but lava has flowed from some. Rather commonly in this case the crater is horseshoe-shaped, and the lava flowed through the open part. Highest of all is the subsidiary Parazit Cone, which rises to 1023 m above sea level on the southwest side of the cone. A narrow stream of lava runs down from it to the coast, forming Cape Podgornyi. A chain of three craters is well preserved near the northeastern base of the cone. The whole coast from Cape Pravyi to Cape Khitryi, over a distance of about 3 km, is flooded by the lavas from these cones. A very large and well preserved cone, Osobaya Peak, lies in the southeast quadrant of the island. This cone has a basal diameter of about 600 m and a crater diameter of about 150 m. Lavas from this cone extend to the sea coast and form Cape Siandrion. To the southeast of Osobaya Cone

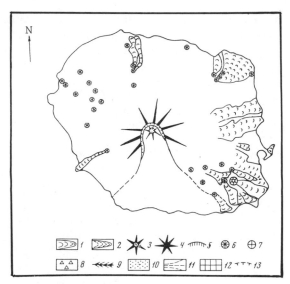

Fig. 9. Sketch of the structure of Alaid Volcano: 1) young lava flows; 2) older lava flows; 3) cone with crater; 4) dome; 5) cliffs; 6) parasitic cones; 7) fumarole fields; 8) moraine; 9) mountain ridge; 10) pyroclastic deposits; 11) slopes; 12) basement; 13) edge of glaciated (U-shaped) valley.

there are two more small horseshoe-like craters that give rise to small lava flows. Many cones stretch out in radial chains of three to five craters.

It is interesting to note that two-thirds of all the subsidiary craters are confined to a zone lying along the long axis of the island. To this zone also belong two submarine craters. One of these is on the submarine flank of the cone to the northwest of the coast; its summit lies at a depth of about 60 m below sea level. In all probability the zone in question marks a tectonic line perpendicular to the general trend of the arc.

Judging from the existing data, the rocks of Alaid Volcano are rather uniform and belong to the augite-olivine andesite-basalts and basalts. Kuno (1935), who studied the lavas of Alaid, called all the rocks basalts, but when the chemical analyses are recalculated by the Zavaritskii method, the $b$ parameter is approximately 20, which corresponds to a typical andesite-basalt. As the chemical analyses indicate, the rocks have a somewhat higher than usual alkalinity.

The analyzed rocks from the summit of the volcano (Suzuki
and Sasa, 1933) are the most leucocratic and judging from the results of recalculation, they are andesites. The lavas of the subsidiary Otdel'nyi Cone are the most basic and are olivine basalts.

The presence of olivine as phenocrysts and microlites, along
with the absence of hypersthene, distinguishes the lavas of Alaid
from those of the other volcanoes of the Kurile Islands.

At present, the summit crater of Alaid is completely dormant,
but in more remote times it rather often displayed violent activity.
In the 1730's and 1740's the volcano showed fumarolic activity, then
for a long time it was quiet. Beginning in 1790 it started to emit,
at varying intervals, warm "vapors" with ash, and in February 1793
it was in a state of violent eruption. Ash "like coarse gun-powder"
fell in a layer 10 cm thick in southern Kamchatka at a distance of
120 km from Alaid. Strong eruptions also occurred in June 1854
and in July 1860. The last eruption from the summit crater occurred in 1894.

In various summaries dates of eruptions of this volcano are
given as 1770, 1789, 1821, 1828, 1829, 1843, 1848, and 1858. However, they are all wrong. The dates of the eruptions of the parasitic craters are not known. The development of only one crater,
Taketomi, in 1933-1934, has been observed. This eruption and its
products were studied in detail by Japanese volcanologists.

The eruption was preceded by strong shocks that began on
October 20, 1933 and that were felt on Paramushir and Shumshu at
a distance of more than 50 km from the volcano. On November 17,
1933 the inhabitants of Shumshu, 50 km away, saw a colossal column
of black "smoke" near the east coast of Alaid. This was the beginning of the submarine eruption. On January 14, 1934 a small volcanic islet about 20 m high was noted. On January 26 investigators
reached the site of the eruption by ship. It was found that the new
islet occurred about half a mile (nautical) from the east shore of
Alaid, where the depth earlier had been 20-50 m. The diameter of
the islet was about 200 m and its height, 50 m. At its summit was
a horseshoe-shaped crater open to the northeast. For an hour or
two at a time strong Strombolian eruptions occurred which ejected
clouds of ash to heights of up to 3 km. In April the height of the
new cone reached 130 m above sea level. A lava flow poured through
the lower, northeastern part of the crater; where the lava reached

the sea, there rose a thick, white cloud of steam. In June a new lava bocca was formed near the base of the cone; from it a lava flow was emitted to form a flat plateau 200 × 250 m wide and 10-15 m above sea level. At this time the crater was filled with blocky lava, and a small flow of it poured out onto the northeast flank of the cone and spread onto the surface of the plateau. At this time the cone had a diameter of 800 m and a height of 145 m. The summit crater, still horseshoe-shaped, had a diameter of 300 m. In August 1934 the eruption stopped and the cone of the new parasitic crater, composed of ash and scoria, began to be destroyed rapidly on the north and east by the sea. The loose material was redeposited near the western and southern shores of the islet, forming two sand bars. In the winter of 1935/36 the eastern bar reached the island of Alaid, converting Taketomi into a peninsula. At the end of 1946 Taketomi was joined to the main island by the second bar. A salt lake was left between the bars, but in 1959 the southern bar was breached by the sea and a small bay was formed.

The lavas and scoria of Taketomi are completely similar to the lavas of the main volcano in mineralogic and chemical composition.

## Shirinki Island

Shirinki Volcano forms an isolated island that rises from the waters of the Sea of Okhotsk 15 km west of Fuss Peak. The island is almost perfectly circular in outline, with a diameter of 3 km. The strongly truncated cone of the volcano rises 761 m above sea level. The summit crater, approximately 750 m in diameter, is surrounded by a crown of cliffs and is broadly open to the south. Approximately in the center of the crater is a broad mound with cliffy exposures; apparently this is the remains of an extrusion dome. In the southern part of the crater lies a small dome (about 450 m across) from which a small tongue of lava extends down along the side. In some places along the sides of the cone flows of lava run down from the crater and reach the coast. Most of the flanks are covered by pumiceous pyroclastics. In the eastern part of the island are found the remains of an older structure with monoclinal dips of the rocks to the east* (Figs. 10, 11).

---

* Judging from aerial photographs, the remains of the eastern structure, contrary to the published data (Markhinin and Stratula, 1965), have no traces of a crater.

Fig. 10.  Shirinki Volcano

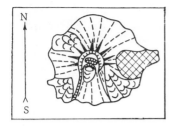

Fig. 11. Sketch of the structure
of Shirinki Volcano (symbols as
in Fig. 9).

Somewhat alkaline hornblende andesites are dominant among the young rocks. In the andesite pumice from the submarine slopes of the volcano, normal green hornblende predominates as phenocrysts. Labradorite, $An_{50-69}$, is of subordinate importance. Sometimes the plagioclase is zoned, with cores of $An_{58-66}$ and outer zones of $An_{49-55}$. The groundmass is hyaline. In the hornblende andesites from the subaerial lava flows labradorite, $An_{50-52}$, predominates; then comes hornblende, mostly brown; in addition, there are traces of augite. The groundmass has a hyalopilitic structure. At the base of the cone hornblende-free augite andesites are found.

The age of the present edifice is post-glacial; the summit dome is certainly very young, even historic, though the direct information on eruptions of Shirinki are lacking.

The presence around the island of a 140-meter submarine terrace and the presence in the eastern part of fragments of a more ancient structure allow us to say that the present cone rose on the remains of an older, preglacial volcano.

Makanru Island

The island of Makanru is situated in the Sea of Okhotsk, 30 km northwest of the north shore of Onekotan, from which it is separated by the Fifth Kurile (or Evreinov) Strait. The island is elongate in a north-south direction; its dimensions are $6 \times 9$ km, and its area is 49 km$^2$.

Makanru has been subjected to twofold glaciation; the southern and lower part of the island is composed of a moraine, apparently of the second glaciation, that issues from a U-shaped valley. In all probability, volcanic activity had already ended here in preglacial times. Thus distinct volcanic forms are not preserved. All the slopes are strongly eroded by ice and water, and a small portion of the original slope is preserved only at one place. Thus it is very difficult to judge the original form of the volcano. Nevertheless, an eruptive center is located at the point indicated by the

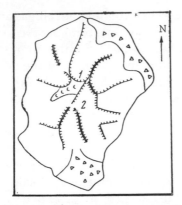

Fig. 12. Sketch of the structure of
Makanru Volcano (symbols as in
Fig. 9).

number 1 on the map (Fig. 12).
One other center very probably
occurs near point 2. We do not
exclude the possibility that activity here had a "line of clusters"
character* and that several closely
spaced centers existed.

The rocks of Makanru are
dominantly basalts and andesite-
basalts. Among the phenocrysts
labradorite ($An_{60-70}$) and bytownite
($An_{70-89}$) predominate; augite is
present in very marked amounts,
but hypersthene is less common.
Olivine is always present, but in
small amounts. The structure of the groundmass is hyalopilitic or,
less commonly, intersertal. In the andesites of Cape Poludennyi
hornblende is present.

Avos' Rocks

At a distance of about 20 km southwest of Makanru lies the
cone-like rock of Avos' (35 m above sea level). Another rock to
the southeast and two more to the northeast, joined together and to
the first one by a subsurface reef, form a semicircle. Apparently
these are the remains of an old destroyed crater. Judging from the
presence of a rather large 140-meter submarine terrace, Avos'
Volcano is preglacial, and no activity has occurred in later times.

The volcano rises from the floor of the trench that marks the
northern end of the Kurile Basin of the Sea of Okhotsk.

Ekarma Island

Ekarma island is situated 8.5 km northwest of Shiashkotan.
In plan the island has an oval form, its long axis east-west, whose
dimensions are $5 \times 7.5$ km; its area is 30 km$^2$. The island consists
of two volcanoes united at their bases. The eastern volcano shows
evidence of intense erosion, in particular of glaciation. It has the
form of a broad ridge up to 800 m high, at whose summit there is

─────────

*Cf. the description of Paramushir island below.

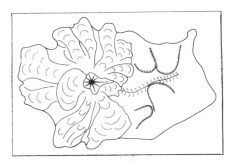

Fig. 13. Sketch of the structure of Ekarma
Volcano (symbols as in Fig. 9).

Fig. 14. Sketch of the
structure of Chirinkotan
Volcano (symbols as in
Fig. 9).

an area of bleached rock; apparently this is the region of the former crater, where the rocks were altered by postvolcanic activity. This volcano has shown no activity in postglacial time.

Ekarma Volcano, which forms the western part of the island, is young (postglacial) with a fresh volcanogenic topographic form. The difference in age of the two volcanic structures is well emphasized by the disposition of the 140-meter submarine terrace. The eastern part of the island is tied to the northern part of Shiashkotan by the terrace, but in the western part of Ekarma the terrace is absent.

Ekarma Volcano forms a slightly truncated cone 1171 m high. In all directions from the summit run off lava flows up to 3 km long. The summit crater is sealed by an extrusion dome that completely overlaps the rim of the crater and gives the summit its pointed form.

The shoreline of the volcano locally has a many-lobed form: the descending "tongues" are composed of lava flows, with pyroclastic materials occurring in the embayments between them (Fig. 13).

The lavas of Ekarma Volcano are two-pyroxene andesites and andesite-basalts with phenocrysts of labradorite ($An_{50-65}$), augite, and hypersthene. The structure of the groundmass is microlitic and hyalopilitic. The upper dome and the later viscous lava flows are composed of more acid two-pyroxene andesites with important amounts of hornblende among the phenocrysts.

An eruption of Ekarma is known to have occurred in 1767–1769; apparently the upper dome was produced at the time of this eruption. In the first half of the nineteenth century the volcano displayed fumarolic activity; at present it is quiescent, but the possibility of its catastrophic awakening is not excluded. Near the north slope of the cone warm mineral springs occur.

## Chirinkotan Island

The island of Chirinkotan lies 29 km west of Ekarma. The island is a single volcano, having in plan view a circular form 2.5-3 km across. Vegetation is almost completely absent.

Although Chirinkotan Volcano is of small dimensions (its area is 6 km$^2$), its structure is rather complex (Fig. 14). The structure rising above the water is essentially only the summit of the volcano. No submarine terrace is present near the island, which points to a Holocene age for the structure. The submarine slopes, perhaps complicated by parasitic craters, run all the way down beneath the waters of the Sea of Okhotsk to depths of almost 2500 m. The actual height of the cone comes to about 3000 m. Chirinkotan Volcano actually rises from the floor of the Kurile Basin.

At the summit of the volcano, on the west, traces are seen of a large explosion crater about 1 km across. In this crater rise the remains of an inner cone (or possibly, a dome) that at one time filled or almost filled the crater. Subsequently an important part of the inner cone was blown away, and the summit is presently occupied by the large amphitheater of the crater, open to the south and about 0.8 km in diameter. The floor of this crater is more or less flat; it is cut up by the small valleys of mud flows, and locally there are small mineralized ponds. Along the edge of the crater, near the wall of the amphitheater, the vents of strong fumaroles are noted.

Near the north foot of the inner cone, approximately on the rim of the old crater, there is a secondary cone which produced lava flows which flowed down as far as the north shore of the island. Fresh lava flows have also poured down the southeast flank of the volcano.

Effusion of just these flows was observed in the 1880's by Captain Snow. At that time the inner cone was still whole. According to the descriptions by Snow, "this is a double volcanic cone, the outer one of which is destroyed on its southeast side. Smoke issues from the crater, and sometimes lava issues through a breach in the crater and flows down the side of the mountain to the sea." Snow sailed in the Kuriles between 1878 and 1889, but unfortunately he gives no dates for the eruptions.

Thick lava flows also are exposed in the coastal cliffs of the volcano and are composed of basic two-pyroxene andesites and andesite-basalts, almost all of which contain small amounts of hornblende. Farther up on the flanks more acid pyroxene-hornblende andesites with clinopyroxene and orthopyroxene are found. The groundmass structure is microlitic.

Pumice-like pyroclastics dredged from the submarine slopes of the cone are of hornblende andesite.

On the floor of the present crater Bogoyavlenskaya in 1961 found a small fresh scoria cone with a lava flow. This cone probably was formed in 1955, because we did not see it in 1954. The lava flow is composed of pyroxene-hornblende andesite, with phenocrysts of labradorite ($An_{50-60}$), augite, and hornblende. The groundmass has a hyalopilitic structure.

Chirinkotan Volcano displayed active fumaroles and possibly explosive activity in the eighteenth century. The strong explosive eruption which formed the present crater has not been recorded. Most probably it occurred at the end of the nineteenth or at the very beginning of the twentieth century. In later years the volcano produced no fumarolic activity that could be noticed from afar. Even in 1953 we did not notice any fumaroles while passing the island on a ship. However, in 1954 strong fumarolic activity resumed: fumarolic steam was seen at distances of up to 50 km. In the fall of 1955 we received a short telegram about very strong gas activity of the volcano. Unfortunately, we could not make out what gave rise to the message: whether it was an eruption or a distinct increase in fumarolic activity. As noted above, apparently the new scoria cone, which produced a lava flow, was built in the crater at this time. During later visits to the volcano in 1961-1963, the usual gas activity was observed.

All the islands of the western zone are relatively small. The largest of them, Alaid, is the ninth largest of the Kurile Islands. Outcrops of the Tertiary basement are never found on these islands.

Moreover, two of the islands (Alaid and Chirinkotan) were formed only in Holocene times; Ekarma Volcano was formed at this same time. Of all the islands of the western zone, only Avos' and Brouton Volcanoes have displayed no activity during the Holocene. A distinctive feature of the lavas is the common presence of hornblende.

## Volcanoes of the Main Volcanic Zone

### North Kurile Islands

Six islands occur in the northern group: Shumshu, Paramushir, Onekotan, Kharimkotan, Shiashkotan, and Lovushki Rocks. Of these, Shumshu is the only one among all the islands of the Greater Kurile chain on which there are neither active nor extinct volcanoes. On all the other islands there are volcanoes or their remnants. In this whole group as many as 60 volcanoes can be counted, of which more than half (34) are of Holocene age. At present nine volcanoes are active.*

The first island of the Greater Kurile chain, Shumshu, is separated from Kamchatka by the First Kurile Strait, whose width is all of 11 km, but whose depth does not reach 40 m. The Second Kurile Strait, separating the islands of Shumshu and Paramushir, is scarcely 2 km wide, with a depth of up to 50 m. Even in late Würm time both of these islands were joined to Kamchatka. Actually, Shumshu is a torn-off piece of Paramushir.

### Paramushir Island

The island of Paramushir is the largest of the North Kurile Islands and the second in area, after Iturup, in the whole Kurile chain. The island is elongated in the direction of trend of the arc, with a length of 100 km and an average width of 20 km. The area of the island is 2042 $km^2$.

The island is composed of Neogene, dominantly sedimentary-volcanogenic rocks, which are thrown into gentle folds usually trending parallel to the trend of the arc. A fauna of middle Miocene to Pliocene age is found here. The underlying unfossiliferous series is presumably lower Miocene and upper Paleogene. Some outcrops of granitoid intrusions are of presumably middle Miocene age.

---

*In all instances, Fuss Peak has not been included in these figures.

Quaternary volcanic rocks and modern volcanoes are concentrated in the northern and southern parts of the island, forming
the northern part of Vernadskii and Karpinskii Ridges. On Paramushir, as on the other large islands of the Kurile chain (Urup and
Iturup), an important role is played by eruptions similar to the
fissure type; on Paramushir these eruptions played a dominant role.
Of all the volcanoes on this island, only two are structures of the
central type, Fersman and Fuss Peak Volcanoes.

The old $(Q_1-Q_2)$ lavas were erupted onto a slightly leveled
surface of upper Tertiary deposits from closely spaced centers
that were restricted to the near-summit parts of the ridges. As a
result of this activity broad lava plateaus were formed that sloped
more or less gently in all directions from the ridges. The first
glaciation, which was very strong on Paramushir, to a marked degree destroyed the volcanic plateaus and the volcanic centers themselves; only isolated patches of the margins of the plateaus and
"ribs" of the near-summit parts remained. In most cases the processes of erosion went so far that now, without special studies, it
would be almost impossible to recognize the centers of early
Quaternary volcanic activity. In some cases the distribution of
the preglacial centers is represented by the remnants of necks or
by rocks of the vent facies. Sometimes the periclinal attitude of the
volcanic rocks is recognized, and so on.

The centers and flows of interglacial times are also often destroyed (by the second glaciation), but in many cases the surfaces
of the flows have preserved their general morphologic outlines, and
the loci of almost all the eruptive centers can be determined. Postglacial centers, flows, and pyroclastic deposits are generally very
well preserved and are only slightly affected by erosion.

On the strength of what has been said, we will take up the description of the volcanoes of Vernadskii and Karpinskii Ridges in
"reverse geologic" order, from postglacial to preglacial.

Vernadskii Ridge − Modern Volcanoes. The
modern volcanic centers of Vernadskii Ridge have a "linear-clustered" disposition. The separate eruptive centers are rather closely
grouped in "clusters" located at some distance from one another
along the trend of the ridge. There are three groups of modern
volcanic centers; from north to south these are the Ebeko, Bogdanovich, and Vernadskii groups.

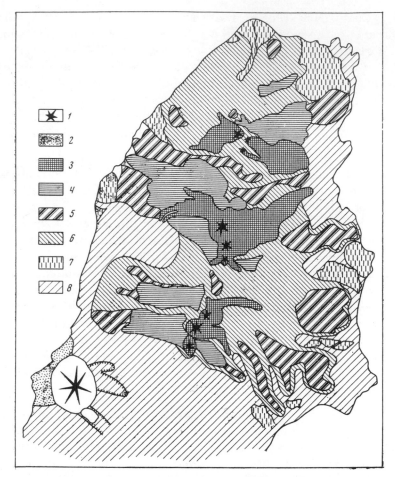

Fig. 15. Map showing the distribution of the volcanoes of Vernadskii Ridge:
1) volcanic edifices; 2) pyroclastic deposits of Fersman Volcano; 3) Holocene
lavas; 4) interglacial lavas; 5) areas of preglacial lavas preserving their
original topograph; 6) eroded portions of Quaternary structures; 7) moraines;
8) basement and marine terraces.

**I.** In the Ebeko group (Fig. 16) there are three well defined
and two already strongly eroded cones.

**1.** Ebeko Volcano is a cone 200-220 m high (absolute height,
1037 m); it is strongly elongate in an east-west direction, with three
adjoining craters at the summit that stretch like a chain from north
to south. The dimensions of all three craters are approximately the

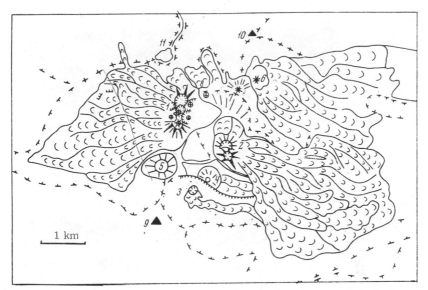

Fig. 16. Sketch of the structure of the Ebeko group of volcanoes: 1) Ebeko Cone;
2) Neozhidannyi Cone; 3) Nezametnyi Cone; 4) explosion crater; 5) southern cone;
6-8) interglacial eruptive centers; 9, 10) preglacial centers; 11) Zelenaya Mountain (symbols as in Fig. 9).

same: the rim diameters are about 300–350 m, the floor diameters
about 200 m. The aggregate dimensions of the summit craters are
350 × 800 m. On the eastern and western sides there are two open
amphitheaters that probably are lateral explosion craters, strongly
broken down by subsequent erosion; a similar crater of smaller
size and containing a small lake lies at the north slope.

The floor of the southern summit crater is uneven; its depth
reaches 70 m. A strong group of solfataras that are depositing sulfur occurs in the eastern part of the crater, and a large boiling
spring that (in 1952) was throwing a column of hot (93°C) water up
along with the gases lies in the center of the crater. On the north
this crater is partly cut off by the next, middle crater, which is
filled by a lake whose level lies 40–50 m below the floor of the first
crater. In the western half of the crater, on the shore and from
the floor of the lake, numerous strong, sulfur-depositing solfataras
occur. Here there are many steaming pits, niches, and tunnels in
which liquid sulfur boils and bubbles. Prior to the 1963 eruption,
the water of the crater lake had a beautiful turquoise color with a

Fig. 17.   Crater of Ebeko Volcano; photo by N. K. Klassov.

milky appearance. After the eruption the water became muddy from the masses of ash that fell into it. The fumarolic gases rising from the floor of the lake give the illusion that the water is boiling, although in 1953 the temperature of the water in the hottest areas did not exceed 62°C, and the average temperature of the lake at that time was 30-35°C. In 1956 the temperature of the water began to fall gradually, and in 1959 it stood at about 20°C; at the same time the level of the lake fell. The depth of the lake* is about 20 m. The last crater to the north is in close contact with the middle one, so that the barrier between them is almost nonexistent. The floor of this crater slopes gradually to the north, and here near the north wall of the crater lies a small cold lake with the shape of a half moon. (Fig. 17).

Solfataras and boiling springs occur on the outer slopes of the cone and on the floors of the lateral explosion craters. The maximum temperature of the solfataric gases near their vents in 1953 was 144°C, but temperatures of 98-120°C are most common. Steep sulfur cones or hollow pipes are usually formed at the solfatara vents; gases escape under great pressure from their tops, and molten sulfur boils within them. The outer parts of the sulfur "pipes" consist of common, bright yellow, rhombic sulfur; the inner part of the pipes in the higher temperature solfataras has an orange color and apparently consists of the monoclinic form. Rivulets of molten sulfur rather commonly run out of the solfataras.

Carbon dioxide (up to 92% of the gas other than water vapor) predominates in the make-up of the fumarolic gases; $H_2S$, $SO_2$, and HCl are present in large amounts. The water of the crater lake has a strongly acid reaction (pH = 3), and it contains a significant amount of Cl' and $SO_4$" ions.

Numerous flows of blocky lava run down to the west from the foot of Ebeko Cone. These flows have completely flooded the cirque and upper U-shaped valley of Gorshkov River, extending 3.5 km from the cone. The area here flooded by lavas has a crudely triangular form, with a base about 3 km long and a height of 3-3.5 km. One of the lava flows spilled over into the U-shaped valley of Yur'ev River. The lava outpourings took place mainly from boccas at the

---

*In the fall of 1965 the lake level fell suddenly and the floor of the crater was exposed, then the level gradually rose again.

base of the cone, and effusions over the lip of the crater have occurred only from the northern crater.

These lavas have the composition of pyroxene andesite-basalt
or, more rarely, andesite. Phenocrysts of labradorite, $An_{54-61}$, and
of pyroxene (diopsidic augite) are dominant; hypersthene, sometimes
as relics in the augite, is found in smaller amounts, and olivine is
even rarer. The structure of the rock is seriate-porphyritic with
a hyalopilitic (to locally microlitic) groundmass. Plagioclase
(usually acid labradorite, $An_{50-55}$) and augite predominate among
the microlites; orthopyroxene is more commonly found as microphenocrysts and larger microlites. In some andesites the groundmass is a microlite-free glass.

Ebeko is the only active volcano on Vernadskii Ridge. An
eruption is known in 1793. In September 1859, during an eruption
of Ebeko, dense sulfurous steam covered the adjoining island of
Shumshu, causing nausea and headaches among the inhabitants. The
last strong eruption occurred in 1934-1935. In September 1934
earthquakes began to be felt on Shumshu, and on October 5 a dark
column of ash first rose above the crater, which earlier was emit-
ting only white steam. On October 12 ash formation intensified considerably and fallout of ash was noticed on Shumshu. On October 17
clouds of ash with sulfur dioxide affected the whole island of Shumshu. On December 28 a very strong explosion was noted at the beginning of the eruption. From June to August 1935 convoluted clouds
of gas and ash rose to a height of 1.5 km above the crater. Liquid
sulfur poured from fissures on the side of the cone. Explosions occurred from an east-west fissure on the floor of the northern crater,
which was dry during the eruption. Lava flows were not emitted,
but bombs of the "breadcrust" type were ejected in large amounts.
These bombs are composed of gray, porous, two-pyroxene andesite.
The structure of the rocks is seriate-porphyritic, with a hyalopilitic
groundmass. Andesine-labradorite, $An_{45-52}$, predominates among
the phenocrysts; diopsidic augite and hypersthene are present in
smaller amounts. Andesine, $An_{45}$, and orthopyroxene predominate
in the microlites, but clinopyroxene is also present. The eruption,
which belongs to the Vulcanian type, ended in the fall of 1935.

In March 1963 an explosion occurred on the north wall of the
amphitheater cut into the east side of the cone and a small crater,
2-3 m across, was formed. From this crater gases carrying a
small amount of ash were emitted with great force. In the summer

of 1963 the intensity of the gas jet weakened, and in July gases were issuing quietly from the new center.

2.  Neozhidannyi Cone is located 1.5 km southwest of Ebeko. This cone has a relative height of about 100 m (about 1070 m absolute).  On the summit there is a crater about 200 m across and 25-30 m deep; the floor of the crater is occupied by a rather shallow fresh-water lake fed by melting snow; the lake dries up by the end of the summer.  The eastern and southeastern sectors of the base are a plateau of blocky lava; numerous lava flows reach downward for almost 5 km, descending to about the 300-meter level.  In their upper part these flows fill the U-shaped valleys of the second glaciation, but lower down they broaden to about 3 km and fill the whole area between the Kuz'minkaya and Matrosskaya Rivers.

Lavas from the top of the cone resemble in composition and groundmass structure the andesite bombs of Ebeko; they are distinguished by their high density and by the more basic composition of their plagioclase, which is labradorite, $An_{50-68}$.

Near the north foot of Neozhidannyi Cone are the remains of an old, lower cone that now is almost completely buried beneath the rubble of the lavas of Neozhidannyi Cone.  Only the north wall of the cone and the adjoining portion of the crater remain uncovered.  In its time this cone also produced numerous lava flows, which now can be distinguished from the flows of Neozhidannyi proper only with difficulty.  On the whole, Neozhidannyi Volcano is to a lesser degree a two-stage structure.  The activity of both cones was characterized by the dominance of lava eruptions; explosive activity was of subordinate importance.

3.  Just to the south of Neozhidannyi Cone lie the remains of still another badly ruined cone, still older than Neozhidannyi Cone. The possibility exists that this center had already begun its activity  at the end of the second glaciation.  However, it is certain that a large eruption occurred in postglacial times and left a wide (about 5 km across) crater.  At present, this crater is destroyed on the south and has the form of a wide funnel open to the south.  During the final explosion a large quantity of "breadcrust" bombs was ejected.

4.  Still farther south, on the floor of a cirque of the second glaciation at the source of the Snezhnaya River, there is a

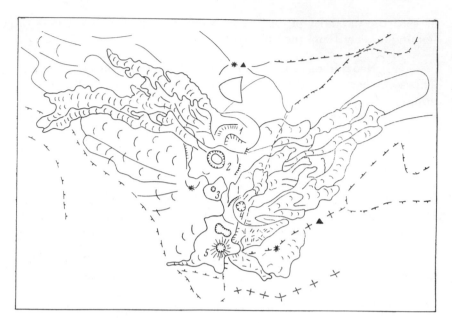

Fig. 18. Distribution map of the volcanoes of the Bogdanovich group:   1) Mt. Krasnukh;
2) Bogdanovich Volcano; 3) Ploskii Crater; 4) Krasheninnikov Crater; 5) Kozyrevskii
Crater (symbols as in Fig. 9).

saucer-shaped crater* which produced a single lava flow that runs
down along the U-shaped valley for about 2 km, at which point it
runs together with the lavas of Neozhidannyi.

   5. The remains of another early postglacial structures lie
south of Ebeko Cone on the ridge between the cirques of the Snezhnaya and Gorshkov Rivers.  This structure is substantially a pyroclastic cone; the remains of a crater are present on it.

   II.  The second group of modern volcanoes, the Bogdanovich
group, occurs at a distance of 5-6 km south of the volcanoes of the
Ebeko group.  There are five eruptive centers in the Bogdanovich
group (Fig. 18).

   1.  Farthest north in this group is Mt. Krasnukh, a rather
badly destroyed volcano that has already lost its crater.  The south-

---

*I suggested the name "Vlasov Crater" for it; the volcanologists of the Sakhalin Com-
   plex Institute have named it Nezametnyi Cone.  The latter name is intrenched in the
   literature.

eastern part of this structure is cut open by a large cirque in whose wall rocks altered by solfataric activity are exposed. Probably this cirque is the remnant of an explosion crater. Two lava flows run down to the northwest from the sides of this crater. One of these extends for about 4 km over lavas of interglacial age, and the other reaches a length of almost 6 km; a tongue of it runs down the large U-shaped valley of the Burnaya River.

2.  Bogdanovich Crater adjoins Mt. Krasnukh on the south. It is a small, very low, and very gently sloping cone with a basal diameter of about 1 km. On the west it is bordered by a bench of interglacial lavas and on the east by the cliff of the broad trough of the Nasedkin River. At the top of the cone is a wide (up to 500 m) crater about 40 m deep. The floor of the crater is covered by a fresh-water lake; along the east shore its depth is not very great (thus it is called Malovodnoe ["shallow"] Lake), but in the west its floor slopes rather steeply. In the west wall of the crater sheets of lava are exposed. A narrow stream of lava, about 6 km long, extends to the northwest from the cone. This flow parallels one of the flows from Krasnukh Volcano almost throughout its length, and at the end of its course it also descends into the valley of the Burnaya River.

Lava from the area of Bogdanovich Crater is two-pyroxene andesite containing dominantly labradorite ($An_{52-62}$) phenocrysts. Diopsidic augite, some with polysynthetic twinning, and hypersthene are found in somewhat smaller amounts. The last-named mineral is more often found as relics surrounded by a rim of augite. The structure of the rock is seriate-porphyritic, with microlites of andesine and clinopyroxene. Cristobalite and possibly tridymite occur in cavities.

3.  Farther south, side by side with Bogdanovich, lies a small cone with a miniature, swampy crater.* This cone probably is purely pyroclastic.

4.  About 1 km south of Bogdanovich Crater is Krasheninnikov Crater. It consists of two nested cones separated by a small depression. The outer cone is badly destroyed; only its western portion is preserved, along with the heads of lava flows that run to the north. The inner cone has a very fresh appearance. On its summit

---

* The volcanologists of the Sakhalin Complex Institute have called it Ploskii Crater.

is a crater 250-300 m across and about 50 m deep. In the western part of the crater is a deep explosion crater. The eastern part of the crater rim has a deep gash cut into it, and from here flows of blocky lava that overlap the older flows of the outer cone descend to the northeast, toward the Pacific Ocean. The other flows cover the whole floor of the large U-shaped valley of the Nasedkin River. Here, apparently, there are still older flows from Bogdanovich Crater or from Krasnukh Cone. The total width of the complex of lava flows in the Nasedkin River valley is 2-3 km; its length is 8 km. The flows overlap the lowland moraines, but do not extend the remaining approximately 2 km to the sea coast.

The lava flow close to the crater is an andesite with clearly dominant phenocrysts of andesine and andesine-labradorite, $An_{43-53}$. In addition, there are phenocrysts of augite and more rarely of hypersthene. The groundmass is vitrophyric.

5. A rather low saddle separates Krasheninnikov Crater from Kozyrevskii Cone. The latter is a rather large, filled crater with a relative height of 100-150 m (1160 m absolute). The crater of the volcano has the form of a completely closed funnel about 100 m deep, with a flat, dry floor. Near the northern base of the cone are two conjoined lateral explosion craters that are elongate in an east-west direction. The western crater is filled with a fresh-water lake. On the eastern edge of the second crater a gigantic bomb with a volume of over 100 $m^3$ and a "breadcrust" surface strikes the eye. A small ribbon of lava descends westward from the foot of the cone; about 1 km downstream it reaches the steep wall of the U-shaped valley at the source of the Burnaya River. The east foot of the cone drops off steeply to the head of the U-shaped valleys of the Severyanka and Ptich'ya Rivers (the headwaters of the Nasedkin River). On the steep side of the mountain east of the explosion craters is a lava bocca from which a small flow ran down along the valley of the Pitch'ya River. A second flow (or perhaps it is a mound that looks like a flow) descends from beneath the cliff at the head of the Severyanka River. This cone, including the parts near and within the crater, is composed exclusively of pyroclastic material, in the make-up of which angular blocks predominate. Most probably not a single lava flow has issued from the summit crater, and its activity was purely explosive.

The lavas that make up the gigantic bomb, the small blocks, and the eastern lava flow were studied under the microscope. All are very similar two-pyroxene andesites. Labradorite, $An_{52-65}$, predominates among the phenocrysts; pyroxene phenocrysts are dominantly augite, which is sometimes twinned, but hypersthene is also common. Rarely, crystals of olivine are found. The groundmass is glassy, with rare microlites of andesine and andesine-labradorite, $An_{40-55}$. Orthopyroxene is found in still smaller amounts, and augite is rarer still. The lava flow contains relic grains of hypersthene with augite rims.

III. The third and last group of modern volcanoes on the Vernadskii Ridge lies 4-5 km south of the preceding group and consists of two cones, Vernadskii and Bilibin, and a small scoria pile. This group was studied only on aerial photographs.

1. The cone of Vernadskii Crater is bounded on the south and west by benches of interglacial lava. It is about 1 km in diameter and about 150 m high. A small flow of andesitic lava extends eastward from the base of the cone and then turns 90° to the south, running for 2.5 km down the glacial trough of the left fork of the Zaozernaya River. The remains of a small crater about 400 m across are preserved at the top of the cone. The obelisk-like remains of an extrusion dome rise above the northwest part of the crater. The northwest part of the cone consists of the agglomerate mantle of this dome, which was formed in the final stage of its activity. As a result, the cone is somewhat elongate in an east-west direction. According to oral information from Vlasov, Vernadskii Cone shows evidence of weak fumarolic activity.*

2. Bilibin Cone adjoins Vernadskii Cone on the northeast and occurs at the very head of the left fork of the U-shaped valley of the Zaozernaya River. A crater about 250 m across is located on the summit of the cone. The north wall of the crater is steep; from here a rather wide ribbon of lava descends, enters the adjacent U-shaped valley of the Levashov River, and runs down along it for almost 4 km.

3. A small pile of scoria without a crater adjoins Vernadskii Crater on the south. It is somewhat elongate parallel to the trend of the ridge.† To the south the scoria rests directly on the eroded Tertiary basement.

* The volcanologists of the Sakhalin Complex Institute did not note any such present-day activity.

† The volcanologists of the Sakhalin Complex Institute called this hill Lineinyi Crater; according to their data, the scoria is basaltic in composition.

Thus, there are 10-12 postglacial eruptive centers in Vernadskii Ridge. Most of these centers were single-stage structures
and were not active for very long, although activity occurred to a
small degree in two stages in some of them (for example, in Neozhidannyi Cone and Krasheninnikov Crater). The northernmost
structure, Ebeko Volcano, from which several eruptions are known,
is more complex in structure and in the character of its activity.
Apparently Bogdanovich Crater has a complex history interrelated
with that of Krasnukh Cone. The longest lava flows were formed in
this central group of volcanoes.

The lavas are predominantly two-pyroxene andesites or, more
rarely, andesite-basalts.

Vernadskii Ridge – Interglacial Volcanoes.
The interglacial centers of eruption generally have been somewhat
destroyed by the second glaciation. Only one center escaped this
fate. It lies right on Severo-Kuril'skaya Mountain, near the very
top of the eastern side of the ridge, at the head of one of the U-
shaped valleys of the first glaciation, between Neozhidannyi Cone
and Vetrenaya Mountain. Strictly speaking, this center is late
glacial rather than interglacial. Most probably it was formed at
the end of the second glaciation; its lavas show only insignificant
effects of glacial activity (along Savushkin Creek and Kuzminka
River). The eruptive center is a craterless pile of lava from which
a wide fan of lava flows descends eastward for 3-3.5 km; downslope
the width of this fan reaches 3 km. A considerable part of the trail
from Severo-Kuril'skaya Mountain goes just along these flows.

These flows are composed of two-pyroxene andesites with
seriate-porphyritic structure. Andesine-labradorite ($An_{51-55}$), isolated large crystals of diopsidic augite, and small crystals of hypersthene dominate among the phenocrysts. The composition of the
large grains of hypersthene is somewhat different from that of the
small grains. Apparently the first-generation hypersthene crystal-
lized before the augite and has a more magnesium-rich character;
the hypersthene of the second generation is smaller and richer in
iron. Grains of olivine are rarely found. The structure of the
groundmass is hyalopilitic, with microlites of andesine and andesine-
labradorite, $An_{40-50}$, and also of hypersthene; it contains augite in
subordinate amounts.

An additional one or two centers were located about 1 km southwest of this center, but at present only a large, half-destroyed cirque of the second glaciation* remains here. Small tongues of lava from these centers descend into the upper glacial trough of the Zelenaya River and to the east along the headwaters of the Kuzminka River.

Between the wall of this structure and Ebeko Cone there is a strong group of solfataras, the so-called "Revushchie fumaroles." About 1 km farther south, near the remains of a destroyed inter- glacial structure, there is another very weak group of fumaroles.

Lavas of interglacial age on the bench above the Revushchie fumaroles are andesite-basalts. Labradorite-bytownite ($An_{68-69}$) and abundant augite and olivine occur as phenocrysts, and hyper- sthene is present in small amounts, mainly as relic grains in the augite. The structure of the groundmass is hyalopilitic, with mi- crolites of labradorite ($An_{58-60}$), augite, and subordinate hyper- sthene.

At a point 2-3 km to the north, not far from Mt. Zemlepro- khodets, a basaltic interglacial flow also runs down northeastward from the summit of the ridge; the source of these flows has been destroyed, but most probably the centers from which these flows issued were active at the end of the glacial period. The whole divide between Sestrichko Creek and Artyushin Bay is covered by a peculiar moraine,† although the topography at the north end of Vernadskii Ridge, which this moraine adjoins, is not very much dis- sected. The impression is created that much of the material in this moraine resulted from the influx of fresh volcanogenic products.

Farther south, in the Bogdanovich and Vernadskii Volcano groups, the eruptive centers generally are not preserved, but many of the lava flows are rather well preserved. Moreover, one can rather commonly find marginal embankments and other features of

---

* Earlier (1954) I said that the glacier enlarged a crater that existed here earlier, and I called this part of the structure the "second somma." However, the most recent ob- servations have led me to the conclusion that only a purely glacial depression is pres- ent here and that the eruptions took place not from one, but from several centers. The same thing applies to the "first somma."

† Here our data disagrees with that of the volcanologists of the Sakhalin Complex In- stitute.

the flow surfaces, and as a result we can recognize both individual flows and the probable sites of their eruption.

The longest flows were erupted on the west side of Mt. Nasedkin, 3 km north of Bogdanovich Crater. These flows, which have the details of their surface relief rather well preserved, run down over preglacial lavas to the west; 8 km from the point of their eruption they reach the shore of the Sea of Okhotsk.

The lavas are of pyroxene andesite, with phenocrysts of labradorite ($An_{55-62}$) and augite; hypersthene is also present, apparently in two generations, the first of which is represented by relics rimmed by augite. The structure of the groundmass is microlitic, with microlites of andesine ($An_{45-48}$) and clinopyroxene.

In this same group is a rather well preserved center 1 km southwest of Bogdanovich Crater, but here part of the lava flows are cut off by the U-shaped valley of the Burnaya River. In the Bogdanovich group of volcanoes several more (at least three) centers were located somewhere in the area of the present craters. Their flows, descending to the west, are well preserved. In all probability, flows also descend to the east, but subsequent glacial erosion has completely destroyed them, leaving the great glacial troughs of the present Nasedkin and Severyanka Rivers.

On the ridge between the headwaters of these troughs are preserved the remains of a structure that apparently is of late glacial age. The northern part of the structure is cut off by the U-shaped valley of the Ptich'ya River, and the southern portion is occupied by part of the head of the U-shaped valley of the Severyanka River. This feature apparently is related to the nonuniform retreat of the glaciers.

Farther south, in the Vernadskii group of volcanoes, there are also no less than three destroyed or half-destroyed interglacial eruptive centers. Their flows descend westward across preglacial lavas for 4-5 km and are partly cut off by the U-shaped valley of the Medveditsa and Alyaska Rivers. One small flow runs down to the west into the headwaters of the Zaozernaya River.

Vernadskii Ridge – Preglacial Volcanoes.
The centers of preglacial activity have been almost completely destroyed by the two successive glaciations. Only the ends of the lava flows are preserved; these form isolated plateaus on either side of

the island. On the Okhotsk coast the remains of the dissected plateau stick out from beneath interglacial flows between the glacial
troughs of the Alyaska and Medveditsa Rivers, and also between the
troughs of the Burnaya and Groshkov Rivers, where they locally
extend to the sea shore. One of the eruptive centers apparently
occurs in the area of Uglovaya Mountain (between Ebeko Cone and
Mt. Nasedkin).

On the shore of the Pacific Ocean parts of the plateau are preserved on the divide between the U-shaped valleys of the following rivers: Zaozernaya—Medvezh'ya, Medvezh'ya—Levashov, Leva-
shov-Severyanka, Severyanka—Ptich'ya (Palernoe Plateau), and
Nasedkin—Matrosskaya. In the last case the lavas extend as far as
Mt. Nasedkin, where one of the centers of eruption apparently is located. In the other cases the upper parts of the flows are cut off
by glacial erosion. In particular, the center of lava eruption on Mt.
Levashov has been destroyed, and only a neck some 200 m west of
the summit is preserved. In many areas the glaciers sliced through
the whole series of Quaternary lavas and exposed the Tertiary basement.

Locally on the east flank of the ridge, at a distance of 0.5
to 3-4 km from its axis, isolated necks and rocks of the vent facies
are exposed. Sometimes dozens of such necks are grouped into
"clusters" over an area of 1-2 km$^2$. The country rock around the
old vents has undergone intensive postvolcanic reworking; in particular, the well-known "Sernoe Kol'tso" deposit of volcanic sulfur
is confined to one of these areas.

On the shore of the Pacific Ocean, where early Quaternary
lavas are exposed in high (170-260 m) cliffs, the fact that the horizontal series of these lavas rest on deformed and beveled Tertiary
deposits (e.g., at Cape Okruglyi) is well displayed.

Farther north the character of the early Quaternary rocks
changes rather sharply. On the Okhotsk coast along the Gorshkov
River a distinct change in the rocks occurs. In the cliffs on the left
(south) bank of this river thick lava flows are exposed resting on
the Tertiary basement, but on the higher right bank a section is exposed through a typical stratovolcano with dominant pyroclastic material. The same changed character is shown by the rocks farther
north of the Yur'ev River. Here the Tertiary basement is present
only locally, along the very shore of the sea. All of this part has

been badly destroyed by glacial activity, and no old centers of eruption at all are preserved.  Here, however, the eruptions had a fissure rather than a central character.  Numerous east-west dikes
and the remains of necks testify to this in the headwaters of the
Yur'ev River* and at the north end of Vernadskii Ridge, from Vetrenaya Mountain to Mt. Zemleprokhodets.†

    Thus, at the north end of Vernadskii Ridge early Quaternary
activity had a mixed effusive-explosive character, whereas farther
south more quiet, massive eruptions of lava predominated.  The occurrence of the Tertiary basement at greater depth in the northern
part of the island is related to the plunge in this direction of the
axis of the Tertiary anticline (Sergeev, 1962).  The possibility is
not excluded that faulting occurs in the vicinity of the valleys of the
Gorshkov and Snezhnaya Rivers.

    The composition of the preglacial eruptive products is also
distinctly different.  In the central and southern part two-pyroxene
andesites and andesite-basalts predominate, analogous to those described above, but the basalts are clearly subordinate.  At the north
end of Vernadskii Ridge the rocks are almost exclusively basalts;
flows of andesite-basalt are found only rarely.

    Hornblende dacites are found at the base of the preglacial
lavas of the Ebeko volcanic group (Rodionova et al., 1963), but these
rocks may be Pliocene in age.

    On the whole, the most characteristic rocks for all of these
volcanoes of the Vernadskii Ridge are two-pyroxene andesites.
Basalts are dominant only in the north.  The andesites cover an
area of 300 km$^2$; the basalts, 60 km$^2$.

    A unique, isolated volcano that occurs in the main volcanic
zone of Paramushir,‡  F e r s m a n  V o l c a n o  rises on the shore
of the Sea of Okhotsk, near the west foot of Vernadskii Ridge.  This
is a very large extrusion dome that rises almost 800 m (1052 m,

---

* The volcanologists of the Sakhalin Complex Institute called this group of necks
  Vlodavets Volcano (Aver'yanov et al., 1964).
† Gorkin, Rodionova, and others (1963) recognize "numerous parasitic craters" here; we
  did not find these.  Apparently the remains of the preglacial centers of the linear-
  clustered type were called "parasitic."
‡ The other isolated volcano, Fuss Peak, belongs to the western zone.

absolute) above the surrounding countryside. The basal diameter
of the agglomerate mantle of the dome is 3.5-4 km, and the talus
extends westward 1.5-2 km farther, reaching the shore of the Sea
of Okhotsk. The upper, monolithic portion of the dome is partly destroyed and to a large degree is covered by talus. If we judge from
its general outline, Fersman Volcano is not a single dome but, as
is often the case, consists of several (three) closely adjoining
domes. No remains of an earlier layered structure are noticeable.

The dome was formed at the very foot of Vernadskii Ridge,
where the glacial valleys open out onto the coastal plain, and it
separates several troughs of the first glaciation, specifically those
of the Sokolik and Shumnaya Rivers. The Sokolik River by-passed
the dome to the north, but the Sumnaya River appears to have been
dammed up and to have formed a lake (Glukhoe Lake) here in the
trough. Traces of the second glaciation are absent in this area, so
that the age of the dome can be said with certainty only to be younger
than the first glaciation. The relatively good preservation of the
flanks of the dome allow us tentatively to assign it a late glacial or
postglacial age.

The dome is composed of rather acid (59% $SiO_2$) two-pyroxene
andesite. Plagioclase predominates among the phenocrysts and the
pyroxenes — monoclinic and particularly orthorhombic — occur in
subordinate amounts. The hypersthene is usually strongly decomposed and is rimmed by a wide opacitic margin. The groundmass
is similar in structure to hyaline, but is devitrified.

Karpinskii Ridge — Modern Volcanoes (Fig. 19).
As in Vernadskii Ridge, the volcanoes of Karpinskii Ridge have a linear-
clustered character. Here also three groups of volcanoes are
recognized: Chikurachki-Tatarinov, Lomonosov, and Karpinskii.

The first group includes two closely adjoining structures,
Chikurachki and Tatarinov, and is composed of six to eight cones
(Fig. 20).

This chain begins in the north with Chikurachki Volcano. Seen
from a distance, from the north or northeast, it appears to be a normal isolated cone rising directly from sea level to a height of 1815
m (Fig. 21). In absolute height, Chikurachki is the highest volcano
on Paramushir and the third highest in all the Kuriles. However,
the present cone of Chikurachki is actually situated on the remains

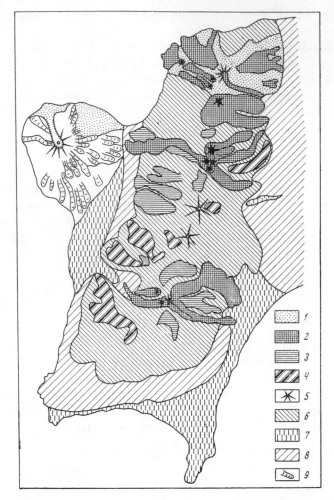

Fig. 19. Distribution map of the volcanoes of Karpinskii Ridge and Fuss Peak: 1) pyroclastic deposits; 2) Holocene lavas; 3) interglacial lavas; 4) areas of preglacial lavas; 6) eroded portions of Quaternary structures: 7) moraines; 8) basement; 9) lava flows of Fuss Peak.

of a high Pleistocene volcano, and the relative height of the cone is only 250-300 m.

The scoria and lava of the active cone have covered all the irregularities in the older surface and, descending as rather thin flows to sea level, they produce the illusion of a high, regular

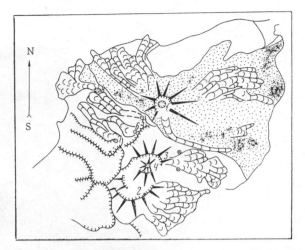

Fig. 20.  Distribution map of the volcanoes of the Chikurachki-
Tatarinov group: 1) Chikurachki Volcano; 2-4) structures of
the Tatarinov volcanic group; 5-7) Holocene structures in the
saddle; 8) cone at the base of the ridge (symbols as on Fig. 9).

cone.  The remains of the older structure stick out only in the north,
where they form a distinct break in the slope.

The crater of Chikurachki Volcano has a diameter of about
450 m and a depth of up to 200 m.  The southeast part of the lip is
destroyed almost to the very floor of the crater.  The wall of the
inner cone stretches along the south edge; within it, low-tempera-
ture (60-80°) fumaroles fumed along a number of arcuate fissures
at the time of our visit (August, 1953).  This whole portion was
covered with a crust of varicolored clay and efflorescences.  On the
floor of the crater was a small snowfield.

In the walls of the crater a series of lavas separated by layers
of cherry-red scoria is exposed.  Scoria covers the whole surface
of the cone and descends to the base of the old structure.  Particularly thick accumulations of scoria are found near the east base of
the old structure, where a lifeless plain has been formed.

In the lower part of the eastern slope the outlines of numerous lava flows "show through" from beneath the sheet of scoria.
Lava flows also descend along the northwest slope and reach as far
as the Sea of Okhotsk where, 4.5 km from the summit, they form
Capes Chikurachki and Svirepyi.

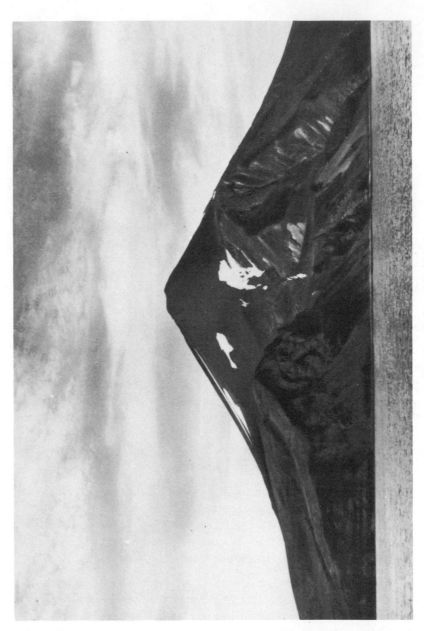

Fig. 21. Chikurachki Volcano.

A large, very young flow of lava runs down across the depressed part of the crater to the south, and then turns around to the southeast. The source of this flow has been covered by bombs and scoria.

The red bombs and scoria that cover the upper part of the volcano's cone give it a distinct cherry-red color. As on Alaid, the color of the scoria derives from processes of posteruptive oxidation and points to a high temperature for the material at the moment of its eruption.

A very strong eruption of Chikurachki occurred in December 1853. It was characterized by the ejection of very large quantities of scoria and pyriform bombs. The ejecta from this very eruption now cover the surface of the cone and are spread over the vicinity of the volcano. In Shelekhovo, 12 km east of the volcano, scoria lies on the surface in a layer 40 cm thick, and still almost no vegetation has begun to grow on it. Fine gravel and sand from Chikurachki covers the Vernadskii Ridge area, 50 km from the volcano, in a thin (1 cm) layer. The total volume of pyroclastic material ejected in December 1853 is estimated at approximately 1 $km^3$. After the December paroxysm weak explosions apparently occurred and built the inner cone; they lasted until 1859. On the basis of a soil profile in Shelekhovo, we may guess that catastrophic eruptions of Chikurachki occur about every 700-900 years.

A weak eruption of the volcano occurred on May 26-27, 1958. In Shelokhovo and at the Podgornyi whaling station (18 km southeast of the volcano) 3-4 cm of ash fell. No substantial changes in the shape and size of the crater occurred as a result of this eruption; only the floor of the inner cone was somewhat deepened (Shilov and Voronova, 1962).

A later eruption occurred on May 2-21, 1961; it was accompanied by very weak ash falls. Separate, very weak explosions also occurred on July 22-23, and August 9-10. The latest, also very weak eruption occurred in January 1964.

The lavas of Chikurachki Volcano, seen in the walls of the crater, are plagiophyric andesites and two-pyroxene andesite-basalts. The phenocrysts in the andesites are almost exclusively of plagioclase of labradorite composition ($An_{57-69}$); clinopyrocene and orthopyroxene are found as scattered grains. The structure of the groundmass is microlitic to pilotaxitic or in some flows, hyaline. The andesite-basalts are distinguished by the abundance of phenocrysts of diopsidic augite and hypersthene, along with which grains of olivine are found. The groundmass consists of idiomorphic crystals of andesine-labradorite ($An_{50-55}$) and of almost as much again idiomorphic crystals of clinopyroxene and orthopyroxene.

Lava from the 1854-1859 flow is basic andesite with predominant phenocrysts of labradorite-bytownite, $An_{70-72}$; pyroxenes occur as scattered grains. The structure of the groundmass is hyaline. The scoria from the 1854 and earlier eruptions is distinguished from the lavas by the large content of mafic minerals; sometimes even olivine is found. The plagioclase also is somewhat more basic, and the groundmass is a strongly vesicular, dark-brown glass.

The Tatarinov Group of Volcanoes, closely adjoining Chikurachki on the south, has a rather complex structure and consists of at least six eruptive centers. The present centers are located on the remains of an ancient, but rather high (up to 1400 m) volcanic structure that has been badly destroyed by glaciation. Glacial topographic forms play an important role in the external appearance of this group, and only on the east slope are the erosional landforms overlapped by the modern lavas.

The central part of the massif is composed of the remains of two eruptive centers (2, 3, in Fig. 20). Of the northern cone the northern half is preserved, and of the southern cone, the southern half. Thus, the impression is given of one large crater* more than 1 km across. However, the clearly asymmetrical form of this depression, the disposition of the lava flows, and the remains of morainal deposits on its floor pointed to the fact that two craters were established here in postglacial or more probably, in late glacial times on the edge of a wide glacial amphitheater.

Both of these centers gave rise to large lava flows that descend along the east side of the massif all the way to its base. Later both cones were partially destroyed. Not only purely erosional processes, but also explosive ones very probably played a role in their

---

*In our first publication it was interpreted in just this way.

destruction. The history of the southern cone probably ended at
this point, but in the area of the northern eruptive center continued
explosive activity led to the succesive formation of two nested explosion cones on the floor of the depression. Of the outer cone,
only the remains of the southern part of the embankment are still
preserved. The miniature inner crater is surrounded by a small
cinder wall and is filled by a small lake. On the northeast edge of
the lake in 1953 there were weak, dying solfataras that had ceased
in 1959 (Shilov and Voronova, 1962). The whole northwest side is
composed of native sulfur, which testifies to the formerly intense
solfataric processes.

Above, on the steep northern side of the amphitheater, the
characteristic black and cherry-red scoria of Chikurachki is found
interstratified with thin deposits of the pale yellow ejecta of Tatarinov Volcano. Clearly, the crater on the floor existed for a rather
long time and from time to time displayed explosive activity. The
last eruption, judging from the character of the ash layers between
the two upper layers of Chikurachki scoria, occurred approximately
at the end of the seventeenth century, i.e., just before the discovery
of the island by the Russian explorers.

On the east flank of the volcano, at a height of about a thousand meters above sea level, are two solfatara fields. Here, along
with the strong solfataras, there are acid springs and boiling springs
that eject columns of hot water to heights of 2-3 m. Earlier there
were probably lateral explosion craters here that are now filled
with the products of postvolcanic activity.

The rocks of this part of the massif are strongly altered by
posteruptive processes. Fresh lavas from the northeast lava flow
are andesite with phenocrysts dominantly of andesine-labradorite,
$An_{48-51}$. Among the mafic minerals diopsidic augite is dominant,
and then comes hypersthene. The groundmass structure is hyalopilitic. Among the microlites andesine, $An_{42}$, is dominant.

The remnants of three craters (5-7 in Fig. 20) are preserved
on the small ridge that joins Chikurachki Cone to the central massif
of Tatarinov Volcano. Each of these craters erupted lava flows to
the northwest.

Finally, near the northwest foot of the massif, a parasitic
crater is located from which extend the lava flows that formed the
fan-shaped platform of Cape Skal'nyi (8 in Fig. 20).

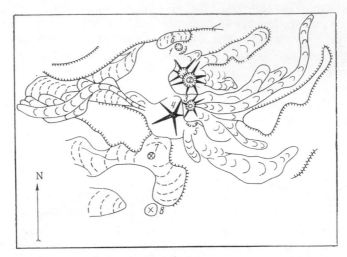

Fig. 22. Sketch of the structure of the Lomonosov group of volcanoes. Explanation of the numbers is in the text (symbols as
in Fig. 9).

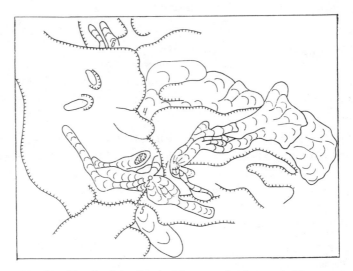

Fig. 23. Sketch of the Karpinskii group of volcanoes. The explanation of the numbers is given in the text (symbols as
in Fig. 9).

The Lomonosov Group of Volcanoes, which adjoins the massif of Tatarinov Volcano on the south, consists of five eruptive centers. The first of these occurs right at the boundary between the two massifs; it is purely explosive, without any lava flows (1 in Fig. 22).

The second cone, Mt. Borisyak, has on its summit a crater from which flows of coarsely blocky lava run down to the east for 3.5 km (2 in Fig. 22).

South of Borisyak Crater lie two large cones. The first of these (3 in Fig. 22) has a small crater from which a long lava flow descends to the east. At 5.5 km from the crater, in the valley of the Tukharka River, it reaches the base of the massif. The other cone (4 in Fig. 22), Lomonosov proper (1681 m), is an extrusion dome. A lava flow more than 7 km long extends westward from beneath the agglomerate mantle of the dome, apparently from an old crater now sealed by the dome; it reaches almost to the shore of the Sea of Okhotsk.

The loose pyroclastics and lava flows from these two cones almost completely overlap a small cone located to the south of them (5 in Fig. 22). This last cone also emitted a lava flow which runs to the southeast and then turns eastward. The flow is 4 km long, and its surface has a very coarsely blocky character.

Approximately 10-12 km south of the Lomonosov group of volcanoes lies the last of this series of "clusters" of Holocene eruptive centers, the Karpinskii Group of Volcanoes, consisting of three independent apparatuses.

The first cone lies on the east slope of Karpinskii Ridge, near its crest and at the top of a wide glacial cirque (1 in Fig. 23). Resting against the side of the ridge, this cone is well formed only in its eastern half. The diameter of the crater, which is open to the northwest, reaches almost 300 m; within it lies a small inner horseshoe-shaped cone. Near the northeast wall of the outer crater a dense cloud of gases with a temperature at its edge of 148° (1953) is emitted from an opening reminiscent of an explosion crater. The activity of the fumarole at the time of our visit to the volcano in 1953 had strongly increased relative to 1946 when we first visited this place. In 1946 there was a cluster of not very strong fumaroles here. Apparently the strengthening of the activity was related to the strong

earthquake of November 5, 1952, following which the inhabitants of
Cape Vasil'ev on Paramushir saw a column of dark-colored gases
rising above the volcano.

To the northeast of the cone a thick lava flow runs down into
the hollow of the glacial valley, then turns to the southeast, and
comes out onto the coastal plain. The wide field of lava covers the
interfluve area between the Trudnaya and Lesnaya Rivers; it is 2
km wide and 7 km long. Here we can recognize at least three groups
of flows of different ages. A smaller flow also descends to the southeast from the cone into the headwaters of the Galochkin River.

The second cone, entirely of lava, lies on the very crest of
the ridge and forms the high point of the massif, 1345 m (2 in Fig.
23). This cone emitted lava flows both to the southeast, into the U-
shaped valley of the Galochkin River (3 km long), and to the west,
into the broad caldera-like depression which we earlier called
Karpinskii Caldera.* The western flow attains a length of 4.5 km
and descends almost 750 m. At a height of about 900 m this flow
overruns a group of large blocks of granodiorite.

The lavas of a young flow from this center (2 in Fig. 23) are
basic andesites or andesite-basalts. If we judge from their chemical composition, the lavas from the early parts of this flow are
basic andesites, close to andesite-basalt, but the lavas from the
later parts of the eruption are still more similar to typical andesite-
basalt. The phenocrysts are labradorite ($An_{50-66}$), diopsidic augite,
and hypersthene; olivine is found in very small amounts (1-2 grains
per slide). The structure of the groundmass is hyalopitic. The
microlites are of sparse andesine-labradorite, $An_{55}$, and of abundant
large laths of ortho- and clinopyroxene.

Near the east wall of the depression in question, at a height
of 1100-1200 m, there is a rather broad explosion crater from which
a flow of lava has descended to the west. Apparently this crater occurs at the site of a cone destroyed by an explosion. Within the
crater hot springs issue and numerous intense solfataras form sulfur cones up to 3-5 m high. Inside these cones sulfur boils and sometimes is sprayed outward. One of the hot springs forms a very

---

* According to our latest observations, this broad depression was formed mainly by glaciation. It is rhomb-shaped in form, and the walls are of different ages, ranging from
early Quaternay to recent.

effective tilted fountain that reaches a height of 2 m; its temperature is 80°C.

Thus, on Karpinskii Ridge we can count at least 14 postglacial
eruptive centers. As on Vernadskii Ridge, some of these centers
are one-stage structures, having formed as a result of activity that
was not repeated. Other centers have a rather long and complex
history of activity. Chikurachki Volcano shows the traits of the
usual central volcano with a large number of distinct eruptions.

Karpinskii Ridge – Interglacial Volcanoes.
Most of the centers of interglacial activity are badly destroyed and
are often overlapped by modern structures and flows. The remains
of six or seven eruptive centers of interglacial age are preserved.

One of these is found at the very boundary between the Tatarinov and Lomonosov groups (6 in Fig. 22); its remains form Srednyaya
Mountain, and the flows on the left bank of the Alenushkin River apparently are related to this center.

Shatskii Cone (7 in Fig. 22) is better preserved; it lies south
of Lomonosov Cone. Its flows often form the summits of ridges
and often descend westward to the lower course of Krasheninnikov
River. Right beside it are located the remains of still another cone
(8 in Fig. 22). Part of the west slope of Karpinskii Ridge between
Krasheninnikov and Fuss Rivers is rather weakly dissected and does
not have a clearly expressed alpine form. In all probability this part
is covered by interglacial lavas. One can suppose that there, in the
area of Arkhangel'skii Mountain, another interglacial center occurred.

Finally, the remains of three interglacial centers occur in the
area of the Karpinskii group of volcanoes. The center located to the
north of the present cone (4 in Fig. 23) had the largest dimensions;
its lava flows descend eastward to the very foot of the ridge. A second center lies south of the young cones (5 in Fig. 23). To the east
of Mount Topor, in a U-shaped valley of the first glaciation, small
interglacial flows (6 in Fig. 23) were erupted, but their sources were
completely destroyed during the second glaciation.

Lavas of interglacial age are also found on the floor of the
caldera-like depression, but they have been subjected to glacial erosion. In appearance these lavas are rather fresh, but under the microscope significant changes are seen in them that are related to

glacial weathering. In transmitted light we find a seriate-porphyritic structure, with phenocrysts of plagioclase, clinopyroxene, and
orthopyroxene. The plagioclase is almost completely converted
into an isotropic aggregate, although in plain light one can see all
the characteristic features of its structure, for example, zoning,
inclusions of the groundmass, and so on. Judging from the relics,
the plagioclase was andesine-labradorite, $An_{52-55}$. The pyroxenes
also have been affected by the alteration, but only along fractures.
The groundmass is completely converted to an isotropic aggregate;
in plain light a relic structure (fluidal glass with a small amount
of microlites) is seen.

Karpinskii Ridge — Preglacial Volcanoes.
Karpinskii Ridge, being higher than Vernadskii Ridge, has undergone even more intensive glaciation, and the preglacial volcanoes
are now wholly destroyed; the preglacial lava flows are almost
never preserved.

Remnants of the ancient volcanoes include the tops of Arkhangel'skii and Belousov Mountains; these have a clearly expressed
alpine topography, and on their steep flanks the periclinal attitude
of the ancient lavas can be seen.

Remnants of lava flows are seen in the Karpinskii group. These
flows belong to several eruptive centers, but not one of these has
been preserved. As is well displayed in the steep walls of the cirque
of the Strela River, the flows rest with angular unconformity on the
Tertiary deposits.

Preglacial lavas in the side of the Karpinskii "caldera" are
andesite-basalts. The rock structure is seriate-porphyritic, with
phenocrysts of labradorite ($An_{53-55}$ in one case, $An_{59-64}$ in another),
rather abundant augite, and somewhat less hypersthene. Olivine is
found in small amounts, sometimes intergrown with the pyroxene.
The structure of the groundmass is hyalopilitic or locally microlitic.
Among the microlites, labradorite ($An_{50-55}$) and both pyroxenes are
represented.

Judging from the character of the structures preserved, preglacial and interglacial activity was of the "linear-clustered" type
and predominantly effusive. Present-day activity has essentially
the same character, but the very northernmost volcano of the Karp-

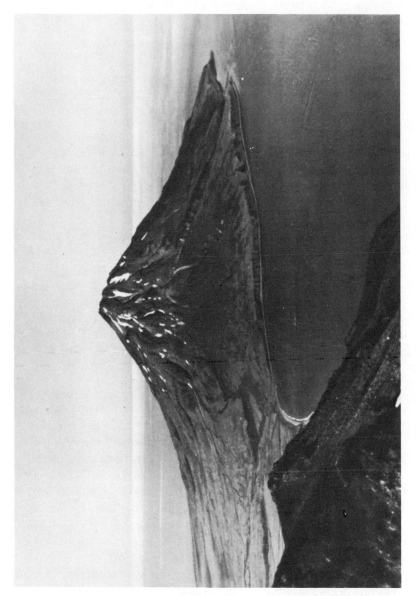

Fig. 24. Fuss Peak, seen from the top of Chikurachki Volcano.

inskii group, Chikurachki Volcano, also displays intense explosive activity along with lava effusion and has the characteristics of an independent central structure.

The lavas of the volcanoes of Vernadskii and Karpinskii Ridges are rather similar and are represented by two-pyroxene andesites and andesite-basalts, sometimes with olivine. However, the rocks of Karpinskii Ridge are somewhat more basic and clinopyroxene is occasionally dominant over orthopyroxene among the microlites.

Fuss Peak is a unique, isolated stratovolcano on the island of Paramushir that, as noted earlier, actually belongs to the western zone. The volcano forms a distinct peninsula near the southwest coast of the island; it is semicircular in form and is joined to the foothills of Karpinskii Ridge by a low isthmus. Fuss Peak forms a normal, beautiful, strongly truncated cone (Fig. 24). At the summit is a crater with the form of a steep-sided funnel up to 700 m across and 300 m deep. On the floor of the crater is a well-like pit, the site of the last eruption of the volcano. The north-northwest edge of the crater is cut through to the very floor by a deep, steep-sided canyon that, cutting through the flank, stretches down to the shore of the Sea of Okhotsk. The opposite, southern edge of the rim is the highest point, 1772 m above sea level. The submarine terrace is absent in the Fuss Peak area, so that the cone rises directly from the floor of the Sea of Okhotsk, and its actual height amounts to almost 2800 m.

In the middle and lower parts of the cone numerous well-preserved lava flows are found. Particularly numerous are the flows in the eastern and southeastern sectors of the cone, where they extend for 5-6 km from the summit and reach the very base of the cone. A series of flows also descends to the north and west. Thus, for example, the extreme western part of the cone, Cape Neproidennyi, is formed by a lava flow. At the mouth of the canyon coming from the crater and in the northeastern part of the cone there are debris cones composed of ejecta with significant quantities of pumice bombs. The west side of the cone drops off very steeply, and in an excellent section it is seen that the lower part of the volcano is composed exclusively of pyroclastic products, while above this the rocks are mainly lava flows.

A characteristic feature of the rocks of Fuss Peak is the rather large amount of hornblende among the phenocrysts. Tuffs and lava flows from the base of the cone are composed of two-pyroxene andesites, sometimes with small amounts of hornblende. The rocks have a seriate-porphyritic structure. Among the phenocrysts labradorite, $An_{53-61}$, and clinopyroxene predominate; hypersthene is somewhat less common, and hornblende (brown, with strong pleochroism) is found only in trace amounts. The structure of the groundmass is hyalopilitic and microlitic; the microlites are of labradorite ($An_{50-57}$), clinopyroxene, and orthopyroxene.

Among the basal tuffs crystalloclastic ones predominate, and vitroclastic ones are also recognized. The upper part of the cone is composed of pyroxene-hornblende andesites.

The lavas and some of the bombs show a seriate-porphyritic structure, with a hyalopilitic structure in the groundmass. Scoriaceous bombs and pumice have a vitrophyric structure, with crystallitic and perlitic structures in the groundmass.

In one of the lava flows the plagioclase is labradorite, $An_{65}$, in the core and andesine-labradorite in the outer zone; the hornblende is brown, with strong pleochroism and a small (about 4°) extinction angle. Pyroxenes (rhombic and monoclinic) are usually present.

In the bombs plagioclase and hornblende usually predominate. The plagioclase is andesine-labradorite in the cores (up to $An_{55}$) and andesine in the margins or in small grains ($An_{40-50}$). The hornblende prisms reach 1-2 mm in length; they are greenish-brown, with pronounced pleochroism and an extinction angle of 10-15°. The pyroxenes (rhombic and monoclinic) are subordinate in amount. In the microlites andesine, $An_{40-50}$, is dominant; orthopyroxene, clinopyroxene, and rarely hornblende are recognized. Hypersthene is absent as microlites.

The amount of hornblende present as phenocrysts increases from the bottom to the top of the section, and in the young pyroclastics this mineral becomes dominant.

Evidence of glaciation is unimportant on the flanks of the volcano, and the absence of the 140-m submarine terrace allows us to give the whole structure a postglacial age.

Thanks to its perfect form, Fuss Peak is easily identified as a volcano. Apparently for this reason, almost all eruptions on Para-

mushir have been assigned to this volcano. Actually, only one rather strong explosive eruption in July, 1854 is authentic. At present the volcano does not show even traces of fumarolic activity.
The dates of eruptions given in some summaries (1737, 1793, 1857,
and 1859) are false.

### Onekotan Island

The island of Onekotan is separated from Paramushir by the
wide (55–60 km) Fourth Kurile Strait. It is a rather large island:
its length is 42.5 km, and its width varies from 7.5 km in the north
to 17.5 km in the south; its area is 425 km$^2$.

On the Soviet (Geologiya SSSR, 1964) and Japanese (Geological
Map, 1959) geological maps, sediments of Tertiary age are indicated along both shores in the middle part of the island. On the west
coast, at the base of the Mount Shestakov massif, there are actually
very thin deposits that may be Tertiary. On the east coast the deposits indicated as Tertiary are actually pyroclastic flows dated by
radiocarbon as Holocene in age (see below).

There are two active volcanoes on the island, Nemo Peak and
Krenitsyn Peak, along with older, rather completely destroyed
structures (Fig. 25).

Nemo Peak is the central cone of a complex volcanic structure that forms the northern part of the island. Here, between the
north end of the island (Mounts Petr and Asyrmintar)* and Platform Mountain 11.5 km away, there extends a wide depression that
is open to the west and that is bounded on the east by Sovetskii Ridge
The outlines of the mountains and ridges surrounding the depression
as well as the attitude of the rocks allows us to suggest the existence here of two old calderas that are partly nested (Fig. 26). All
the small ridges and summits show evidence of glacial erosion, and
if we judge from the low heights (the floor of the depression ranges
from 50 to 100 m; Platform Mountain, 590 m; other heights, from
200 to 400–500 m), this is evidence of the first, more intense glaciation.

---

*Mount Asyrmintar does not seem to be a presently active volcano, as we indicated
earlier, but rather an old erosional summit with monoclinal lavas.

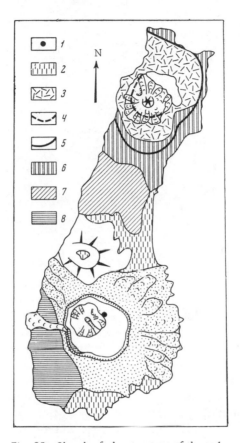

Fig. 25. Sketch of the structure of the volcanoes of Onekotan: 1) 1952 dome; 2) pyroclastic deposits related to the formation of
Tao-Rusyr Caldera; 3) ignimbrites of the
caldera of Nemo Peak; 4) assumed outline
of the inner caldera of Nemo Peak; 5) outline of the older caldera of Nemo Peak; 6)
older (in part Tertiary) rocks of the outer
structure of Nemo Peak; 7) Mount Shestakov
massif; 8) remnants of ancient Mednyi Volcano
(other symbols as in Fig. 9).

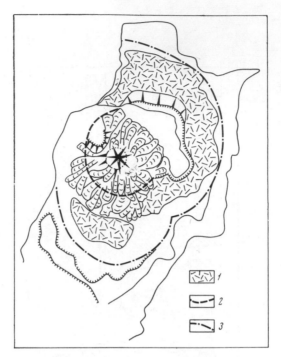

Fig. 26. Sketch of the structure of Nemo Peak: 1)
ignimbrite deposits; 2) edge of the inner caldera;
3) outlines of the old caldera (symbols as in Fig. 9).

On the floor of the depression, to the east of Nemo Peak, are
preserved the remains of another, younger structure that was enclosed in the depression approximately at the boundary between the
two supposed calderas. This structure has been blown apart; of it
there remain the remnants of the somma (250 m high) and part of
the caldera (Nemo Caldera), in which there is a lake (Chernoe Lake)
with no outlet and with the shape of a half moon. A fragment of this
structure also forms part of the Okhotsk shoreline from Nemo Bay
to the mouth of the Ozernaya River. The crest of the caldera has
an irregular form. A low scarp that is overlapped by much younger
lavas of Nemo Peak leads from the south shore of Chernoe Lake.
Perhaps the crest ends at the point where a break in the relief occurs on the south slope of Nemo Peak.

The whole flat floor to the north and east of Chernoe Lake
(Shirokaya Valley) and to the south of Nemo Peak, as well as at the

mouth of the Ozernaya River, is covered by a uniform layer of ignimbrites of supposedly Holocene age. In the coastal cliffs of Nemo Bay the ignimbrites are seen to rest on top of supposedly morainal deposits. Farther south the ignimbrites are covered by lavas of Nemo Peak. Undoubtedly these ignimbrite deposits are related to the Nemo Caldera.

The rocks of the Nemo somma are dominantly andesite-basalts analogous to the rocks of the central cone, Nemo Peak (see below). The ignimbrites are acid andesites or andesite-dacites with 64% $SiO_2$.

The central cone, Nemo Peak, rises somewhat eccentrically with respect to Nemo Caldera, overlapping its southern edge. In addition, the site of Nemo Peak falls in the area of overlap of the two older calderas.

The height of the cone reaches 1019 m (its relative height, above Chernoe Lake, is 947 m) and its basal diameter is about 5 km. Nemo Peak has the form of a beautiful, weakly truncated cone. The rather even, regular slopes are covered by numerous flows of lava that locally are covered by a surface of scoria.

In the southwestern part of the cone, at a height of 750–800 m, there is a scarp, the rim of the old crater whose diameter apparently was as much as 800–900 m. This old crater is completely filled by an inner cone whose crater (350 m across) in turn is sealed by an extrusion dome. The dome completely overlaps the rim of the crater and gives it a pointed form. The height of the dome is about 100 m; on its summit there is a collapse crater about 150 m across. On the surface of the 800-m bench there is an explosion crater 40 m across and up to 30 m deep, the site of intense solfataric activity.

The regular appearance of the central cone is disturbed only on the northwest, where the remains rise of another structure that is overlapped by the young flows of Nemo Peak. Apparently this was an eccentric, young cone whose evolution was brought to an end by a northwest-directed explosion that destroyed the structure. Apparently the earlier central cone was double, and later only Nemo Peak continued to be active.

Nemo Peak is composed of andesite-basaltic and basaltic rocks. Their phenocrysts are of labradorite, $An_{50-70}$; augite; hyper-

sthene, which sometimes is found only as relic grains in augite;
and olivine, which occurs in small, but ubiquitous amounts and which
is almost always corroded and replaced by pyroxene. The structure of the groundmass is hyalopilitic or andesitic. In the microlites are found iron-rich hypersthene and smaller amounts of augite.

The upper short flows of lava and the summit dome are composed of augite andesite (59.3% $SiO_2$) similar to the lavas just described. The phenocrysts here are labradorite ($An_{50-59}$), augite
with thin fused margins, hypersthene as large grains and relics (in
the augite crystals), and small amounts of olivine, locally rimmed
by magnetite and augite. The structure of the groundmass is microlitic, with dominant plagioclase and subordinate hypersthene.

The history of Nemo Peak is very complex. Initially, in preglacial times, a double caldera was formed on some ancient volcanoes. Later the island was glaciated; the caldera walls were
abraded and the floor of the depression apparently served as an ice-
collecting basin. Moraines were deposited in the basin.

The somma of Nemo Peak was formed in interglacial times;
its cone was superimposed on the glacial topography produced by
the first glaciation. The formation of the somma was brought to an
end by a gigantic explosion, which almost completely destroyed the
cone. The explosion was accompanied by the eruption of incandescent pyroclastic flows and the formation in the depression of an
ignimbritic topography. The diameter of the caldera formed was
about 5 km.

Then, in postglacial times, the central cone was formed. In
its initial stages of development the cone was double; later, eruptions continued only from one crater (Nemo Peak proper). The
eruptions were characterized by the emission of large numbers of
lava flows that alternated with Strombolian explosions. The cone
filled almost the whole basin of the caldera and even overlapped its
rim on the south and west; part of the atrio, filled by the waters of
the lake, was preserved only to the north of the central cone. In
the final stage an extrusion dome was squeezed out from the summit crater of the central cone.

Eruptions of Nemo occurred in the eighteenth century, and
an eruption (with the formation of the extrusion dome?) is known to
have taken place in 1906. The volcano displays constant fumarolic

a

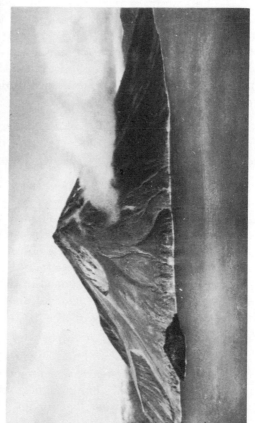

b

Fig. 27. a) Central cone (Krenitsyn Peak) in the caldera of Tao-Rusyr. View from the north rim of the caldera. b) Krenitsyn Peak: on the slope to the left is the 1952 explosion crater and at the base is the 1952 dome. View from the northeast crater rim.

activity from the explosion crater located on the edge of the old summit crater (on the southwest slope at a height of about 800 m).

In the middle part of the island lies the strongly eroded massif of Mount Shestakov. This massif was formed in preglacial times and then underwent glacial abrasion and postglacial erosion to such a degree that at present it is extremely difficult to restore its original form. As has been indicated, sedimentary rocks, apparently of Tertiary age, underlie the lavas in the western part, near the shoreline.

One of the most beautiful volcanoes in the Kurile Islands, Krenitsyn Peak, is the central cone of the large caldera of Tao-Rusyr, which is situated in the southern, broad portion of Onekotan (Fig. 27).

The somma of the volcano forms a gently sloping, shield-like volcano with slopes ranging from 7° in the east to 14° in the west. The basal diameter of the somma reaches 16-17 km. At the top of the Tao-Rusyr somma is a completely closed caldera, 7.5 km across; within it there is a deep caldera lake (Kol'tsevoe) with a diameter of 7 km whose surface lies at an altitude of about 400 m above sea level (Fig. 28).

The depth of the lake is not known, but it is very great. Near the north shore even a 150-meter line does not reach its floor. The lake waters are dark blue of a very deep tone.

In the southwestern part of the island, next to the rim of the caldera, there are glacial troughs and cirques. In this part the remains of an older, preglacial volcano (Mednyi Volcano) adjoin Tao-Rusyr and are partly overlapped by its lavas. In the cliffs along the Okhotsk shore the structure of a complex stratovolcano cut by dikes is exposed; its height reaches almost 870 m. The remains of this basaltic volcano are also exposed in the southwest wall of the caldera, where we can see the nonconformable overlap of the Tao-Rusyr lavas on the older, eroded surface.

The remains of a second preglacial volcano lie to the northwest of the Tao-Rusyr somma, where the amphitheater of the caldera-volcano Kryzhanovskii is preserved. This volcano shows clear evidence of glaciation. The diameter of the caldera is 3 km, and it is widely open to the west, but its flat floor is probably covered by

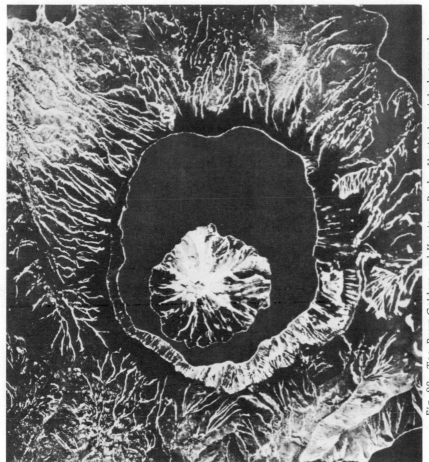

Fig. 28. Tao-Rusyr Caldera and Krenitsyn Peak. Vertical aerial photograph.

pyroclastics from the caldera-forming explosion of Tao-Rusyr. In the east the rim of the caldera reaches a height of 550 m. The rocks of Kryzhanovskii are mainly basalts.

The southern and western parts of the somma of Tao-Rusyr itself also show evidence of glaciation in the form of cirques and small U-shaped valleys; moreover, the head of one of the valleys is cut off by the caldera. However, the eastern and northwestern slopes of the somma are smooth, without any clear traces of glacial action, and here there are numerous lava flows with well preserved traces of surface features. In addition, near the western foot one of the flows, which forms Cape Angibya, clearly flows down the U-shaped valley of the Angibya River.

The facts cited allow us to state that the shield volcano of Tao-Rusyr arose in interglacial or late glacial times on the ruins of preglacial volcanic structures and continued to erupt lavas into post-glacial times. Then there occurred a gigantic explosion that beheaded the volcano and formed the caldera. This explosion was accompanied by pyroclastic flows, the deposits of which occupy an important part of the coast of the island, but which gradually diminish in thickness from south to north.

These deposits also fill the area between the Tao-Rusyr somma and the Mount Shestakov massif along Fontanko and Ol'khovaya Creeks, and most probably they occur on the south shore of the island in the region of Capes Terrasnyi and Krenitsyn.

In the walls of the caldera alternating lavas and pyroclastics are found, with the lavas clearly dominant. All the lavas are very similar, but we can still distinguish aphyric basalts (at the base of the section), predominant olivine basalts, and subordinate two-pyroxene andesite-basalts. At the very top of the section interlayers of dacitic pumice covered by thin flows of basalt are found.

The structure in the aphyric basalts is intersertal; labradorite ($An_{60}$), olivine, hypersthene, and augite occur as microlites and microphenocrysts.

In the olivine basalts labradorite ($An_{58-70}$) and olivine, often resorbed, are the dominant phenocrysts. More rarely augite is present, and hypersthene is even rarer. The structure of the groundmass is intersertal and microdoleritic. Augite is dominant over plagioclase among the microlites.

The andesite-basalts make up about 15% of the section; both aphyric and porphyritic varieties are found. Labradorite ($An_{57-68}$), augite, and hypersthene are present as phenocrysts; the structure of the groundmass is pilotaxitic and hyalopilitic.

The deposits of the pyroclastic flows have an agglomeratic character, with large, black, rounded blocks of pumiceous two-pyroxene andesite (58.7% $SiO_2$). Considerable quantities of rounded blocks and chunks of crystalline diorite and gabbro-diorite are very characteristic. At the base of the pyroclastic flows are found the remains of carbonized scrubby vegetation. Radiocarbon age determination allows us to date the caldera as having formed 7040 years ago.

Krenitsyn Peak is somewhat eccentric; the central cone rises from the waters of the lake in the western part of the caldera, to a height of 900 m (1325 m above sea level). The diameter of its base at lake level is 3.5-4 km; along the shore tongues of lava alternate with mounds of pyroclastics, and in the cliffs is seen the alternation of these and other rocks that is characteristic of volcanoes.

The summit crater, about 350 m across and up to 100 m deep, is broadly open to the southeast at the head of a deep ravine that extends to the lake shore. There is also a small notch in the crater on the northwest, and a second ravine runs down the slope from this point. On the east edge of the crater rim is a large "tooth" of monolithic dark lava, shaped like a pointed cone. This protuberance looks like the remains of a devastated dome that once sealed the vent of the crater.

On the northeast side of the volcano, at a height of about 900 m, an old, obliterated lateral explosion crater 400-450 m across is exposed. In line with it, somewhat farther south, is the 1952 lateral crater. It is pear-shaped in plan view, widening downslope; its dimensions are 350 × 450 m. Below this crater, on the steep eastern flank of the cone, the dark, flat top of the 1952 sublacustrine extrusion dome sticks out of the water. The diameter of the dome is about 300 m and its height is about 30 m. The dome apparently grew within a large subaqueous explosion crater whose western edge is cut into the flanks of the dome a little above lake level. Reckoning in terms of its size relative to the main cone, the dome formed a miniature peninsula.

The lavas of Krenitsyn Peak are very uniform andesites with phenocrysts of labradorite, $An_{52-62}$, and augite, sometimes with traces of hypersthene; glomeroporphyritic aggregates of plagioclase and pyroxene are often found.

The summit extrusion is characterized by a more acid plagioclase (andesine-labradorite, $An_{46-52}$).

The 1952 eruption gave rise to products of varied form: andesitic ash, lapilli of pumiceous andesite, and a lava dome. In the lapilli the phenocrysts are labradorite ($An_{52-62}$), augite, and hypersthene. The structure of the groundmass is hyaline and contains abundant oriented vesicles.

The dome lavas are pyroxene andesite, with phenocrysts of andesine and andesine-labradorite ($An_{40-55}$), augite, and more rarely, hypersthene. The structure of the groundmass is microlitic, with recrystallized glass and with microlites of plagioclase and very iron-rich hypersthene.

A broad view from the top of the mountain of the gigantic bowl of the caldera presents an unforgettably fine picture: the deep-blue lake from which the cone rises, covered with green grass and varicolored volcanic rocks, sparkles in a framework of gloomy cliffs.

Krenitsyn Peak showed weak solfataric activity in 1846 and 1879; then for a long time it was quiet and was even regarded as extinct. However, in November 1952 a violent eruption took place unexpectedly. It began with explosions through a newly formed explosion crater on the east flank of the volcano. Then the site of the explosions migrated to the base of the dome, where the lava dome was later produced. After the eruption the volcano renewed fumarolic activity at three points: on the east edge of the summit crater, in the lateral crater, and near the base of the cone at the edge of the subaqueous crater. Here, near the new dome, a column of hot mineralized water is being ejected.

## Kharimkotan Island

The island of Kharimkotan, $8 \times 12$ km, lies 15 km southwest of Onekotan across the Sixth Kurile Strait. The island, which is elongate in a north-south direction, is a single volcano of rather complex structure. Its lower parts are covered by sparse grasses and scrub, and its summit is bare. Its area is 68 km$^2$.

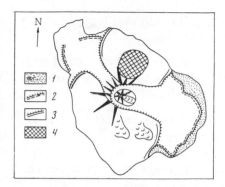

Fig. 29. Sketch of the structure of Khar-imkotan Volcano: 1) part of the island formed as a result of the 1933 eruption; 2) old cliff; 3) still older cliff; 4) remnant of the preglacial structure (other symbols as in Fig. 9).

Seen from afar, the volcano seems to be a rather gently sloping, strongly truncated cone with smooth slopes. However, on closer study its structure proves to be very complex (Fig. 29).

The basal (old) cone has a crudely oval outline, elongate in a north–south direction; its basal dimensions are 7×10 km. This cone is divided into two segments, northern and southern, by two wide, deep depressions, one of which runs to the east and the other to the northwest from the summit. The sides of the old cone are rather smooth, but on the northeast an area with complex, strongly dissected relief reminiscent of alpine forms sticks out at a height of about 700 m above sea level. Most probably the remains of a still older volcanic structure of preglacial age protrude from beneath the cone at this point.

In the steep slopes of the old cone the layered structure of a stratovolcano is exposed, with a great predominance of pyroclastic material. A layer of welded tuff is found in the steep slopes near the top; the lavas and pyroclastic rocks of the somma are hypersthene and two-pyroxene andesites.* The phenocrysts are plagioclase ranging in composition from andesine ($An_{42}$) to labradorite ($An_{70}$), hypersthene, and sometimes augite. The structure of the groundmass is vitrophyric and hyalopilitic.

The depressions through which the streams run were formed mainly by explosive means, during violent directed explosions. At the mouth of the eastern depression is a vast debris cone that forms a peninsula 6 km across. There is a debris cone of similar type at the mouth of the northwest depression; it is shaped like a square 3 km on a side.

_____
* Markhinin and Stratula (1965) also note basalts.

These two debris cones break up the original form of the island and "extend" its outline to the northwest.

The sides of the volcanic cone that are not interrupted by the hollows are separated from the debris cones by clearly defined terraces of the old shoreline.

The head of the northwestern depression is filled by the remains of a young cone which joins together both segments of the old structure. Until the 1933 eruption the young cone was whole and also filled the head of the eastern depression; it rose almost 70 m above the old cone (1213 m above sea level).

As a result of the catastrophic eruption of 1933, almost all of the young cone was destroyed and the highest point on the island is now the edge of the old cone, 1145 m. The wide crater formed in 1933, which is open to the east, united with the eastern depression and is now a large amphitheater up to 1.7 km across. The walls of the crater are composed of rocks of different ages: on the north and east lavas and pyroclastics of the old cone are found; on the northwest, pyroclastics of the young cone. Within the crater rose an extrusion dome 2 × 1.5 km across, from which a small flow stretches eastward from the top of the dome to its base.

The deposits of incandescent avalanches, looking like lateral moraines, run down to the east from the base of the dome; the whole eastern depression, the adjoining debris cone, and also parts of the old cone are covered with the deposits of a thick pyroclastic flow and directed explosion. Similar, still older deposits are seen in exposures, testifying to repeated explosions in the past.

The northwestern debris cone, which was not affected by the 1933 eruption, was formed in the past in at least two stages. Another pyroclastic flow has pushed the shoreline out 1.5 km, leaving behind it the prominent cliff of the old shoreline.

A strong eruption occurred in Kharimokotan in 1713; later eruptions are known in 1846, 1848 (?), 1883, and 1931. As already indicated, a very strong eruption took place in 1933. There are two winter quarters on the island, the huts of which are situated in the northern part of the island on the shore of Severgin Bay. The course of the eruption is described from their data (Miyatake, 1934). The eruption was preceded by frequent earthquakes in the fall of 1932.

The eruption began at dawn on January 8, apparently with a roar, and a fiery column shot into the sky. At this same time some tsunamis attacked the island; their height reached 20 m. Tsunamis were also experienced on Onekotan and Paramushir. A southwest wind carried the products away from the huts, but at 10 PM on January 8, the wind changed to the south, and ash, lapilli, and pumice bombs up to 30 cm in diameter fell in the vicinity of the huts of the winter quarters until 4 AM on January 10. The glass in the windows and tiles on the roofs were broken by them, and the docks were also broken up. With a change in the wind once again the products of the eruption were carried off to one side. The explosions ended on January 12. In the vicinity of the huts the thickness of the pyroclastics amounted to 40 cm. The pumice from this area was studied by Nemoto (1934) and was indicated as being a two-pyroxene andesite with 60.5% $SiO_2$.

Explosions were also noted on January 30 and April 14. In the summer of 1933 the island was visited by Miyatake, an engineer in the Ministry of Rural Economics. According to him, the top of the volcano had been demolished and reduced in height by 200 m, the outline of the east coast was much altered, and the shoreline had been extended by up to 100 m (Miyatake, 1934).

Comparison of the Japanese maps of 1916 with present ones allows us to determine the scale of the extension of the shoreline (Fig. 29). Behind the new shoreline a distinct cliff still stands at heights of up to 70-80 m.

The extension of the shoreline was not produced by the simple fallout of pyroclastics. It is very clear that the culminating eruption of 1933 was a directed explosion that carried away a large part of the young cone. Following the explosion, a pyroclastic flow took place to the east, and its products have extended the shoreline. Then an extrusion dome began to grow in the newly formed crater; growth of the dome was accompanied by incandescent avalanches. In the final stage the top of the dome was ruptured by more fluid lava that produced small flows.

The pumice of the 1933 pyroclastic flows is a dacite with 67.5% $SiO_2$. Under the microscope it is seen that the phenocrysts of plagioclase and hypersthene are immersed in a clear, transparent, vesicular glass in which augite and hypersthene occur as microphenocrysts. The dome and flow are composed of hypersthene andesite (59% $SiO_2$) with phenocrysts of labradorite, $An_{55-56}$, and hypersthene; olivine occurs in small but constant amounts. Augite is also present as microphenocrysts and as tiny crystals. The structure of the groundmass is hyalopilitic.

In character the 1933 eruption is referred to the "Bezymyannyi type" (Gorshkov, 1962), but its scale is vastly smaller than the explosion on Bezymyannyi in 1956.

At present the volcano shows intense solfataric activity. A strong solfatara field with deposits of sulfur is found near the west foot of the dome in the depression between the dome and the crater wall.

## Shiashkotan Island

The island of Shiashkotan is located 29 km southwest of Kharimkotan, across the Shiashkotan Strait. The island consists of two separate volcanic massifs joined by a low (about 150 m) ridge that scarcely reaches a width of 1 km. This ridge is composed of Tertiary rocks, which are also found in the west and in the basement of the volcanoes. The island reaches 25 km in length, and its area is 122 $km^2$.

The active Sinarko Volcano is on the northern massif. This $9 \times 11$ km massif has a rather complex structure. Our earlier idea of a large somma broken by faulting probably is not correct. Possibly the northern massif was constructed as a "linear-clustered" type, but it has been strongly eroded, in particular by glacial processes, and it is not possible to reconstruct the preglacial centers without special study.

Gently sloping, smoothed-out cirques and U-shaped valleys on the east and west slopes of the ridge were formed mainly at the time of the first glaciation. On the north of the massif there is a wide depression separated from the massif by rather steep cliffs. Earlier we regarded this depression as a sector graben; however, more detailed examination of its configuration inclines us to the opinion of its dominantly erosional origin. The tectonically weakened zones that are undoubtedly present on this island may have played a certain role. However, the formation of a broad explosion crater that apparently was open to the northwest and subsequent large-scale

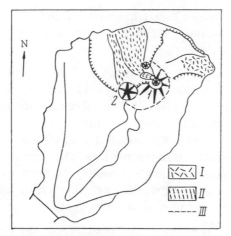

Fig. 30. Sketch of the structure of the Sinarko
Massif: I) Ignimbrites; II) deposits of incandescent avalanches; III) assumed outline of
the old crater; 1) Sinarko Volcano; 2) Zheltokamennyi Dome (other symbols as in Fig. 9).

collapse played the main role in initiating this depression. At present, this crater or caldera (its diameter is about 2 km) is completely buried beneath a young cone and has been reconstructed on
the basis of the rock sequence and breaks in the relief (Fig. 30).

In all probability the deposits of welded tuff in the northeast
sector of the massif, particularly those that make up Cape Krasnyi,
are related to the formation of just this ancient, broad crater.

The central cone of postglacial age completely overlapped the
caldera and filled the heads of the adjacent valleys. The northwest
flank of this cone has been destroyed (whether by explosion or by
collapse is not clear), and a short, broad tongue of lava runs down
along a ravine here. From the crater a steep, dark-colored extrusion dome rises; its agglomerate mantle almost completely overlapped the rim of the crater, and in many places it stretches down
the sides of the cone. The top of the dome is flat and is covered by
a chaotic accumulation of rocks; this is the highest point on the
massif, 934 m above sea level.

A second extrusion dome, Zheltokamennaya Mountain, 898 m
high with a well preserved form, lies 1.5 km southwest of Sinarko
Dome. A badly destroyed structure that is surrounded by a thick

field of altered rocks lies 1 km north of Sinarko. Apparently this is still another dome. All three domes are confined to the edge of the supposed large crater.

In the lavas of the old preglacial structure two-pyroxene andesites predominate, sometimes with olivine among the phenocrysts. The sintered tuffs of Cape Krasnyi and on the left bank of Sernaya Creek are of andesitic composition (57-59% $SiO_2$) and contain visible pieces of plagioclase, hypersthene, and more rarely, pyroxene andesites. Hypersthene usually predominates among the mafic phenocrysts; its large phenocrysts show evidence of corrosion, and along their edges they are fringed by augite. Hypersthene is also dominant in the microphenocrysts.

Eruptions of Sinarko are known from the first half of the eighteenth century, in 1846, and in 1855. In 1872 an Ainu village was destroyed by an eruption on Shiashkotan. We thought earlier (Gorshkov, 1954) that the village was near the southern volcano of Kuntomintar. However, during later investigations it was shown that no trace of a recent eruption is to be found there, and that the form thought to be a lateral explosion crater is actually erosional. According to all the evidence, this village was on the shore of the Shiashkotan Strait, near the north end of the island. It may be suggested that in 1872 a directed blast occurred and demolished the northwest part of the young cone and the village. Then a thick flow of viscous lava was erupted, and finally an extrusion dome was formed. In 1878 Snow visited this place and saw fresh evidence of incandescent avalanches and burned vegetation on the sea shore. Apparently the eruption which began in 1872 had continued for all these years in the form of a lava extrusion. It is possible that the village was destroyed by incandescent avalanches. Even now the deposits of the incandescent avalanches are well exposed. The old, now overgrown deposits of the incandescent avalanches extend from the Zheltokamennyi Dome.

At present columns of fumarolic gases rise from the contact between the crater and the dome and from the surface of the dome on the northwest.

Kuntomintar Volcano occupies the southern, broader part of Shiashkotan. In plan it has an oval form 6 × 7 km across, with its long axis parallel to the axis of the island, in a northeast direction (Fig. 31). The highest point on the volcanic massif reaches 828 m above sea level.

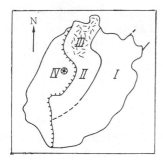

Fig. 31. Sketch of the structure of the Kuntomintar Massif: I) Preglacial structure; II) inner structure; III) ignimbrites; IV) fumaroles.

The Kuntomintar Massif is essentially the eastern half of a complex volcanic structure, the other half of which has been destroyed.

This is a double volcano of the "Somma-Vesuvius" type; however, the structure is largely concealed. The somma makes up the eastern part of the massif; its southeastern part shows evidence of intense glacial erosion. In the northeast the outline of the old caldera can be traced; judging from this area, its diameter was about 4-4.5 km. The inner cone completely filled this caldera, and in the southeast the outline of the caldera is concealed. On the western edge of the massif an amphitheater-like caldera is situated, open to the west and 2-2.5 km across. On the northwest stands high, solitary Bashne Rock, which is also a part of the caldera rim. Apparently the inner caldera was a reservoir basin during the second glaciation. The glaciers descended through the low area to the north and west, leaving Bashne Rock projecting as a nunatak. The floor of the caldera is uneven, with a large number of trenches and closed basins. The northern part of the massif is covered by a mantle of ignimbrites that are undoubtedly related to the formation of the inner caldera. Cape Grotovyi, which sticks out into the Sea of Okhotsk, is composed of ignimbrites.

The structure of the stratovolcano is well exposed in great cliffs; pyroclastic rocks consisting mainly of two-pyroxene andesites are clearly predominant. The plagioclase phenocrysts are of labradorite, $An_{50-70}$; among the mafic minerals hypersthene predominates, and augite occurs in smaller grains. The groundmass has a hyaline, hyalopilitic, or microlitic structure.

In the andesite-dacite bombs phenocrysts of andesine and andesine-labradorite ($An_{42-55}$) and of hypersthene are immersed in a hyaline groundmass; isometric grains of quartz are present.

Along the west shore of Kuntomintar Volcano massif there are distinct faults along which the western part of the massif has been dropped down.

In postglacial times eruptions of the volcano are not recorded, but it shows continuous solfataric activity near the east wall of the caldera. A hot sulfur spring issues here from a steep-sided valley; springs and hot mineralized ponds occur on the floor of the caldera.

## Lovushki Rocks

The rocks are situated in the Kruzenshtern Strait, 20 km south of Shiashkotan. There are four rocks (up to 42 m high) that are disposed in the form of a horseshoe open to the west; they undoubtedly are the remains of the top of a submarine volcano. In the crater of the volcano, inside the semicircle of rocks, the water is about 20-25 m deep. Around the rocks stretches a wide submarine terrace, whose presence defines the age of the volcano as preglacial.

In the northern Kuriles we saw a large variety of volcanic forms and types of eruptions. In the complex volcanoes, as a rule, a slight increase in the silica content of the lavas can be noted in the transition from Pleistocene to Holocene. The predominant rocks are two-pyroxene andesite-basalts and andesites. In contrast to the adjoining western zone, hornblende usually is not found even in the acid extrusions and pumices.

### Central Kurile Islands

There are six islands in the central group: Raikoke, Matua, Rasshua, Ushishir, Ketoi, and Simushir. In addition, between the islands of Rasshua and Ushishir is the group of Srednii Rocks (the remnants of a submarine volcano) and between Raikoke and Matua two submarine eruptions have been recorded. In all of this group there are twelve terrestrial volcanoes, of which only two have not been active in the Holocene; eight are referred to as active at the present time.

## Raikoke Island

The island of Raikoke is separated from Shiashkotan by the 50-km-wide Kruzenshtern Strait. Its closest neighbor to the south, Matua, is 16 km away, across the Golovnin Strait. In plan the island is oval, somewhat elongate in a north-south direction, measuring $2 \times 2.5$ km. Its slopes are bare, with almost no traces of vegetation.

Raikoke Volcano is a strongly truncated cone with a height of 551 m above sea level. The 130-m submarine terrace is absent in the western half of the island.

It may be that the modern cone is set on the edge of an older submarine structure that rises directly from the floor of the Sea of Okhotsk. In this case the height of the volcano amounts to about 2500 m.

On the top of the cone is a large closed crater about 700 m across and up to 200 m deep. The southeastern part of its rim is somewhat higher than the northwestern part, and thus a view of the crater is available only from the Sea of Okhotsk. The broad, steep-sided crater in a relatively low cone presents a very unusual and effective picture. The walls of the crater are extremely steep, and the characteristic structure of a stratovolcano can be seen in the cliffs. The southern flank is covered by pyroclastics, the eastern flank by lava flows.

The rocks are augite basalts. The phenocrysts are mainly labradorite, $An_{65-70}$, and augite; olivine and hypersthene are found in some rocks. The structure of the groundmass is intersertal or hyalopilitic.

Raikoke Volcano erupted in the middle of the eighteenth century, and all the vegetation on the island was burned off. A catastrophic eruption occurred in 1778; the eruption broke out suddenly and 15 persons under the command of Capt. Chernyi, who was returning from Matua to Kamchatka, died under the hail of bombs. In 1780 Capt. Sekerin was sent to Raikoke "to describe and to indicate on a plan the extent to which the island has been built up by the eruption of burning volcanic peaks." This was the first specifically volcanological expedition to the Kurile Islands. According to Capt. Sekerin's description, the upper third of the island was blown away and its outline was changed beyond description; unfortunately, we did not succeed in finding his sketches. Later, apparently, the walls of the crater crumbled and became gently sloping, and its depth was decreased. A hundred years later, in the 1880's, according to Capt. Snow's descriptions, the crater was 30-60 m deep. On February 15, 1924, a strong eruption occurred that significantly deepened the crater, and the outline of the island was again changed. At present the volcano is very quiet and its crater serves as refuge for a multi-

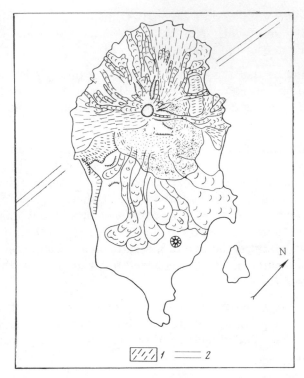

Fig. 32. Sketch of the structure of Sarychev Peak. 1) Incandescent avalanche deposits; 2) fault line (other symbols
as in Fig. 9).

tude of sea birds. The dates of 1777 and 1780, mentioned in some
summaries for eruptions, are mistaken.

Matua Island

The island of Matua lies 18 km south of Raikoke. In plan
Matua has the appearance of a poorly drawn oval, 6 × 16 km across
and elongate in a northwest direction; its area is 50 km$^2$. The
southeast part of the island has a rather flat surface, 50-100 m
above sea level, while the active cone, Sarychev Peak, rises in the
northwestern part. A small, flat-topped island, Toporkovyi, lies
1 km east of Matua. The southeastern part of the island is covered
by scrub vegetation, but there is almost no vegetation on the flanks
of the active cone.

Matua Volcano displays a very complex structure; although the young pyroclastics and lava flows of Sarychev Peak overlap and obscure the older parts, the basic outline of this volcano can be determined rather confidently (Fig. 32).

Tertiary rocks occur in the basement of Matua Volcano. On the Japanese geologic map (Geological Map, 1959) Tertiary rocks are shown on Toporkovyi islet and in the extreme southeastern part of Matua. The presence of presumably Tertiary rocks at the latter site was also confirmed by Markhinin (1964).

The remains of Matua Volcano, which is a somma relative to Sarychev Peak, form the southeastern half of the island. The remains of the old volcano are covered by recent cinders and are thickly covered with vegetation. However, a tremendous cliff in the southwestern part of the island exposes the inner structure of the somma. In an excellent section the layered structure is well displayed with a predominance of lava flows.

The surface of the southeastern part of the somma is covered by thick flows of lava which, particularly in the lower part, show through quite well beneath the cinder layer. Some flows even preserve traces of their surface structures.

One can recognize flows of different ages; the westernmost of these flows descends along a steep gorge that is probably a glacial trough, but the head of this flow is cut off by the rim of the caldera. Thus, this rather young flow is related to the building of the somma. An adjacent, very long flow issues from above the rim of the somma and undoubtedly belongs to the central cone. The next flow to the east is also related to the central cone. The remaining flows most probably are related to the somma and in part to the subsidiary craters, one of which, on the side of the cone, has produced lava flows and a second of which, at the base, shows no lava emissions. To a considerable degree the flows fill in the irregularities of the somma cone; however, evidence of intense, probably glacial erosion shows through beneath the flows.

In the southwestern part of the island, at a height of about 850 m, there is a small area where we can follow the boundaries of the caldera. Farther to the east and north the caldera is hidden beneath the deposits of the central cone, though its outlines can be fixed by a break in slope. The diameter of the caldera is 3-3.5 km.

A 140-m submarine terrace adjoins the eastern part of Matua; right there, where the structure of the somma ends and that of the central cone begins, the submarine terrace ends. This terrace is absent in the area of the central-cone structure. Special echo-sounding work carried out on the R/V "Geolog" supports the hypothesis that we have put forward, that the northwestern part of the island has been dropped down along a fault of great displacement. This fault cut off part of the somma and of the caldera.

The central cone, Sarychev Peak, occupies the northwestern part of the island. Seen from the northwest, the cone has a very regular form with a weakly truncated summit; seen from the southwest or northeast, its summit seems to be somewhat elongate in a northwest direction. The western slopes of the cone descend directly to the Sea of Okhotsk, but the eastern slopes butt against the caldera, to a considerable degree overlapping it.

The diameter of the crater of the volcano is about 250 m; its southeastern rim is considerably higher than the other parts and reaches 1497 m above sea level. The walls of the crater are vertical, locally even overhanging (see Fig. 33). During an ascent in 1946 we did not succeed in making out the structure of the floor through the masses of rising steam; in 1954 a weakly convex shield of congealed lava, covered by a network of characteristic fissures, was easily visible on the floor of the crater at a depth of about 200 m. From the highest part of the crater rim, on the southeast, a small ridge with two "shoulders" extends to the southeast flank; here we can recognize traces of an older crater. Apparently the old central cone whose flows run down to the southeast was located somewhat farther to the east. Later the crater apparently migrated to the northwest. As a result of this movement, the central cone is somewhat elongate from southeast to northwest.

The upper part of the cone immediately adjacent to the crater is "ringed" by the sources of well-preserved lava flows (Fig. 33). Below the lava "collar" the surface of the cone is covered largely by deposits of pyroclastic flows and nuées ardentes. However, in many places lava flows occur right at the surface or just beneath its pyroclastics. Many capes are composed of lava tongues. A number of flows have begun to overlap the still preserved northern part of the summit. Below the lava "crown" there are many deep ravines; young flows descend along some of them, and all of them served as routes for the descent of the 1946 pyroclastic flows.

Fig. 33. Crater of Sarychev Peak.

The lower, more gently sloping part of the flanks is composed mainly of a coarsely layered series of incandescent avalanche deposits. The material in these was very hot, and in deposits that were sufficiently thick the processes of posteruptive oxidation continued for a long time and led to a reddening of the rocks. Numerous steam columns were still rising from the surface of such deposits in 1954, eight years after the violent 1946 eruption. The thickness of these deposits, which are chaotic mixtures of fine ash and sand with blocks of a great range of sizes, exceeds 20 m in the coastal cliffs.

The lavas of the volcano are composed of basic two-pyroxene andesites and andesite-basalts. The phenocrysts are of labradorite ($An_{53-68}$), augite, hypersthene, and isolated grains of olivine. Rarely one of the pyroxenes is absent. The structure of the groundmass is hyaline, hyalopilitic, or microlitic. Sopochka Kruglaya (an old subsidiary cone) is composed of aphyric basalt.

In cliffs in the lower part of the cone interlayers of light-colored lapilli of hornblende andesite are recognized among the pyroclastics; the hornblende is green with strong pleochroism, and the groundmass is a vesicular fluidal glass. It may be that these lapilli are related to the caldera-forming explosion.

Eruptions of Sarychev Peak are rather frequent. They are expressed in a variety of ways, and rather commonly they are of considerable violence. A very strong eruption occurred in the 1760's. In the winter of 1878/79 a quiet emission of lava occurred on the northeast flank and ran down to the shoreline. Flows descending to the southwest apparently are even younger, but the date of their emission is unknown. Explosions with the ejection of ash and scoria occurred from January 17 to 22, 1923 (Kamio, 1931). An explosive eruption occurred on February 14, 1928. A short, but very strong eruption took place on February 13, 1930, when a vast amount of pyroclastic material accumulated near the foot of the cone.

One of the strongest eruptions occurred from November 9 to 19, 1946. It began with a rather weak explosion of the Vulcanian type; then a light phenomenon was seen above the crater and incandescent bombs began to be ejected. On the morning of November 13 the eruption became stronger, and bombs that fell up to 7-8 km from the crater completely destroyed the scrub vegetation on the east slope of the somma and the coast. During the day of November

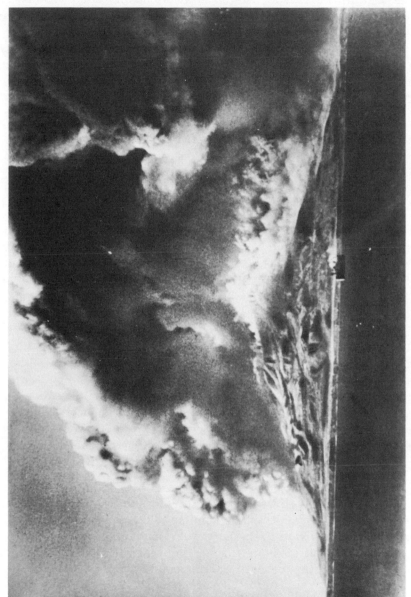

Fig. 34.  Eruption of Sarychev Peak in November 1946.  Photograph by B. Prokakhin.

13 numerous incandescent avalanches, above which rose ash clouds of the "cauliflower" type, flowed down along the northwestern half of the cone. The whole island seemed to be enveloped in flames (Fig. 34). A mass of loose material accumulated near the base of the cone and considerably changed the shape of the coastline. Later sea waves eroded the loose material of the newly formed headlands and redeposited the sand in small bays, producing further changes in the topography of the island.

At the end of the summer and in the fall of 1954 the volcano renewed its activity somewhat, rather rare and weak ejections of ash occurred, and at times light effects were displayed above the crater. A single explosion that raised an ash cloud 4.5 km above the crater occurred on August 30, 1960 (Shilov, 1962).

The history of development of the volcano may be summarized in the following way. A volcano began and was formed in early Quaternary times. Along with all the Kurile Islands Matua was subjected to the first glaciation; the glaciers left troughs and cirques on the flanks of the cone. Later activity has "healed" the scars of glaciation. The genesis of some of the subsidiary craters contributed to this. The second glaciation left no evidence of erosion; here in the Central Kuriles it was weaker than in the North Kuriles. At this time the 140-m submarine terrace was produced. In the beginning of the Holocene a series of faults with very considerable displacement was established along the west edge of the Central Kuriles. Perhaps the beginning of tectonic movement provoked the development of the caldera on the top of Matua Volcano. The northwestern half of the island was dropped beneath the surface of the sea, and in the amphitheater of the remaining half of the caldera arose the central cone, which not only completely filled the basin of the caldera but almost everywhere overlapped its rim.

The individual flows of lava, after crossing the caldera rim, flowed out onto the flanks of the somma. The central cone crater migrated once or twice to the northwest, and as a result the cone took on a somewhat elongate form. The migration of the crater most probably was related to continued movement along the fault plane.

Eruptions of the central cone are characterized both by the rather quiet emission of voluminous lava flows and by violent explosions with the development of pyroclastic flows. Thick deposits of various pyroclastics are seen to the east of the cone where they cover the flanks and base of the somma as a continuous mantle.

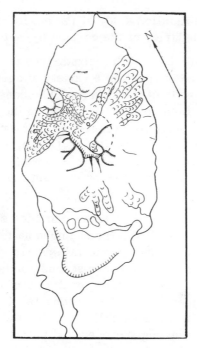

Fig. 35. Sketch of the structure of Rasshua
Volcano (symbols as in Fig. 9).

## Rasshua Island

Rasshua is separated from Matua by Nadezhda Strait, 28 km
wide. In plan the island has an oval form and is strongly elongate
in a north-south direction. It is 6 × 13 km across, with an area of
62 km². An important part of the island is covered by low, brushy,
but difficult to penetrate vegetation.

Rasshua Volcano has a complex structure of the Somma-
Vesuvius type (Fig. 35). The remains of the somma form the northern and southern ends of the island, but at the very south end the
basement, mainly of Tertiary age, crops out. The old somma has
been subjected to intense erosion, apparently including glacial. The
northern massif is a rough high area about 800 m above sea level;
the southern is a crescentic ridge up to 500 m high. Judging from
the disposition of the preserved parts, the somma had a caldera up
to 6 km across. The western edge of the caldera is near the western coast of the island and the eastern edge lies beyond the present

shoreline. More detailed study of the coastal sections allows us to
establish the boundaries of the caldera more closely.

In the middle part of the island rises the central cone, complex and composed of rocks of a variety of ages, which almost completely fills the basin of the caldera. The cone has a somewhat eccentric position, displaced to the north; on the south an area of the
flat atrio, 1-2 km wide, is preserved. The atrio has free access to
the Sea of Okhotsk and to the Pacific Ocean. On its floor are two
fresh-water lakes; from one of these flows a small stream that
empties into Nepristupnyi Bay on the east, by way of a spectacular
25-m waterfall.

The central cone has a rather complex structure. The main
cone, whose slopes make up the southern half of the central massif,
is rather strongly eroded. Some valleys are reminiscent of cirques
and apparently are related to the second glaciation. In the southern
half the rim of the broad summit crater, strongly altered by erosion, sticks up; judging from the remnants, its diameter was as
much as 2 km. Regardless of the fact that the upper part of the
slopes are rather strongly eroded, individual lava flows are recognized near the base of the cone in the south.

On the northwest shore lies a small isolated cone, Mount
Razval (736 m), which is apparently of the same age. Its crater,
up to 300 m across, is open to the Sea of Okhotsk.

In the ruined crater of the old central cone rise two Holocene
cones. One of them is the highest point on the island (956 m); on
its summit a rather shallow crater is preserved, elongate in an
east-west direction; its dimensions are $75 \times 120$ m. Numerous lava
flows extend westward from the crater. Some of the flows have
flooded the floor of the old crater, others run down still farther,
to the coast; one of the flows descends between Razval Cone and the
remnants of the somma. Some flows also extend to the northeast,
toward the Pacific Ocean. The eastern, now active, cone has a wide
(more than 500 m) explosion crater that is broadly open to the southeast, like an amphitheater. Light-colored, altered rocks that stand
out clearly against the dark background of the brushy vegetation extend from the crater to the seashore as a broad front. A thick flow
of very fresh-looking lava extends down the west flank of the cone
as a short, but broad tongue. Older flows run down to the east and
northeast as far as the ocean.

The somma in the southern part is composed of two-pyroxene andesite-basalts and andesites. The phenocrysts are of labradorite ($An_{55-65}$), augite, and hypersthene; scattered grains of altered olivine are found.

The structure of the groundmass is intersertal or microlitic.

Pumice of hornblende-dacitic composition (the hornblende green with a small extinction angle) is found in the atrio. The groundmass is hyaline with perlitic fracture.

Bombs from the somma rim apparently are related to eruptions of the central cone. They are two-pyroxene andesites with phenocrysts of labradorite ($An_{59-70}$), hypersthene, and augite. The groundmass is of dark glass.

A strong eruption of the volcano was noted in 1846, and it may have been during this eruption that the eastern cone was ruined. On November 4, 1946, before the eruption of neighboring Sarychev Peak occurred, Rasshua Volcano markedly increased its fumarolic activity. In October 1957 an increase in activity again was noted, possibly with weak explosions. The volcano displays continuous fumarolic activity in the eastern crater and in the saddle between the eastern and western cones.

The history of development of the volcano is given in the following summary. The volcano began and went through a long cycle of evolution, leading up to the formation of the caldera, in preglacial times.

At the time of the first glaciation, the walls of the volcano and its caldera underwent intensive erosion. In interglacial times a large high cone appeared on the floor of the ruined caldera; its eruptions were brought to an end by a large explosion and the development of a wide crater. At the same time a second eccentric cone was formed near the northwest base of the central cone. The second glaciation eroded the upper slopes of the central cone and somewhat damaged the summit crater.

In postglacial times two more small cones arose in the basin of the broad crater of the central cone; their lava flows have covered the northern part of the old cone and radiate as far as the shore of the Sea of Okhotsk and the Pacific Ocean. In the last stage the summit of the eastern inner cone has been destroyed.

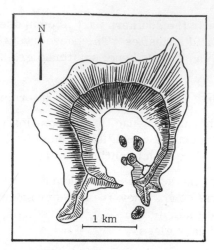

Fig. 36.  Sketch of the structure
of Ushishir Volcano.

To the south of Rasshua stretches one of the most dangerous straits, Srednii Strait.  The rocks in this strait, S r e d n i i   R o c k s , are, in all probability, the remnants of a submarine volcano.  The wide submarine terrace that joins Srednii Rocks to the islands of Ushishir and Rasshua fixes the age of Srednii Volcano as preglacial.

It may be that the rather flat-topped Khitraya Rock is a rather young extrusion dome (Holocene) and that the adjoining rocks (Chernye, Botsman, and Pugovka) are situated on the rim of an old caldera.  In this case the diameter of the caldera is about 3 km.

## Ushishir Islands

Ushishir is located 17 km southwest of Rasshua, across the Srednii Strait.  Two islands are included in the Ushishir group: Ryponkich on the north and Yankich on the south.  The former has the shape of an elongate rhombus, $1 \times 3$ km; the latter, the form of a ring about 2.5 km across and with a bay that cuts deeply into the island from the south.  The only vegetation on the island is grassy.

In the eastern part of Yankich, Markhinin notes outcrops of presumably Tertiary basement and of quartz diorite.

Ushishir Volcano (Fig. 36) has a structure of the Somma-Vesuvius type.  The somma of the volcano has been considerably

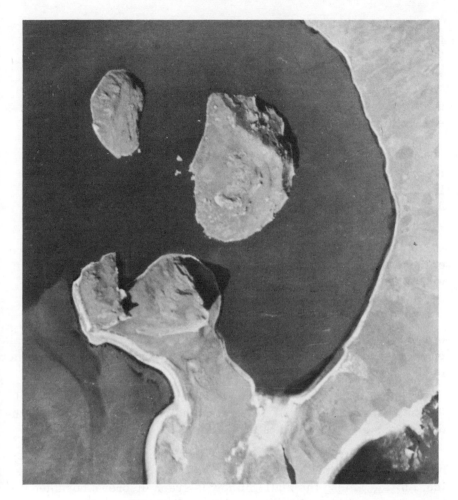

Fig. 37. Domes in the crater of Ushishir Volcano.

destroyed on all sides by the waves of the Pacific Ocean and the Sea
of Okhotsk. Of the former gently sloping shield-like volcano, which
was at least 10 km across, there remain two small islands whose
relief (they are very steep sided) poorly reflects the original form
of the volcano.

Ryponkich is a preserved portion of the base of the somma.
The rather even surface of the island rises gradually from 20 m on
the north to 130 m on the south. This part of the somma, from which
we can judge the original dimensions and form of the volcano, is cut
off on all sides by steep, high cliffs.

Yankich, a remnant of the near-crater part of the somma, forms a steep-sided ring-like ridge enclosing a caldera up to 1.6 km across. The southern part of the caldera wall is broken, and waters of the Pacific Ocean fill its floor, forming a caldera bay 1 km across and, according to Capt. Snow's measurements, 58 m deep. At present the entrance to the bay is 300 m wide. The eastern wall of the caldera is also badly destroyed; the width of the barrier between the bay and the open ocean here amounts to only 200-300 m. The height of the north and west walls of the caldera reaches 250 m above sea level; the highest point on the rim is in its northwestern part, where it rises to 400 m above sea level. Judging from the presence around the island of a wide submarine terrace, the somma is preglacial in age.

Within the caldera rise four large monolithic rocks. The two of these in the southern part of the crater have a pointed form and are joined by a low sand bar. These rocks are monolithic and apparently are remnants of old extrusion domes.

Two flat-topped well-preserved extrusion domes that are bean-shaped in plan stick out approximately in the center of the caldera bay. One of them is $100 \times 200$ m across and 12 m high; the other is $200 \times 300$ m across and 32 m high (Fig. 37).

These four rocks (domes) are set in the form of a ring about 0.5 km in diameter. In all probability their distribution reflects the dimensions and position of a submarine conduit.

These four domes are composed of rather uniform pyroxene-hornblende andesites. The plagioclase of the phenocrysts is andesine, $An_{40-45}$; among the mafic minerals green or brown hornblende is dominant, with strong pleochroism and a rather small (10-12°) extinction angle. Sometimes the hornblende is partly decomposed and surrounded by an opacite rim. Pyroxenes are present in small amounts; hypersthene is more abundant than augite. In some sections large crystals of hypersthene are surrounded by rims of hornblende; the hypersthene is yellowish with distinct pleochroism. Quartz is sometimes present. The structure of the groundmass is cryptocrystalline.

The only rock sample that we studied from the eastern part of the caldera was called a hornblende dacite with phenocrysts of andesine ($An_{48}$), quartz, and green hornblende. Ortho- and clinopyroxenes were noted as 1-2 grains per thin section. The structure of the groundmass is microfelsitic.

According to Markhinin's data (1965), the rocks of Ushishir are augite andesites and andesite-basalts.* In addition to amphibole-pyroxene andesites, Nemoto (1938) also noted two-pyroxene andesites and dacites with traces of olivine.

Near the southeastern part of the caldera wall a cluster of intense fumaroles and hot springs lies on a low sandy shore near the water level; in the eighteenth and early nineteenth centuries this was a sacred place to the Kurile Ainu.

A colorful description of the cult ceremony is provided by Cossack Captain Chernyi, who also gives an excellent description of the crater bay. The bay made a great impression on Snow, who devoted several pages to its description. However, even a visit to the crater bay does not give one a complete impression of the form and beauty of this unique volcano. The volcano is seen as a whole only from the air: on the dark-blue, even surface of the ocean a distinct ring emerges, tinted by the varied color of the volcanic rocks and the green of the vegetation, and in its center shines the crater bay.

The fumaroles and hot springs on the shore of the bay continue with unabated strength at the present time; true eruptions, however, have not been noted. [Radiocarbon dating of carbon from the bottom of the calderic pumice (done in 1968 in the Institute of Geology) set the age of the caldera at $9400 \pm 50$ years, which corresponds to the early Holocene. Thus, the author's supposition as to the age of the caldera was corroborated.] Some phrases from the "Journal" of Captain Chernyi allow us to suppose that weak explosions sometimes occurred here at the beginning of the eighteenth century. A weak explosion in the area of the fumarole field, apparently purely phreatic, occurred in July 1884. In spite of the absence of information on eruptions, one can assume that the development of the two northern extrusion domes took place after 1769, when Captain Chernyi was on the island. He described the island in detail and even noted the individual rocks and stacks near the

---

* According to Markhinin and Stratula (1964), the domes are composed of augite andesites and andesite-basalts; we did not find any such rocks in a single one of the numerous thin sections of the dome lavas.

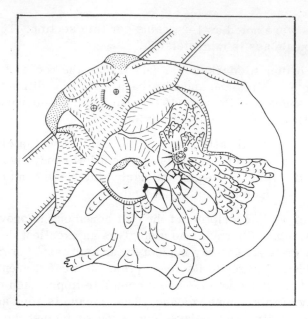

Fig. 38. Sketch of the structure of Ketoi Volcano
(symbols as in Fig. 9).

coast of the island. In his description there is mention of two coni-
cal rocks in the bay, whereas at present there are four rocks in the
caldera. Clearly the two fresher-looking domes arose after the
visit of Captain Chernyi, i.e., after 1769.

### Ketoi Island

Ketoi is separated from Ushishir by the Rikord Strait, whose
width is up to 26 km. In plan the island has the form of a rather perfect circle 10 km in diameter and with an area of 71 km$^2$. Vegetation on the island is mainly scrub, though there are some low birches;
here for the first time the Kurile bamboo appears, locally forming
almost impenetrable thickets.

Ketoi Volcano is one of the most complex structures among
the Kurile volcanoes, and in what follows only a preliminary sketch
of its structure can be given (Fig. 38) until a special, detailed study
has been made.

The greater part of the island is occupied by remnants of an
old caldera volcano. The rim of the caldera is preserved only in

the northeast sector, where its height reaches 720 m.  In this same
area the gentle slope of the old volcano is preserved, bounded on
the ocean side by a 100-m cliff.  Other parts of the old cone either
were destroyed by later processes (on the northwest) or were overlapped by younger flows and protrude as basement from beneath the
young flows; in places they are exposed in the coastal cliffs.

The diameter of the old caldera is estimated at 5 km.  Judging from the blocks of ignimbrite (not found in place), the formation of the caldera was accompanied by incandescent pyroclastic
flows with the formation of ignimbrites.  In all probability these
rocks may be found in place in the eastern part of the island which
we did not investigate.

The younger central structure forms the northwestern part
of the island.  Its height reaches 1172 m.

The eastern part of the central structure has been intensely
eroded, in part by glacial action, and no volcanic forms are preserved; the crater also is completely ruined, and the northwestern
part of the cone, along with the old structure, is broken up by a
series of longitudinal faults (relative to the trend of the arc). These
have produced a deep graben and an adjacent horst (or perhaps a
somewhat less strongly depressed block).

In the lower part of the horst remnant an almost horizontal
series of shallow-water tuff-conglomerates containing a mass of
small, well-rounded pebbles is exposed; this is overlain by normal
subaerial lavas and pyroclastic rocks.

The fault tectonics are strongly underscored by the disposition of the submarine terrace.  Deep sea (several hundred meters)
adjoins the island on the northwest, but all the rest of the island is
surrounded by a 140-m submarine terrace.

In the graben there are two solfatara fields and the remains
of an enclosing pyroclastic cone.  This cone has been destroyed by
intense erosion, but between the solfatara fields there is a thick
patch of pyroclastic material with a monoclinal dip to the northeast
that is unconformable against the walls of the graben.  In all probability there were two or perhaps more craters in the graben.

In the central part of the island lies a rather broad, steep-
sided depression whose floor is occupied by a fresh-water lake

Fig. 39. Pallas Peak.

about 1.5 km across. This is a young explosion crater that was probably formed as a result of the explosion of one of the young intracaldera cones. The existence of a former cone here is indicated by some well preserved flows (one of which forms Cape Monolitnyi). The southern edge of the young caldera apparently coincides rather closely with the rim of the old caldera. The young caldera has also bitten into the adjacent part of the old central structure, and high (up to 500 m) steep cliffs extend from the high points of the island down to Ketoi Lake in the caldera.

On the east edge of Ketoi Lake there is a young extrusion dome and in line with it the remains of a second intracaldera cone with large flows of lava stretching to the southeast.

The old central edifice partly overlapped the rim of the old caldera, but on the east a rather wide atrio remained in which, judging from the deposits, a lake existed for a long time. The contemporary intracaldera active cone, Pallas Peak, lies right in this atrio (Fig. 39).

The structure of Pallas Peak resembles in miniature a structure of the Somma-Vesuvius type. The southwestern, older and higher (about 1000 m) part of the volcano forms an amphitheater open to the south, in which is located the young, still higher cone. The northeastern flanks of the latter descend to the atrio of the old caldera, while the southeastern flank, which rests against the wall of the amphitheater, is very poorly developed; between the rim of the inner cone and the wall of the outer one there is a small depression — a miniature atrio.

In general the cone is somewhat elongate from southwest to northeast, and the rim of the crater in plan has the form of a weakly concave figure eight. The relative height of the cone above the level of Lake Ketoi is 320 m. The crater size is large for such a small cone; its total width along the axis is 550 m. The diameter of the crater of the inner cone is about 400 m. Its walls slope steeply downward, and on its floor is Glazok (crater) Lake, 300 m in diameter. The color of the lake water is turquoise of a very beautiful shade.

Old, strongly overgrown lava flows that are apparently related to the outer structure run down to the southwest from the cone to the shore of the young caldera lake. Numerous young lava flows extend eastward all the way to the sea coast, 4.5 km away. A series of flows from the crater of the inner cone extend northward, filling the atrio of the old caldera. Pumice deposits related to the explosion of the young caldera are exposed beneath these lavas, and still lower are lake beds of the old caldera.

On the northeast side of the cone, in one of the small canyons, is a linear group of very strong fumaroles.

The rocks of Ketoi Volcano vary from olivine basalts to acid (andesitic) ignimbrite and pumice. In the building of the old caldera and of the old central cone basalts and andesites are dominant. The modern lavas of Pallas Peak are normal two-pyroxene andesites.

In the first half of the eighteenth century the volcano showed no traces of activity, even fumarolic. The first information on the activity of Pallas Peak dates from 1843: "In July and August the whole island seemed to be enveloped in flames" (Perrey, 1864). The eruption was very violent (apparently large effusions of lava took place) and lasted until 1846. An eruption is known from 1924, but details on it are lacking. Ejection of pumice was noted on September 27, 1960.

At present the volcano shows fumarolic activity on the outer slopes of the cone. Apparently there are fumaroles in the crater near the north shore of the crater lake. At the time of our visit to the crater sporadic bubbling noises were heard to the north, but it did not seem possible to get there because of the steepness of the crater walls.

The history of the formation of Ketoi Volcano may be summarized as follows: In the early Quaternary a broad shield volcano was formed on the Tertiary basement. Then a gigantic explosion occurred, with the formation of incandescent pyroclastic flows and ignimbrite deposits.

As a result of this explosion a caldera, up to 5-6 km across, was developed. It may be that the caldera is eccentric relative to the outline of the island (is displaced to the northwest) and the island took on a half-moon shape, with a caldera bay that had a rather narrow connection with the Sea of Okhotsk. The first glaciation left no obvious traces; clearly, the low, gently sloping sides of the caldera cone were not suitable for the development of glaciers.

In interglacial times the central structure was formed, asymmetrically with respect to the caldera; thus, the northwestern part of the central cone extended beyond the caldera, partly overlapping its flanks, whereas on the east, between the central structure and the wall of the caldera, a wide atrio with a lake was left. The lake sediments are exposed in the atrio beneath the lavas of Pallas Peak and the pumice of the young caldera. The stream flowing north from the atrio does not cut a canyon near the exit from the caldera, but percolates through the lake sediments and comes out on the other side of the break as numerous high-volume springs that unite farther down to form a single channel.

The gravel that occurs in the basal part of the northwest "horst" probably is related to the lower levels of the caldera lake, but the possibility is not excluded that they were deposited in a shallow caldera bay or in an adjoining part of the sea.

The second glaciation considerably eroded the central structure (particularly its upper part). Evidence of the second glaciation may be buried (along with that of the first?) beneath the young lavas on the outer slopes of the old caldera edifice.

At the beginning of postglacial time the northwestern part of the island was broken by a system of faults. A large part of the central structure and a small part of the old volcano were involved in the subsidence, forming a long, narrow graben.

In this graben a chain of cones arose whose deposits filled the graben almost to the top. These cones were later rapidly destroyed by running water, but remnants of the pyroclastic deposits blown out into the graben are still preserved. Here intensive solfataric activity also continues.

However, the main postglacial activity was concentrated to the east of the central structure, approximately at the edge of the old caldera.

At first a rather large cone that adjoined the old central structure on the north and whose lava flows on the south spread down the outer slopes of the somma to the sea coast arose near the southern edge of the old caldera. Capes Monolitnyi and Okruglyi are composed of its flows. The development of this cone was ended by a large explosion that formed a young explosion caldera about 2 km across.  The pumice from this explosion covers the lake sediments in the atrio and the lower parts of the flanks of the central cone.

At present this caldera is occupied by Ketoi Lake whose level is at about 600 m.  The waters of the caldera lake flow south-eastward through Stochnyi Creek into the Pacific Ocean.  The rim of the caldera retains its form best in its southern half, where it cuts off the heads of some fresh lava flows.  The height of the rim here is about 800 m; the inner wall of the caldera falls steeply for 100–200 m to the surface of the lake, but to the east its outlines are disrupted by later volcanic structures.

After the formation of the young inner caldera a large extrusion dome was formed on its eastern edge.  Then near the dome a cone rose that produced lava flows 2.5–3 km long.  Finally, Pallas Peak was formed somewhat farther to the north; its lava flows ran farther to the east as far as the coast, and to the north they flooded the remaining part of the atrio where they overlapped the pumice deposits of the inner caldera.

Simushir  Island

The island of Simushir is separated from Ketoi by the 19-km wide Diana Strait.  This is the southernmost and largest of the Central Kurile Islands.  It stretches over a length of 59 km, with a width of 4–15 km and an area of 340 km$^2$.  There are six volcanoes on the island; these form four mountainous massifs that are separated from one another by low isthmuses (Fig. 40).  On these isthmuses and locally in the bases of the volcanoes occur rocks of the Tertiary basement.

Uratman  Volcano  forms the northeastern end of the island.  This volcano is constructed according to the Somma–Vesuvius type.  Of the somma, only a narrow arcuate ridge is preserved, bordering a broad caldera with a rim-to-rim diameter of 7.5 km.  The rim of the caldera is not closed on the southeast, and that is where the central cone is located.  In addition, the wall of the caldera is breached on the north, facing Diana Strait.  A narrow (about 200 m) passage joins the strait with Brouton (caldera) Bay.  The bay, in the form of a half-moon, is about 2.5 × 5.5 km across and up to 250 m deep.  The height of the caldera walls reaches almost 450 m, so that the total depth of the caldera is as much as 700 m.

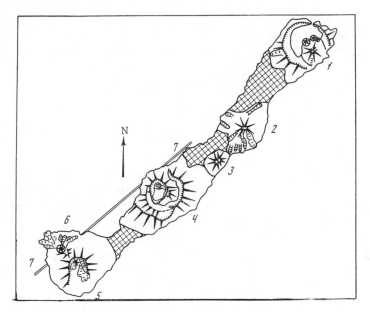

Fig. 40. Sketch of the distribution of the volcanoes on Simushir: 1) Uratman Volcano; 2) Prevo Peak; 3) Ikanmikot Volcano; 4) Zavaritskii Caldera; 5) Milne Volcano; 6) Goryashchaya Sopka (other symbols as in Fig. 9).

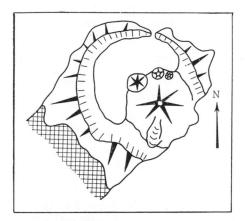

Fig. 41. Sketch of the structure of Uratman Volcano (symbols as in Fig. 9).

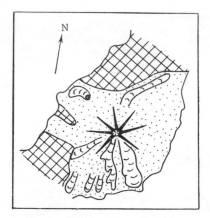

Fig. 42. Sketch of the structure of Prevo
Peak (symbols as in Fig. 9).

To the southwest the caldera is closely joined to the Olenii Ridge; in other places it drops off directly into the sea.

The central cone, Uratman, is located somewhat eccentrically, in the southeastern part of the caldera. It is a strongly overgrown stratovolcano with a ruined circumcrater area. Its basal diameter reaches 4 km and its height, 679 m. Between the foot of the cone and the wall of the caldera are two portions of a narrow atrio which is open toward the caldera bay and toward the Pacific Ocean. To the north of the cone, on the shore of the caldera bay, there are three subsidiary structures. They are strongly overgrown and are almost unexposed. Apparently the largest (westernmost) one is a dome and the other two are cinder cones (Fig. 41).

The lavas of the Uratman somma are two-pyroxene and hypersthene andesites and andesite-basalts. In the lower parts of the section aphyric and sporadically porphyritic varieties occur, with very sparse phenocrysts of plagioclase and pyroxene. The groundmass is pilotaxitic. Olivine-free pyroxene basalts with intersertal groundmasses are also found.

The central cone and the subsidiary craters are composed of two-pyroxene andesites. The phenocrysts are of labradorite ($An_{56}$), orthopyroxene, and clinopyroxene, often in the form of glomeroporphyritic aggregates. The structure of the groundmass is cryptocrystalline or hyaline. The plagioclase microlites are andesine-labradorite, $An_{48-50}$.

The subsidiary dome is composed of amphibole-pyroxene andesites. The phenocrysts are of labradorite ($An_{56-58}$), brown hornblende with a small extinction angle, hypersthene, and augite, with an occasional grain of olivine. The structure of the groundmass is andesitic. The plagioclase microlites are andesine, $An_{46}$.

Fig. 43. Crater of Prevo Peak.

The age of the construction of the somma is thought to be preglacial; the caldera was formed after the first glaciation (tentatively, in interglacial times). The central structure is Holocene. At present the volcano shows no signs of activity.

Prevo Peak lies in the middle part of the island, 16 km southwest of Uratman Peak, near the southern end of Olenii Ridge. This volcano forms a beautiful, very regular, isolated, truncated cone, the profile of whose slopes has the characteristic form of a logarithmic curve. The Japanese placed Prevo Peak in the same class with regard to beauty as Fujiyama, calling it Simushiru-Fuji, i.e., the Fuji of Simushir.

The cone rises near the eastern foot of Olenii Ridge, so that the eastern flanks of the volcano, whose growth is unimpeded, stretch for 11 km along the shore of the Pacific Ocean. The lavas and pyroclastic products of Prevo Peak began to spread westward only when the cone had reached an already rather considerable height and "overflowed" through the spurs of Olenii Ridge which here reaches a height of 400-500 m. Thus, along the shore of the Sea of Okhotsk the rocks of Prevo Peak stretch only for a distance of 4 km and are interrupted here and there by protrusions of the basement rock (Fig. 42).

The crater of the volcano is somewhat elongate in an east-west direction and measures $450 \times 660$ m. Its rim is lower on the south, and on the northeast it rises to a high point at 1361 m.

In the crater is a small, closed, inner cone in the form of a rather low ring-wall composed of pyroclastic material. On its summit is a deep funnel-like crater 350 m across that passes downward into a perpendicular shaft 200 m across (Fig. 43). In the vertical walls of the shaft alternating layers of lava and pyroclastic material are seen, and on its floor there is a small crater lake.

The upper part of the cone is covered by a mantle of scoria from beneath which the tops of lava flows protrude. In the lower part of the slopes the lava flows show through more clearly and locally, especially in the south, they have a fresh appearance. Cape Vasin, the other capes on the Pacific coast, and also Cape Polyanskii on the Okhotsk coast are composed of the lavas of Prevo Peak; the flows are up to 4-5 km long. On the western foot of the volcano are two small subsidiary cones with strongly overgrown tongues of

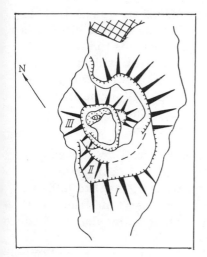

Fig. 44. Sketch of the structure of Zavaritskii Caldera: I) first somma; II) second somma; III) central cone with present caldera (symbols as in Fig. 9).

lava somewhat more than a kilometer long. These cones are not very young, and the lavas from Prevo Peak flow around them. The age of the whole structure is postglacial.

The lavas and pyroclastics of Prevo Peak are basalts. The phenocrysts are of basic labradorite ($An_{65-72}$), olivine (often with opacite rims), and coarse augite crystals (also rimmed). The structure of the groundmass is intersertal or hyaline.

At present the volcano shows no signs of activity, but it has been very active in the recent past. Thus, a violent eruption occurred in the 1760's, apparently with incandescent scoria flows that killed off all the vegetation at the foot of the cone. Weaker eruptions took place in the first half of the nineteenth century. In more recent times the volcano has held its peace; fumarolic activity was noted only in June 1914.

At present Prevo Peak does not show even traces of fumarolic activity, but there is still a good possibility of a strong eruption here.

Ikanmikot Volcano is on the shore of the Pacific Ocean right next to Prevo Peak. This is an extinct, now badly ruined cone. Its height reaches 645 m; on its top is a deep basin, and the southeastern part is cut off by a cliff several hundred meters high. Ikanmikot Volcano arose after the first glaciation; it belongs mainly to interglacial times.

Zavaritskii Caldera is situated in the southern half of the island, 15 km from Prevo Peak; it forms the third mountain massif from the north. It is a very complex volcano, in the structure of which two sommas and a partly blown-away central cone with a lake-filled caldera are involved (Fig. 44). The modern eruptive apparatus is located in the inner caldera. Tertiary rocks are exposed in the basement of the first somma.

The first somma is only partly preserved as a semicircle in the southern part of the volcano. The rim of the somma rises to a height of 450-500 m, and on the northeast it reaches to the rim of the second caldera. The northern part of the first caldera has not been preserved; it may be that the high ground between the northeastern part of the second somma and Ikanmikot Volcano is a remnant of the northern part of the first somma. The diameter of the first somma very probably was as much as 10 km.

The outer slopes of the first somma are well preserved on the southwest. Their upper parts slope at about 15°; their lower parts, where they descend gently to the narrow Kostochko Isthmus that separates Zavaritskii Caldera from the massif of Milne Volcano, have slopes of only 2-3°. The upper half of the slope is drained by numerous small narrow valleys that combine downslope into several large wide ones. Here there is evidence of some strongly modified troughs of the first glaciation. The western part of the first caldera is cut off by a fault, and a steep cliff 520 m high descends to the Sea of Okhotsk. In the lower half of this cliff the remains of a Tertiary volcano are exposed, while at a height of about 300 m above sea level the old eroded stratovolcano is unconformably overlapped by a gently dipping suite of the thick lava flows of the first somma. The base of the southeast flank is badly destroyed by marine abrasion, and steep cliffs up to 200 m high drop away to the Pacific Ocean.

The inner slopes of the somma are steeper than the outer ones. In the extreme western part the difference in height between the rim of the somma and the depression (atrio) that separates the first and second sommas amounts to 350 m. To the east the difference in height gradually decreases. Here in the atrio there are indications of glacial activity; the glacier flowed off to the west.

The first somma is composed mainly of lava flows ranging in composition from andesite to basalt and of subordinate pyroclastic material.

The pyroxene andesites contain phenocrysts of zoned plagioclase (labradorite-bytownite, $An_{70}$, in the cores; andesine-labradorite, $An_{50}$, in the outer zone) and augite. The groundmass is microlitic, with a small amount of glass; the microlites are of andesine, $An_{40}$, and augite.

The andesite-basalt has a seriate-porphyritic structure, with phenocrysts of zoned plagioclase (labradorite-bytownite, $An_{70}$, in the cores and labradorite, $An_{55}$, in the outer parts) and hypersthene. The microlitic groundmass is composed of andesine-labradorite, $An_{53}$, and of large amounts of augite.

In the basalts there are phenocrysts of bytownite, $An_{75}$, augite, and, rarely, hypersthene. Labradorite, $An_{60}$, and abundant augite occur in the microlitic groundmass.

A dike of aphyric basalt is exposed near the high point on the western rim of the somma. The structure of the rock is pilotaxitic; olivine, plagioclase, and augite are found as microlites.

The whole surface of the first somma is covered by a sheet of blocky material that is coarser on the rim and inner slopes and finer on the outer slopes. According to a number of data, this loose material is related to the relatively recent explosions of the central cone and does not have any genetic relation to the first somma.

The second somma is much better preserved and is an almost complete ring, broken only in the northwest. The rim of the second somma reaches its lowest point (425 m) in the west and gradually rises to the east; the high point of the rim and of the whole volcano (624 m) is located in the southeastern part. Farther to the north the rim again declines somewhat (down to 500-580 m). In the south the rim of the second somma is cut by the valleys of three streams. At least two of the three valleys are old, modified U-shaped valleys, most probably of the second glaciation.

The diameter of the caldera of the second somma, measured between the present-day rims, is 7-8 km. The depth of the eastern part of the caldera, where a part of the old atrio exists, is up to 375-400 m.

The outer slope on the southwest is almost unaffected by erosion; it slopes gently (10-12° in the upper part) to the atrio between the first and second sommas. The relative height of the caldera rim does not exceed 225 m here. The northeastern sector, on the opposite side, is adjacent to a Tertiary ridge that extends from here to Prevo Peak (and perhaps to the badly ruined remnants of the first somma). The eastern flank is terminated by a high (200-300 m) erosional scarp.

As noted earlier, the northwestern part of the second somma is in ruins. Both edges of the ruined rim of the second somma and the northwestern edge of the rim of the first somma are located along the same northwest-trending line, a fault line. In all probability a crater bay existed in the caldera of the second somma up to the time of formation of the central cone.

A very remarkable peculiarity of the second somma is the peculiar "roof" of tightly welded tuffs (ignimbrites) that cover its rims and the adjoining flanks. A layer of blocky material lies on top of the ignimbrites, as in the first somma. The ignimbrite sheet rests unconformably on the underlying layers as a result of the great irregularity of the topography. The layer of ignimbrites covers not only the outer slope of the somma, but locally also the inner slope of the caldera, where it dips inward. Moreover, on the southeastern part of the second somma, the sides of one of the valleys are completely armored by the ignimbrite capping from the caldera rim to its valley floor, with periclinal dips away from the high point on the rim and outward and inward with regard to the caldera. This relationship undoubtedly indicates that the ignnimbrite capping was formed not only after the formation of the caldera of the second somma, but even after the original topography of the somma had been considerably altered by erosion.

The ignimbrite layer is also present in the southwestern part of the second somma, but here it is overlapped by a thick suite of layered pyroclastics of varied composition that are related to the explosion of the central cone within the caldera. Structurally the second somma is a stratovolcano, with lavas dominant below and pyroclastics dominant above. Specimens of lava that were studied from the middle part of the section are pyroxene andesites with phenocrysts of zoned andesine (up to $An_{48}$ in the core and up to $An_{35}$ in the marginal parts), hypersthene, and subordinate augite. Sometimes the clinopyroxene is absent and the rock may be called a hypersthene andesite. The groundmass has a pilotaxitic or hyalopilitic structure in which microlites of augite and andesine, $An_{35}$, are present in various proportions.

The pyroclastics in the upper part of the second somma are mainly dacite pumice. Rare phenocrysts of acid plagioclase and even rarer ones of pyroxene and hornblende are enclosed in a vitrophyric porous groundmass. In some of the pumice horizons a weak welding together of the fragments is observed.  Very probably the dacitic pumices are related to the processes by which the second caldera was formed.

The central cone that formerly rose within the second caldera was to a large degree destroyed at some time during its development; only its northern portion, which encloses the adjacent part of the inner, third caldera in the form of a semicircle, has been preserved.

The western and northwestern flanks of the central cone descend to the Sea of Okhotsk coast through a break in the outer caldera, forming a capriciously broken coast line with finger-like projections composed of lava flows.  The northeastern and southwestern flanks extend to the wall of the inner caldera, forming the narrow corridor of an atrio that to the northwest runs down to the sea. To the southeast the atrio widens to 1.5-2 km, and here on the east, at a height of 250 m, there is a flat plain 2 × 4 km across.

The flanks of the cone have the profile of a logarithmic curve that is normal for stratovolcanoes, with slopes of over 30° near the top and as low as 8-10° near the base.  The flanks are covered by a radial system of numerous, but very shallow gullies, incipient barrancas.  A not very high (10-20 m) erosional scarp drops to the Sea of Okhotsk; lavas overlain by several pyroclastic layers are exposed in it.

The walls of the inner caldera, cut into the central cone, in their northern part expose the layered structure of a typical stratovolcano, with pyroclastic material somewhat dominant over lavas. The layered series is cut by numerous vertical dikes of variable cross section, with bulges and apophyses.  One of the dikes is terminated by the remnants of a subsidiary scoria cone that in turn is overlapped by the later layers of central-cone lavas and pyroclastics.

In cliffs in the southwestern part of the caldera, farther from the central cone, pyroclastic material is clearly dominant over lava. A many-layered sequence of postcaldera pyroclastics is even exposed here.

In the inner part of the caldera a layer of strongly welded tuff is exposed almost everywhere; locally it passes over into a lava-flow-like rheoignimbrite.

The composition of the rocks of the inner caldera varies through an extremely wide range: from basalt to dacite.

Among the basalts both aphyric and porphyritic varieties are noted. The former are very vesicular rocks with hyalopilitic structure. In the porphyritic varieties phenocrysts of labradorite ($An_{60}$), olivine, and augite are present; the groundmass has an intersertal structure, with augite and glass in the interstices between the plagioclase microlites.

More widespread are the pyroxene andesites and andesite-basalts: augite-, hypersthene-, and two-pyroxene-bearing varieties. The plagioclase phenocrysts usually have the composition of labradorite, $An_{55-65}$; in some lavas rare olivine is found. The groundmass of the rocks has a hyalopilitic or, in the more basic varieties, a microlitic structure. The dikes are of varied composition. One of them (in the northeastern part of the caldera) is an aphyric andesite. The groundmass has a pilotaxitic structure, with microlites of plagioclase and augite; phenocrysts of andesine-labradorite ($An_{50}$) are found only very rarely.

The postcaldera rocks are mainly dacite pumice. Bomb and scoria layers of andesitic and andesite-basaltic composition are also found.

The inner caldera included not only the upper part of the former central cone, but also its southern base and the adjacent part of the atrio as far as the flank of the outer caldera. The rim of the caldera stretches southward from the highest point on the central cone (484 m) to a point 230 m high, where it reaches the atrio floor and disappears. The southern half of the caldera is cut right into the flat plain of the atrio (Fig. 45).

In plan the inner caldera has the shape of a right triangle 3.5 km long from north to south, 2.5 km wide in the north and 1-1.5 km wide in the south. The inner walls of the caldera are very steep; locally they are vertical or even overhanging in the form of cornices.

In the western part of the caldera a portion of the rim about 1 km long, along with the adjoining zone of the outer slope, has been dropped down about 100 m by a relatively fresh slide, forming a long narrow terrace on the steep inner wall of the caldera. On the opposite flank there is another, older-looking slide with the form of a semicircle about 1 km across and with a displacement of about

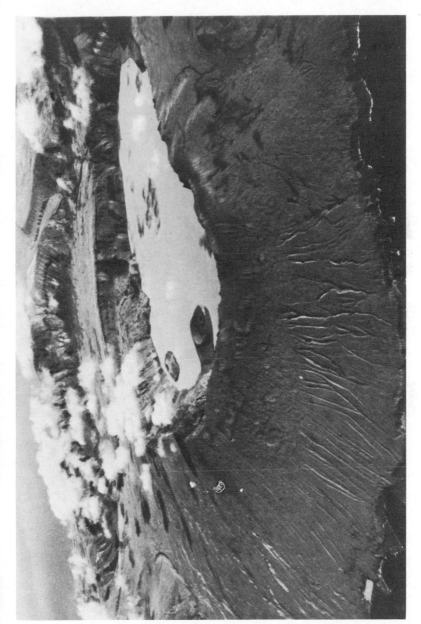

Fig. 45. Zavaritskii Caldera (before the 1957 eruption).

150 m. It is adjacent to the traces of still another notable slide that extends to the south edge of the caldera.

It is impossible not to find echoes of the caldera-forming processes in these manifestations of sliding of different parts of the wall. One can think of the Zavaritskii Caldera as being "alive," as still in the process of formation.

On the floor of the caldera lies Biryuzovoe Lake. The surface of the caldera lake, according to our barometric data, stood at 40 m in 1958, and its depth was 70 m according to measurements by Zelenov. Thus, the floor of the caldera lies at 30 m below sea level, and its depth below the highest point on the rim is 515 m.

Morphologic analysis, a number of geologic and tectonic data, the distribution of fumarole fields, the way in which depths vary in the caldera lake, and the results of magnetometric measurements lead to the conclusion that the inner caldera was formed as the result of an explosion along an explosion trench of north-south trend, and that it was later widened by collapse of its walls.

The history of formation of this complex caldera is rather clearly depicted, especially in its late stages. In the late Pliocene or early Miocene* a broad, shield-shaped volcano rose on the Tertiary basement; a caldera about 10 km across was then formed on its upper part. Subsequently the flanks of this caldera volcano were strongly eroded and the northern portion of the massif was completely or almost completely destroyed. An inner cone was formed in interglacial times in this old, ruined caldera; it passed into a state of intense explosive eruptions at the end of the interglacial stage and was in turn converted into a second caldera volcano (the second somma) with a caldera 7-8 km across. The second glaciation left obvious traces; in particular, U-shaped valleys were formed on the inner slopes of the caldera.

At the beginning of postglacial time the northwestern part of the structure was dropped below sea level along a fault, and a caldera bay was formed in the second caldera. Then the central cone arose in the basin of the caldera. Between the eastern wall of the caldera and the central cone a flat atrio was left, whereas the west-

---

*Remnants of the first somma, according to Bernshtein, have reversed magnetism.

ern flank of the central cone reached beyond the caldera, all the
way to the coast. Continued tectonic movement produced a weakened zone beneath the central cone; into this a thick, dike-like body
was injected that caused a gigantic explosion. This explosion destroyed the southern part of the central cone and formed an elongate
caldera. The initial explosion ejected very hot material which, fall-
ing in the vicinity of the explosion trench, was loose but still plastic
and weldable. The peculiar "roof" of ignimbrite, which we mentioned in the description of the second somma, was produced from
this.

After the formation of the caldera eruptions from the explosion trench continued and a thick layer of mainly pumiceous pyroclastics was deposited in the vicinity of the caldera, including both
the first and especially the second caldera.

At some stage in the development of the caldera a permanent
caldera lake was formed, though its level differed considerably from
that of the present lake. On top of the layers of pumiceous pyroclastics, at a height of 190 m above the present lake level, lies a
coarse alluvial layer, while closer to the lake, at a height of 140
m above its surface, a suite of lacustrine tuffites rests against the
pumice. These rocks are almost flat-lying in the upper part, but
in the lower part they dip steeply, at angles of up to 60°, toward the
lake, though the angle of dip decreases near the present lake level.
These data indicate that the level of the first caldera lake lay at 200
m above sea level. All stream valleys leading into the lake are
hanging and drop into the lake from heights of 150-200 m.

The floor of the lake was situated approximately at the level
of its present surface. The formation of the first lake coincided
with a rather extended period of weakened volcanic activity; a
stream network was formed, controlled by the base level of erosion (the lake level), and a thick series of lake beds was produced.

After a long hiatus violent eruptions again occurred, drying
up the lake and greatly deepening the caldera (by 50-60 m) along the
explosion trench. The caldera again passed into a state of violent
activity and a new series of pyroclastic layers was deposited on top
of the old alluvium and the lake sediments.

After the second stage of activity of the caldera, a new decline in its activity set in. The present caldera lake was formed

Fig. 46.  Northern part of Zavaritskii Caldera (before the 1957 eruption).

and volcanic activity continued, not in all areas of the caldera, but rather confined only to its northern part. The arena of present-day eruptive activity practically coincides with the position in the caldera of the ruined central cone. The almost annihilated central cone has been reawakened, so to speak.

The volcanic activity in the central caldera has changed the outline of the caldera lake, though not to the extent that it has done so in former times. Unfortunately, recent eruptions here have gone unrecorded, and we can judge them only from indirect data.

On the Japanese topographic maps of 1916 a small scoria cone, about 500 m across, with a completely closed crater, was indicated as a peninsula near the north shore of the caldera lake. On maps made at the end of the 1930's the earlier scoria cone in the caldera was noted as having a ruined crater, and to the east of it was a new islet. This allows us to suppose that a new eruption took place in the caldera 'after 1916 and formed a new dome. Recently we became aware of a paper by the Japanese author Harada (1934), who noted this change in 1931. Thus the eruption most probably occurred in the 1920's. We had a chance to see its results in 1954. The southern part of the old scoria cone was found to be in ruins; the diameter of the crater was 350 m and its floor was 250 m across. Approximately in the center of the ruined crater a miniature extrusion dome had arisen, oval in plan view and 100 × 140 m across. The top of the dome was occupied by a rather shallow collapse crater with gently sloping sides and bounded by a narrow ring-like fissure. Two low sand bars joined the dome to the crater walls, separating a small crater lake from the crater bay that existed earlier. A second dome (Harada's "new island"), 200–300 m across, forms an island 500 m to the east of the scoria cone. This dome is separated from the lake shore by a strait 40 m wide (Fig. 46). Later study indicated that this dome is effusive. Thus, in the 1920's, part of the scoria cone was destroyed and then a small dome was extruded into its crater; a second new dome was formed in line with it.

In November 1957 the first recorded eruption occurred in Zavaritskii Caldera. Judging from eyewitness accounts and also from the results of our 1958 studies, a general picture of this eruption can be presented in the following way: The eruption began on November 12, 1957, in the scoria cone or in its immediate vicinity.

Fig. 47. 1957 dome and flow.

For the first day or two the explosions were of Vulcanian character. Apparently the scoria cone was completely destroyed during
this time. During the night of November 13-14 the eruption took
on the character of strong Strombolian explosions. During the period
November 14-18 a large quantity of loose material was ejected and
filled the northern third of the caldera. Possibly the new scoria
cone was formed at this time; its remains are preserved as a wide
ring wall. Then a decline in activity set in and continued for a week.
On November 28 a series of very strong explosions occurred, during which the new cone was destroyed. At the end of the main,
Strombolian phase of the eruption lava of similar composition, but
more viscous, rose to the surface and was squeezed out as an extrusion dome. Then the southeastern part of the dome was carried
away by a flow of block lava. With this the eruption came to an end
on December 1.

As a result of the eruption, a layer of ash and fine scoria (up
to 10 cm) was deposited on the surface of the central cone and in the
atrio. The vegetation (thin bush) was destroyed in the eastern part of
the atrio at distances of up to 500 m from the rim of the caldera (2.5 km
from the point of eruption). A very large amount of water from the crater
lake was thrown out onto the flanks of the cone, where it formed
thick flows (lahars). The northern third of the caldera lake was
found to be covered by the products of the eruption, and the outline
of the lake was radically changed (see Fig. 47). The lake's length
was shortened by a whole kilometer. The eastern dome was joined
to the shore by a belt of fresh pyroclastics. Columns of steam rose
all along the newly formed northern shore for a distance of more
than 1.5 km; the temperature of the water near the shore was found
to be higher than usual, and around the new eruptive center it approached the boiling point. Here hot streams flowed into the lake.
The flat, gently southward-sloping plain, about 2 km long and 1 km
wide, that was formed within the caldera is covered by numerous
scoria bombs up to 1-2 m across. At the site of the former scoria
cone a new extrusion dome was formed, 350 m across and 40 m
high; southeast of the dome extends a short, 300-m tongue of lava.

All the products of the 1957 eruption are pyroxene andesites
that differ among themselves only in details.

The bombs are composed of scoriaceous two-pyroxene andesite with phenocrysts of labradorite ($An_{56}$), hypersthene, and aug-

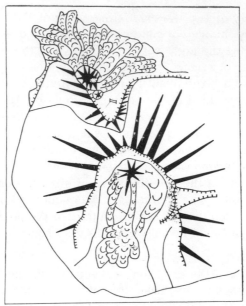

Fig. 48. Sketch of the structure of Milne Volcano [I] and Goryashchaya Sopka [II] (symbols
as in Fig. 9).

ite; the pyroxenes are often glomeroporphyritic. The groundmass
has a hyalopilitic structure, and the plagioclase in the microlites is
andesine, $An_{47}$.

The dome lava is distinguished from the bombs only by the
absence of vesicles and by a microlitic groundmass with a small
amount of glass. The plagioclase microlites in the groundmass are
andesine-labradorite, $An_{50}$.

The rocks of the lava flows are strikingly different from the
preceding ones. They are pyroxene andesites with phenocrysts
mainly of labradorite ($An_{58-65}$) and augite, along with small amounts
of hypersthene and scattered grains of olivine. Judging from the optical properties, the iron content here is less than in the dome rocks.
The groundmass has a hyalopilitic structure, with microlites of andesine-labradorite, $An_{50}$.

During the 1957 eruption some increase in the basicity of the
rocks was observed; this was expressed by a change in the mineralogic and chemical compositions (from 58.8 to 56.5% $SiO_2$).

At present fumarolic activity continues on the dome, and the hydrosolfataras that were active before the eruption near the south shore of the lake are still active.

Milne Volcano forms the southwestern part of Simushir (Fig. 48); it is a volcano of the Somma-Vesuvius type. In the basement of the volcano there are strongly altered, leucocratic lavas which, judging from their reversed magnetization (according to Bernshtein), belong to the upper Tertiary. The lavas of the somma rest on them with angular unconformity.

The somma is a rather broad stratovolcano with a basal diameter of up to 10 km. At the top is a caldera, about 3 km across and open to the south. The rough "level" rim of the caldera amphitheater rises to a maximum height of 1490 m.

The flanks of the volcano are drained by narrow, deep gullies and are strongly overgrown by impassable scrub, especially in the lower parts. The surface of the lava flows is not preserved; however, the distribution of stream valleys and gullies apparently is controlled mainly by the position of the lava flows. On the basis of this evidence an old, strongly eroded structure is recognized that has been strongly beveled by later activity.

At the top of the northeast flank there is a U-shaped valley that was decapitated by the caldera, and on the southwest flank there is a deep cirque. The remaining flanks show no evidence of glacial erosion, but apparently the caldera was an ice accumulation basin, and the whole southeast flank was subjected to intense glacial scouring. Moraines lie at the mouth of the amphitheater.

The northwestern part of the caldera has been cut off by a fault that extends to here from Zavaritskii Caldera. Great steep cliffs up to 200-300 m high bound the somma on this side. As in other cases, the fault is well recorded in the distribution of the submarine terrace.

A small central cone rises in the caldera of Milne Volcano (Fig. 49). Its lava flows descend to the southeast for 4-5 km and reach the coast. The crater of the cone is sealed by an extrusion dome, which is the highest point on the volcano and on the island (1540 m).

The lavas of the somma and of the central cone of Milne Volcano are composed of pyroxene andesite-basalts and andesites. They

Fig. 49. Milne Volcano.

Fig. 50. Goryashchaya Sopka.

are rather similar, but are distinguished by the composition of the plagioclase and the structure of the groundmass. In the andesite-basalts the plagioclase is labradorite, $An_{63-65}$, and the structure of the rock is microlitic. In the andesites the plagioclase is somewhat more acid ($An_{60}$) and the structure is hyalopilitic. Andesine-labradorite, $An_{48-55}$, occurs as microlites.

The summit extrusion is composed of pyroxene andesite. The phenocrysts are of labradorite,($An_{55-58}$), abundant hypersthene (usually with opacite rims), and augite (which is also marginally altered). The structure of the rock is andesitic; the plagioclase microlites are andesine-labradorite, $An_{48-50}$.

The evolutionary history of the volcano is as follows: the volcano arose in early Quaternary times on the Tertiary basement. It underwent erosion during the first glaciation, but later activity has "healed" the traces of the scouring. The volcano was active in interglacial times and throughout the second glacial period. Thus, evidence of the second glaciation is not important on the volcano's flanks. The caldera was formed during the second glacial stage. Volcanic activity stopped after the formation of the caldera, and the caldera served for a while as an important center of glaciation.

In postglacial times activity was renewed in the caldera, forming the central cone with its flows and dome, while the northwestern part of the caldera was down-faulted. It may be that this faulting resulted in the stopping of the central cone's activity. Unfortunately, there are no facts at our disposal that allow us to determine the age of the fault and of the final activity of the central cone. Perhaps Milne Volcano is only "sleeping" and eruptions of it are still possible.

Goryashchaya Sopka is the final volcano on Simushir; it closely adjoins the somma of Milne Volcano on the northwest. In the works of our predecessors, Goryashchaya Sopka is considered to be a parasitic crater on the latter cone; actually, it is an independent complex volcano of the Somma-Vesuvius type.

The somma of Goryashchaya Sopka arose on the fault line that cuts part of the somma of Milne Volcano, to which we referred earlier. Thus the cone of Goryashchaya Sopka could expand without interference only to the northwest, while on the southeast it closely adjoined the flanks of the somma of Milne Volcano. Later on, the northern part of the original cone of Goryashchaya Sopka was destroyed, apparently by a strong directed explosion. At present only the western sector of the cone and the part of it directly adjacent to Milne Volcano are preserved. The amphitheater of the old caldera, which has been strongly modified by later, intense erosion, has very steep, cliffy, and locally perpendicular inner walls. The high point on the somma, Igla Mountain, rises to 1320 m above sea level. It is separated from the flanks of the somma of Milne Volcano by a narrow, not too deep saddle about 100 m deep (Fig. 50).

A large extrusion dome rises eccentrically (it is shifted to the northwest) in the somma amphitheater; its basal diameter is about 1 km and its absolute height is 890 m (its relative height is about 300-400 m). The double summit (of two united extrusion domes)', which is divided by a small saddle, is surrounded by the crown of a rocky rim that passes downward into the loose agglomerate mantle. A small explosion crater is cut into the northern part of the summit, and a small, very fresh lava flow from the last eruption descends to the northeast from the eastern part of its rim. One more miniature dome is situated to the northeast of the main one, beneath the cliffs of the Milne Volcano somma.

Numerous, in part very fresh, lava flows from the now-sealed lava crater stick out from beneath the mantle of the dome. Several of the flows have followed the same paths, leaving three to five and sometimes more parallel marginal ridges. Some flows that went to the northeast descend via very effective lava falls down the steep bench, and a few of them extend for another 3 km to reach the sea near the Skalistyi whaling factory. A series of flows extend westward for 3.5 km, forming rather high lava capes (Capes Aront, Ptichii, and others). The whole north coast of the volcano is made up of older flows; at least two different ages of flows can be recognized among these.

The dome and flow lavas are very similar. They are pyroxene andesite with phenocrysts of labradorite ($An_{55-65}$), augite, and hypersthene. The last-named is often found with opacite or augite rims. Olivine grains are sparse. The groundmass is hyalopilitic.

We have no information on the activity of Goryashchaya Sopka before the middle of the nineteenth century. The first recorded eruption occurred in June 1842; it marked the beginning of a new cycle

of activity and probably was very violent. An eruption in 1849 is
also known. In September 1881 Captain Snow observed the emission of numerous lava flows. Most probably the dome had not yet
been formed. Perhaps it formed in 1883 as the final stage in the
1881 eruption. The formation of the last lava flow and the explosion crater on the dome summit occurred later, but when it is not
known. There are only mentions of weak explosions in 1914 and
1944 (?). At present rather weak fumarolic activity continues in
the explosion crater, at the dome summit, and near its base in
the south.

The structure of the volcanoes of the Central Kurile Islands
varies considerably, from single cones to complex caldera structures with multiphase central structures. In some areas volcanoes
of very different structure are located side by side.

Among the rocks two-pyroxene andesites and andesite-basalts
predominate, but there are also cones of pure basalt (Raikoke,
Prevo Peak). On some volcanoes (especially on Ushishir) hornblende andesites are found. In the history of the individual volcanic
centers a slight increase in acidity is found, but such a trend often
is broken after the formation of a caldera, and the central structure
again becomes basic.

## South Kurile Islands

Boussole Strait, the widest (65 km) and deepest (more than
2000 m) of all the Kurile straits, separates the Central and the
South Kurile Islands. Two small islands (Chernye Brat'ya) and three
large ones (Urup, Iturup, and Kunashir) make up the southern group.

Approximately half of all the volcanoes of the Kurile chain lie
in the southern group: all of 80-85 volcanoes, of which 55 are
Holocene and 16 are active.

Brouton Island will be described in this section; it belongs to
the western zone, but it is geographically close to the southern islands. We will begin our descriptions with this island.

## Brouton Island

Brouton Island is located about 20 km northwest of Chirpoi.
In form it resembles a rhombus $2.5 \times 4$ km across and 8 km$^2$ in area.

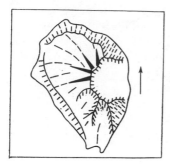

Fig. 51. Sketch of the structure of Brouton Volcano (symbols as in Fig. 9).

It is an extinct volcano, with remnants of a crater at its top. The western half of the crater rim is preserved; its highest point rises to 800 m. The crater diameter is approximately 750 m. The eastern part of the island is steeper and more cliff-like, and the crater has been destroyed on this side. The western flank is also strongly eroded and drops to the sea by a high scarp (Fig. 51).

The rocks of the island are basalts with phenocrysts of labradorite, augite, and hypersthene; strongly altered olivine is also present. The groundmass structure is hyalopilitic.

Around the island is a broad submarine terrace, which establishes the age of the volcano as Pleistocene.

## Chernye Brat'ya Islands

In the Chernye Brat'ya group there are two islands, Chirpoi and Brat Chirpoev. The northern island, Chirpoi, has the shape of an irregular parallelogram, 3.5 × 6 km. Bystryi Strait, 2.7 km wide, separates Brat Chirpoev from Chirpoi. The former has a triangular shape, and is 4 × 5 km across. In spite of their small size, both islands are rather complex volcanic structures (Fig. 52).

Before going into a description of the volcanoes on these islands, we must say something about the submarine topography. Southeast of both islands a submarine terrace extends for 7–8 km, but to the northwest depths increase to several hundred meters close to the islands. Obviously a fault runs through here, as on some of the other islands; this fault "grazes" the Chirpoi volcanic massif. Bystryi Strait, which is rather narrow (less than 3 km) between the island groups, has considerable depth: it reaches 200 m or more. In all probability the islands are separated by a graben transverse to the general trend of the arc. On the Chernye Brat'ya islands five volcanoes can be counted: four on Chirpoi and one on Brat Chirpoev; three of these are Holocene and active.

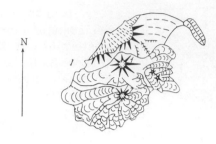

N

Fig. 52. Sketch of the structure of Chirpoi (I) and
Brat Chirpoev (II): 1) Chernyi Volcano; 2) Snow
Volcano (symbols as in Fig. 9).

Chirpoi island is a volcanic massif consisting of four
closely adjoining volcanoes. A narrow sand bar ties long, narrow
Lapka Island to it. Lava and pyroclastic rocks, the remnants of
some older volcano, are exposed in the high (up to 150 m) cliffs.
A narrow and rather shallow (less than 100 m deep) belt of water
stretches to the south of the Lapka Peninsula; this, as it were, marks
the extension farther to the south of a narrow rim that is separated
from the main island by a very deep (170-300 m) area. The Lapka
peninsula, along with this shallow rim, marks a part of the large
caldera whose southern and western parts are cut off by faults.
The diameter of this caldera can be roughly estimated at 6-8 km.
Chirpoi itself should perhaps be thought of as a complex central
cone. Special sounding work is necessary here. Near the Lapka
peninsula pebbles of granitic rock have been found on the shore (by
Bogoyavlenskaya). It may be that the Tertiary basement lies at
shallow depth here and that numerous fragments of it were ejected
during the caldera-forming explosion just as occurred at Tao-Rusyr
Caldera (on Onekotan).

Fig. 53. The volcanoes of Chirpoi island; from left to right: Chirpoi, Chernyi, and Snow.

On Chirpoi itself, as noted above, there are four closely spaced cones. The northernmost cone, Chirpoi (691 m), is cut by a fault on the northwest; the near-crater part on the east is preserved, as is the eastern half of the cone, but the flanks of the cone are rather badly eroded. The second cone, which adjoins it on the southwest, is broken up even more strongly by a fault, and only a small portion of its southern flanks is preserved. The fault that cuts the volcano also drops the 140-m submarine terrace (which is related to the second glaciation) down here, that is, the fault is early Holocene and the two cones about which we just spoke are probably interglacial in age.

Two other cones, Chernyi and Snow, are recent (Fig. 53).

Chernyi Crater is located in the center of the island and is separated from Chirpoi Cone by a small saddle. Chernyi forms a smooth, truncated cone with a well-developed crater at the top. The crater is 330 m across; its rim is smooth, without marked low or high points other than a general decrease in height to the north. The highest point on the rim rises to 624 m above sea level. The inner walls of the caldera have the form of a steep funnel, with slopes of about 60°; the flat floor, 150 m across, lies 150 m below.

The flanks of the cone are smooth, without any prominent ravines. The surface of the cone is covered with light-colored pyroclastics, and lava flows are seen locally, heavily covered with ash at their sources. Some flows may have come from boccas on the flanks. A large, wide flow descends from a crater along the west flank and reaches the shore of the Sea of Okhotsk. A number of flows ran down the east flank. A very long flow, up to 3 km long, runs down the north flank and then along the saddle between this cone and Chirpoi to the east; reaching the sea coast, it forms a very prominent cape.

Near the base of Chernyi Cone on the east are two parasitic craters; they have the shape of miniature amphitheaters from which stream the lava flows that form Cape Udushlivyi, which sticks far out into the sea.

Eruptions of Chernyi Volcano have been recorded in 1712 or 1713, 1854, and July 1857. It may be that the 1854 eruption which is ascribed to Chernyi Cone actually belongs to the adjacent Snow Cone. At present Chernyi Cone shows very active solfataric activity. The vents of the solfataras are located on the floor and walls of the summit crater. A strong, elongate group of solfataras occurs on the west flank near the summit, indicating the presence of a sealed radial fissure.

Small Snow Cone lies near the foot of Chernyi Cone and forms the southern part of the island. It takes the form of a strongly truncated cone which rises to 400 m above sea level; it contains a crater 300 m across. The depth of the crater is not great, about 10 m, and its inner walls slope very gently. The crater has the form of a shallow saucer. In its northern part there is a deep, well-like crater 130-140 m across.

The flanks of the cone are composed almost entirely of lava. Erosional landforms are absent on the sides, and the irregularities present are due only to the superposition of the individual lava flows onto one another. The dark rough cone of Snow is in striking contrast to its light-colored, smooth companion, Chernyi.

Snow Cone is very rich in lava emissions. Numerous thick lava flows descend in all directions from the summit crater and probably from boccas near the cone base. Many flows reach the sea coast and the whole southern part of the island, an area $1 \times 3.5$ km, consists of very fresh-looking lava flows that have run together. The last flow (1879?) ran down to the southwest from the summit crater; at its source it is 300 m wide, and at a distance of 2.5 km from the crater it reaches the sea shore. The chaotic pile of block-lava flows gives the southern part of the island a gloomy, wild appearance, emphasized by the absence here of any vegetation.

Snow Cone apparently was formed after 1770. Cossack Captain Chernyi, who described all of his observations very carefully, noted only one "burning mountain" on Chirpoi. Chernyi was undoubtedly on this island for a long time, because he writes, "There on the highest point I placed a wooden cross inscribed with the year, month, and day, in memory of my having been there." This means that he could not have helped but notice the volcano. The first information on the existence of this volcano and of its eruption in June 1811 we find in Captain Golovnin. An eruption on the island of Chirpoi was in June 1854, but it is not clear which of the two cones erupted. In May-June 1879 Captain Snow saw large lava flows being emitted here; these reached the sea coast and formed a new cape. An eruption occurred on October 20, 1960, and ash fell on the deck of a ship passing 12 km from the volcano. At present rather weak solfataric activity is seen in the crater.

The present lava cones of Snow and Chernyi are of normal two-pyroxene andesites (59-60% $SiO_2$). Various rocks are present on the Lapka Peninsula; andesite-basalts predominate among the lavas. The rim of this peninsula is covered by a layer of dacite pumice that is apparently related to the caldera explosion.

Brat Chirpoev Volcano also has a rather unusual structure. The southeastern part of the island is the remnant of a somma, the rest of which has been destroyed by tectonic and erosional processes. A monoclinal series of lavas and pyroclastics is seen in the cliffs, dipping to the southeast. The northwestern part of the island is made up of a central structure consisting of a large series of closely spaced cones and craters. All of them are located on the arc of a circle that expresses to some degree the outline of an old caldera about 4-5 km across.

The extreme southwestern cone, Brat Chirpoev itself, has the largest dimensions and height (752 m). The summit crater, about 250 m across, resembles a steep-sided, deep funnel. Lavas stick out from the eastern and western walls; elsewhere there is steep talus. The rim of the crater is completely closed, but there are rather low points in the northern and northwestern sectors. The flanks of the cone are covered by a continuous mantle of cherry-red scoria. No lava flows are exposed at the surface; they are found only in sections and are seen where the shoreline juts out in a peculiar manner. Along the west flank a deep, wide ravine of the "sciarra" type extends from the crater to the shore.

On the east this cone is closely adjoined by a lower (about 400 m) double cone on whose summit no crater is preserved. However, in the saddle between the two summits a rather small explosion crater is seen; a second, wider crater is located in the saddle between Chirpoi Cone and the double cone. This whole part of the island is covered by scoria, but the eruptions in the eastern part of the island were not limited to explosions. Lava flows are seen in the coastal cliffs and in the configuration of the coast line.

The small islet of Morskaya Vydra emerges 100 km northeast of the shore of the island; it is 0.5 km across and 150 m high. This island apparently is an old extrusion dome.

The lava and scoria of Brat Chirpoev Volcano are olivine-pyroxene basalt. The phenocrysts are of labradorite-bytownite and bytownite ($An_{65-75}$), olivine, and augite. The groundmass is microlitic or intersertal in the lava, hyalopilitic or hyaline in the bombs. The plagioclase microlites in the groundmass are of andesine-labradorite, $An_{48}$.

No eruptions have been recorded of this volcano in historic times. However, in his "Journal" Captain Chernyi called this island a "burning mountain." Chernyi gave this name only to volcanoes that displayed some evidence of activity in his day (the 1760's). Apparently Brat Chirpoev Volcano showed solfataric activity in the middle of the eighteenth century. The very fresh character of the crater and flanks of the cone do not contradict the possibility of recent Strombolian eruptions.

## Urup Island

The island of Urup lies 30 km from the Chernye Brat'ya Islands, across a strait of the same name. This is one of the largest islands of the chain. In plan it is spindle-shaped, with a length of 120 km and a maximum width of 20 km; its area is 1430 $km^2$. Tertiary sedimentary rocks containing plant remains, Tertiary volcanogenic rocks, and intrusions of granodiorite are well developed on the island. The vegetation on the island is mainly scrub, and very thick, impassable thickets of Kurile bamboo are widespread. Rock birch and alder are present locally.

A belt of linear-clustered eruptions of preglacial age stretches along the entire axis of the island, from Desantnaya Mountain on the northeast to the end of Krishtofovich Ridge on the southwest.

This belt is displaced somewhat to the northwest relative to the geometric axis of the island. Apparently this is why the Quaternary lavas are not exposed anywhere along the coast of the ocean but do occupy a rather considerable part of the Okhotsk coast.

The belt of eruptions consists of five " clusters" separated by saddles in which rocks of the Tertiary basement are usually exposed.

These clusters are as follows: (1) Desantnaya Mountain, (2) northern Shokal'skii Ridge (from Trekhglavaya Mountain to Mount Irin), (3) southern Shokal'skii Ridge (from Mount Kaban to Mount

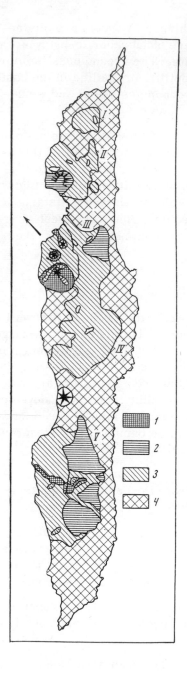

Fig. 54. Sketch of the distribution of volcanoes on Urup: I) Desantnaya Mountain massif; II) north part of Shokal'skii Ridge; III) south part of Shokal'skii Ridge; IV) Petr Shmidt Ridge; V) Krishtofovich Ridge; 1) Holocene lavas; 2) interglacial lavas; 3) eroded portions of Quaternay (mainly Holocene) structure; 4) basement (other symbols as in Fig. 9).

Kavraiskii), (4) Mount Petr Shmidt, and (5) Krishtofovich Ridge (Fig. 54).

The island underwent two-stage glaciation, as a result of which all the older volcanic structures were strongly eroded. In most cases the old centers of eruption have not been preserved.

Late Pleistocene and Holocene activity has been more limited; it developed in the same regions as in the early Pleistocene or in areas aligned with them; with only one exception the late activity has to a certain degree "inherited" the character of the earlier.

On the island as a whole 25 eruptive centers can be counted, of which 14 are Holocene, but only two are active.

The characteristic features of each of the clusters are given below:

1. Desantnaya Mountain rises at the northeast end of the island. It is a Quaternary volcanic massif, round in plan view and about 5 km in diameter. The massif is strongly eroded, and its original form is not preserved. Perhaps, as on Paramushir, this extreme member of the volcanic row manifested activity more of a central than of a "linear-clustered" character. The absolute height of the mountain is 867 m; its relative height above the Tertiary basement is 400-500 m.

2. The north part of Shokal'skii Ridge stretches for about 8 km from Mount Trekhglavaya (1137 m) to Mount Irin (703 m). An eroded remnant of Quaternary rock, Mount Litvinov (590 m), lies somewhat apart from the axis of the "cluster"; in addition, a narrow eroded ridge (Kompaneiskii Ridge) extends to Desantnaya Mountain. The summits and peaks named here, and many others, may be badly destroyed products of central eruptions, but no clear indications of these centers are preserved.

In this same group, to the west of Mount Trekhglavaya, lies the isolated Antipin Volcano (1120 m). It is interglacial in age; its lavas run down into a U-shaped valley of the first glaciation, but U-shaped valleys of the second glaciation are cut into its northwestern and western flanks. There is a moraine in the first of these troughs, and the second one cuts the whole slope from sum-

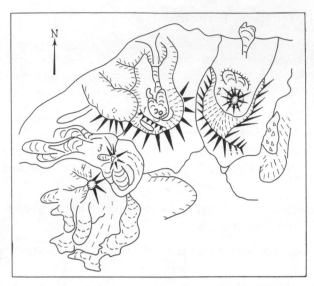

Fig. 55. Sketch of the structure of the Kolokol Volcano
group: 1) Berg Volcano; 2) Trezubets Volcano; 3) Kolokol
Volcano; 4) Borzov Volcano (other symbols as in Fig. 9).

mit to base. The southern and eastern flanks are the best preserved, and here the old lava flows can be followed.

3. The next "cluster," at the southern end of Shokal'skii Ridge,
extends for about 10-12 km from Kaban Mountain to Mount Kavraiskii. The southwestern part of this "cluster" is also strongly eroded
but on the northeast a relatively slightly eroded flank on which one
can trace some strongly abraded and overgrown lava flows descends
southeastward from the vicinity of Kaban Mountain (814 m). Locally, U-shaped valleys and cirques of the second glaciation, containing moraines, have been cut into these flows, but no clear indications of the intensive glaciation of the first glaciation are visible.
It is quite probable that volcanic activity lasted into the interglacial
period in this region.

A group of five Holocene volcanoes adjoins the southern end
of the Shokal'skii Ridge on the west; two of them are active at the
present time (Fig. 55).

The activity of this group, which can be called the Kolokol
Volcano group after the dominant height, began in interglacial times.

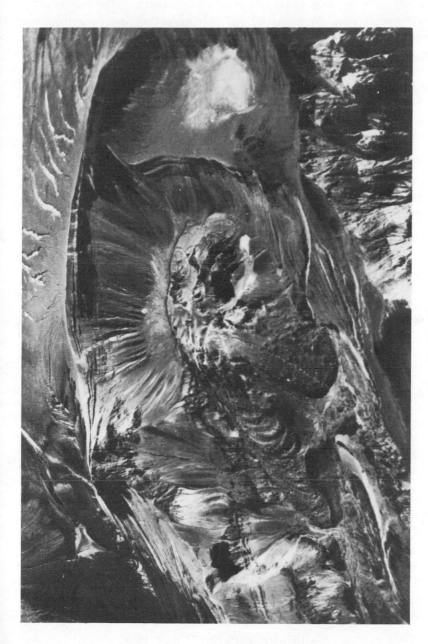

Fig. 56.  Berg Volcano.

Initially there arose a chain of three cones which extended in an approximately east-west direction. At the end of interglacial times a directed explosion occurred from each of the three cones, and on the summits of the cones were formed the amphitheaters of large craters, open toward the sea. At the time of the second glaciation these amphitheaters served as basins for the accumulation of ice. Apparently the craters were widened at this time, and wide U-shaped valleys were developed down as far as the sea coast.

The history of the westernmost cone ended at this point; a portion of the southern flank of the cone and the crater, about 1 km across and open to the west, are still left. The other two cones continued to be active into postglacial times and are still active at present.

The remnants of the northern cone comprise the somma of Berg Volcano; the diameter of its caldera, which is open to the north, is about 2 km. Its rim is generally even; its height is about 1040 m above sea level, and only at its eastern and western ends does it rise to 1100-1200 m, but then it rapidly drops to zero. The southwestern part of the rim, along with a small portion of the flank, is dropped down along an arcuate glide plane, giving the false impression that it is a remnant of an inner somma.

On the southeast the outer flanks of the somma descend to a height of 800-900 m, at which point they run up against the flanks of older ridges that are remnants of the preglacial volcanic rocks. The inner walls of the caldera are steep; the layered structure of the somma can be seen in their upper parts, but lower down this is covered by talus.

There is a rather low extrusion dome in the wide amphi-theater. The dome, which is 700 m across, has a relative height of 200-250 m (approximately 950 m above sea level). The flat sum-mit of the dome is covered by a thick layer of pyroclastics through which protrude scattered spines and ridges of rocks. In the south-ern part of the dome a deep explosion crater is located. Another fresher-looking crater is found in the northwest part of the dome. In the eastern part of the second crater there is a deep explosion pit about 100 m across.

Two rather small flows, apparently of very viscous lava, de-scend to the north of the dome. One of them flows from beneath the

dome and is the older one; the other extends from the top of the
dome and thus is younger (Fig. 56).

Information on volcanic eruptions on Urup is rather incon-
sistent, and we will return to this somewhat later. Reliable information is available only after 1946. There was a weak eruption
in the spring of 1946, and an eruption with the ejection of large
amounts of ash occurred in the winter of 1951/1952. The thickness
of the ash that fell near the volcano reached 10-20 cm. Apparently
the explosion pit on the northwest part of the dome was formed at
this time. The volcano shows continuous solfataric activity; the
solfataras are located in the northwest crater.

The eastern cone forms  the somma of T r e z u b e t s  V o l -
c a n o , which is very similar to Berg Volcano in its structure.
Only the southern half of the somma is preserved in the form of an
obliquely truncated, now semicircular cone that is broadly open to
the north. The diameter of the caldera is about 2 km. The high
point on the caldera rim (1220 m) is in its southern part; from this
point the rim rapidly decreases in height to the northeast and northwest, first down to about the 500-m level and then further until it
finally disappears. The whole northern part of the somma has been
destroyed, and steep cliffs, 400-600 m high, drop down to the sea.
The outer slopes of the somma descend to an altitude of 700-800 m,
where they rest against the eroded remnants of Pleistocene cones.

The inner walls of the caldera are steep and locally perpendicular. A steep, truncated cone, crowned by a ring of almost vertical lava crags, rises to a height of 1017 m above sea level (about
400 m above the floor of the caldera) within the amphitheater of the
half-caldera. This is an extrusion dome whose summit has been
blown away; the diameter of the explosion crater formed is about
300 m, and its depth reaches 100 m. The crater rim is very uneven; its southeastern part is very badly destroyed, forming a passage by which the interior of the crater can be reached. Seen from
the sea to the north three characteristic large crags (after which
the volcano was named "Trident") stick out on the edge of the rim.

The slopes of the dome's agglomerate mantle are rather even,
but on the east slope, at a height of 700-800 m, there is an oval explosion crater 150 × 250 m across and up to 50 m deep. It is suggested that it has formed since 1915 (Nemoto, 1937).

North of the dome lies a gently sloping area in whose slopes
northward-dipping layers of lava and pyroclastic material are exposed. This area is a remnant of an old, layered central cone, now
largely destroyed. A lava flow that extends for 3 km to the coast
where it forms Cape Oval'nyi is also a product of the activity of
the central cone.

The lavas of the somma of Trezubets Volcano are two-pyroxene andesites. Labradorite ($An_{58-70}$), augite, and hypersthene (in
approximately equal amounts) are found as phenocrysts. Olivine
is found in smaller amounts, and quartz is very rare. The groundmass is hyaline. The central-cone lava is also two-pyroxene andesite, with phenocrysts dominantly of labradorite, $An_{58-66}$. Augite
and hypersthene are present in significantly smaller amounts, and
olivine is even rarer. The groundmass of the rock is pilotaxitic.
Finally, the dome lava is also composed of two-pyroxene andesite,
but containing small amounts of hornblende. In the phenocrysts
labradorite, $An_{60-72}$, is clearly predominant; then come augite and
hypersthene and insignificant amounts of basaltic hornblende with
an extinction angle of 8°. The groundmass structure is hyaline,
with very rare microlites.

As already noted, information on eruptions on Urup is not very
clear. All the older Russian and the more recent Japanese sources
speak of only one active volcano on the island; now there are two.
The two volcanoes are situated in a line, and it remains unclear as
to whether one of them was active or both of them, alternately.

Captain Chernyi (in the 1760's) reported nothing at all as to
active volcanoes on Urup, but in the 1770's and 1780's Antipin on
his map (Atlas, Map. No. 159, 1964) noted an active volcano right
in the Kolokol group. Later, in 1844, Voznesenskii (1845), a zoolog-
ist from the Academy of Sciences, also noted one active volcano in
this group without giving its name. Doroshin (1869) reports it in
his summary. At that time the names "Kolokol" and "Trezubets"
were already in existence. Judging from the fact that the name of
the volcano was not given, it may be suggested that the volcano that
erupted was Berg, which had not yet been named. However, it may
have been Trezubets Volcano that erupted.

Japanese sources report an eruption on Urup on July 25-26,
1894 (Omori, 1918). Gubler (1932) assigns this eruption to Kolokol
Volcano, and this eruption has been reported in many summaries,

but we have determined that this volcano could not have erupted in the so recent past (Gorshkov, 1954).

Another eruption on Urup occurred on March 13, 1924 (Kamio, 1932). Nemoto (1937) states that Trezubets Volcano erupted and that a lateral explosion crater was formed at this time. Nemoto studied Urup from 1932 to 1935. During all this period intensive fumarolic activity was noted on Trezubets Volcano, while Berg stayed perfectly quiet. The dome of Berg Volcano has an extremely fresh appearance and it may be that it was formed in the period between 1935 and 1946, when it was first seen.

None of the remaining volcanoes on Urup, judging from their state of preservation, could have had eruptions in the recent historic past.

A second chain of two isolated cones, Kolokol and Borzov Volcanoes, extends to the southwest from Berg Volcano.

Kolokol Volcano has the form of a fine, regular cone* 1326 m above sea level. The crater of the volcano is badly destroyed; its rim is scarcely preserved at all. On its summit remains a flat area in the form of the letter "T," 150 × 200 m across (Fig. 57). The flanks of the cone, up to the very top, are covered with brush and grass. A small lava flow runs down to the west from the crater; farther down there is talus, and several parallel lava flows up to 4 km long extend from the base of the cone almost to the sea coast. A thickly covered flow also extends to the south, into the saddle between Kolokol and Borzov, and then descends northeastward to the base of the cone. A wide flow of very viscous lava "armors" the southeastern flank of the volcano; apparently it was the last outpouring.

A slightly welded pyroclastic flow descends, like a smooth ribbon, southward from beneath the base of the cone as far as the valley of the Rybnaya River. Apparently the large crater that produced this pyroclastic flow is hidden beneath the present cone of Kolokol Volcano.

The cone of Kolokol Volcano is undoubtedly postglacial; there are no traces of any glacial erosion on it, and evidence of erosion of any kind is almost completely absent. There are references to

---

* The Japanese called it Uruppu-Fuji, i.e., the Fuji of Urup.

Fig. 57. Kolokol Volcano group: Kolokol Volcano on the right, Berg Volcano on the left, Trezubets Volcano beyond.

its eruption in 1894; however, there are no traces whatsoever of any so recent an eruption, and these should have survived rather well since 1894.

Borzov Volcano adjoins Kolokol Volcano on the southwest. It is a broader volcano, with a badly ruined crater, at the site of which only a rounded area remains. The slopes are also eroded (especially on the western half of the cone). Farther down, the terminal parts of lava flows are preserved. Judging from the considerably eroded state of the flanks, Borzov Volcano is older than Kolokol; it is early Holocene or perhaps late Pleistocene.

The next "cluster" of early Quaternary volcanics, Petr Shmidt Ridge, extends for about 12-14 km. The ridge is very strongly eroded; locally there are moraines of the second glaciation in the U-shaped valleys near the ridge crest. No traces of the original surface are preserved. Perhaps some of the high points, peaks such as Vitkovskii (982 m), Brigantin (969 m), and Petr Shmidt (1031 m), are remnants of older eruptive centers.

West of the last-named peak, on the shore of the Sea of Okhotsk, rise the remnants of the Tri Sestry [Three Sisters] Volcano (999 m). It also is considerably eroded, but its flanks clearly overlap an old eroded topography of the first glaciation. Tri Sestry Volcano arose and was active either on the very end of the Pleistocene or at the beginning of the Holocene. It is a stratovolcano with a now-ruined summit extrusion dome. On the west flank altered rocks stick out through the continuous cover of scrubby vegetation. Near the Okhotsk coast there is a strong source of thermal waters, apparently the last vestige of activity of this volcano.

A bomb from Cape Neproidesh', near the foot of this volcano, consists of acid (60.5% $SiO_2$) two-pyroxene andesite that contains phenocrysts of quartz and ordinary hornblende in addition to the usual plagioclase, hypersthene, and augite (Nemoto, 1935).

Between Petr Shmidt Ridge and Krishtofovich Ridge the small, isolated cone of Rudakov Volcano (543 m) rises on the Okhotsk coast of the Tokotan Isthmus. Its flanks are covered with deep, wide ravines. The crater rim has also lost its regular form, but the funnel-like crater is preserved and its floor is covered by a fresh-water lake. The diameter of the cone is 3-4 km; the crater is about

Fig. 58. Sketch of the distribution of the volcanoes
of the Ivao group (symbols as in Fig. 9).

700 m across and the lake, 300 m. The flanks of the cone clearly cover a topography produced by the first glaciation. Clear evidence of the second glaciation is lacking, but because of the low height of the cone, none may have ever been produced. The age of Rudakov Volcano is late Pleistocene or early Holocene. Its lavas are of andesite–basalt with mafic phenocrysts of augite, hypersthene, and olivine (Nemoto, 1935).

**4.** The last link in the chain of clusters, Krishtofovich Ridge, about 20 km long, has a most complex structure; the highest point on the island (more than 1400 m) also occurs here.

The ridge is clearly asymmetric in transverse profile. Its northwest flank is deeply eroded; there are many glacial troughs and river valleys, cliffs and spurs here, and the original surface is almost nowhere preserved. On the southeast flank, on the other hand, the original surface of the lava flows is preserved almost everywhere, and only near the axis of the ridge are there a few U-shaped valleys of the second glaciation; downslope, these rapidly become narrower.

The sources of the lava flows are near the ridge axis, but at present the lava cones have been destroyed; only Mount Shabalin represents a preserved cone. The main peaks on the ridge (Sivukh (1326 m), Struve (1320 m), and others, are the remains of cones.

The clear difference in the degree of erosion on the ridge flanks can hardly be explained by differences in abrasion alone. In all probability the cones that erupted lava onto the southeast flank were not only active in preglacial times, but continued to be active and to develop during the first glaciation and during interglacial times.

In the Krishtofovich Ridge area volcanic activity continued into postglacial times (Fig. 58). Initially a chain of small cones arose along the axis of the ridge; from these voluminous flows of rather fluid lava were poured out toward the northwest. At present, the remains of three of these cones are preserved; one of these has the remains of a crater, but the other two are only partial cones. The northwestern portions of these cones were destroyed by large-scale collapses, and a broad area $2 \times 5$ km across is covered by a chaotic accumulation of collapse deposits; beneath these are seen the outlines of lava flows that come out from beneath these deposits farther downslope and locally extend to the coast.

The collapse zone (which may be related to tectonic movements) extends farther to the southwest, beyond the three preserved cones. It may be that the original Holocene cones were larger and that large-scale collapse and later intensive erosion have obliterated their remains.

As mentioned above, the northwest flank of Krishtofovich Ridge is strongly eroded. The cones that we just mentioned are located at the head of a wide depression in whose origin both erosion and tectonics have played a role. The depression extends for 9 km between Sivukh and Struve Peaks. The northern half of the depression is undoubtedly filled by lavas, but in the southern part the presence of fresh lavas is not certain. Above the lavas (and perhaps moraines) the floor of this depression is covered by the products of the collapse; farther downslope, at about the 400–600 m level, there is a steep cliff and still farther down, near the shore, there is a second one. The lava flows that reach the cliffs are converted into lava falls that interrupt the continuity of the flows, but below the cliff the lava again coalesces and continues onward, taking on the peculiar appearance of agglomeratic lava. The character of the surface of a flow and details of its structure, such as the lateral ridges, are very different above and below the benches.

On the southeast side of the ridge, at the head of a large glacial valley (of the Krasotok and Polnovodnaya Rivers), a second chain of cones is formed that is younger than the first one. There are three well-preserved cones here, and they are elongate perpendicular to the trend of the island, i.e., from northwest to southeast. These three cones did not arise at the same time. The oldest of them, Ivao Cone at the northwest end, is the high point on the ridge, 1430 m. Its age is close to that of the first chain; however, its lavas were more viscous and have a smaller extent.

Later the southeasternmost cone arose; its very viscous lavas stretch to the south for 3 km. This cone, which bisects the U-shaped valley, served for a while as the cause of the alpine Lake Ivao.

The third and youngest cone (Krutaya Mountain) is the middle one. It has a distinctly elongate form. A flow of viscous lava runs down to the east from the half-ruined crater to the base of the valley wall.

Thus, as a consequence of what we said earlier, the characteristic feature of volcanic activity on Urup is the predominance of preglacial linear-clustered eruptions. Volcanic activity clearly declined in the Holocene.

## Iturup Island

The island of Iturup is separated from Urup by the Vries Strait, 39 km wide. Iturup is the largest island in the whole Kurile chain; it has a length of 203 km, with a width of 5.5–46 km, and its area is 3200 km$^2$. Rocks of the Tertiary basement underly a considerable area on the island. The island has a very complex form and consists of several mountain groups joined together by isthmuses, some of which are very low. Eight mountain groups and isolated volcanoes can be recognized. In all, 40–48 volcanoes can be counted on the island, of which 31 are Holocene and eight are active.

The vegetation on Iturup, especially to the south of Vetrovoi Isthmus, is richer than on the islands farther north. Here, along the thick brush and Kurile bamboo, there are coniferous and broadleaved forests and groves of rock birches.

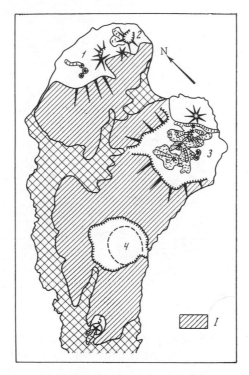

Fig. 59. Sketch of the distribution of the vol-
canoes of the Medvezhii Peninsula group: 1)
Kamui massif; 2) Demon Volcano; 3) Medvezhii
Volcano; 4) Tsirk Caldera; 5) Tornyi Cone; I)
Quaternary lavas (other symbols as in Fig. 9).

Medvezhii Peninsula. The first mountain group ex-
tends for 52 km from the northeast end of the island to the Vetrovoi
Isthmus and forms the Medvezhii Peninsula where three complex
volcanic massifs and several small cones are set on a Tertiary
basement (Fig. 59).

The first of these, the Kamui massif, occupies an oval
area of about 10 × 18 km. The arcuate Kamui Ridge (up to 975 m
high) is a portion of the oldest volcanic edifice in this massif. The
slopes, which descend gently to the south and west, very probably
are remnants of a broad, old cone, while the crest of Kamui Ridge
with its steep cliffs on the east is the remnant of an old caldera.
The lavas of this cone rest directly on the Tertiary basement and
are of preglacial age. Of the old cone only the southwest sector is

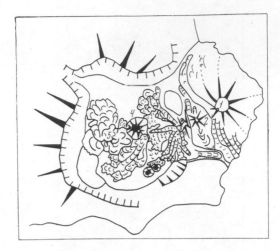

Fig. 60. Sketch of the structure of the Medvezhii
caldera-volcano: 1) Medvezhii Cone; 2) Srednii
Cone; 3) Kudryavyi Cone; 4) Men'shoi Brat Dome;
5) effusive dome (other symbols as in Fig. 9).

still preserved; the remainder has been destroyed, apparently with
the aid of tectonic forces. At the site of the destroyed part north
and east of Kamui Ridge lie the remains of a volcanic structure of
interglacial age. Apparently there were two or three volcanic centers here. One of them, Kamui Cone, is the highest point in this
massif (1323 m). This cone occurs at the head of a large U-shaped
valley of the second glaciation. The remains of the second cone
are located on the rim of the same valley, to the southeast of Kamui,
and the remains of the third cone are to the east. A fourth cone apparently was located still farther to the east. The whole eastern
part of the interglacial edifice has been cut away by a fault.

A group of four small cones is also situated northwest of
Kamui Peak.

The second glaciation left U-shaped valleys along the peri-
phery of the massif. In some of these, moraines of the second glaciation were deposited.

Demon Volcano (1206 m) is located 3 km east of Kamui
Cone. This cone fills the lower part of the glacial trough that extends from Kamui Cone. On the volcano's summit there is a crater
up to 1.5 km across and open to the east. The western part of the

cone is well preserved, but the eastern part has been badly destroyed by erosion. In all probability a fault runs along the easttern flank of Demon Volcano.

The second volcanic massif is the complex caldera volcano M e d v e z h i i (Fig. 60), located to the south of the Kamui massif. The old preglacial edifice of Medvezhii Volcano is represented by the remains of a large caldera volcano. Remnants of the outer slopes, which are cut into by the U-shaped valley (of the first glaciation) of the Slavnaya River, are preserved on the north and west. The southern part of the caldera is badly eroded, while the eastern part is locally overlapped by younger formations and is apparently badly ruined in part. The caldera itself has been strongly modified by erosion, including glacial erosion (during the first glaciation). The caldera rim reaches heights of 530–560 m; its diameter is 8–9 km. The caldera floor lies at an altitude of 175 m. Almost the whole basin of the caldera is filled by younger volcanic formations, and moraines occur at various places on the floor. No clear evidence of the second glaciation is seen.

There are no volcanic formations of interglacial age, though two small, smoothed-off extrusion domes may be of interglacial or late glacial age. On the other hand, postglacial activity was widespread here.

Three closely joined volcanic cones form a small ridge of roughly east-west trend. The easternmost cone, Medvezhii, is the highest (1125 m) and the oldest. It lies outside the caldera, on the site of the ruined eastern part of the somma. The cone's outline is strongly elongate in a north-south direction. This is partly related to the fact that the western flank of Medvezhii Cone is overlapped to a large extent by the younger volcanic structure of Srednii Cone while the eastern flank, which stretches as far as the shore of Vries Strait, is distinctly shorter and correspondingly steeper than the northern and southern slopes. The flanks of the cone are dissected by numerous barrancas; however, no traces of glacial erosion are seen, which fixes the age of the cone as Holocene. A rounded area some 700 m across remains at the site of the crater; in its eastern part traces of the crater rim can be seen.

Approximately 1.5 km west of the summit of Medvezhii Cone, beyond a 200-m-deep saddle and about on the site of the caldera rim, rises the summit of Srednii Cone (1113 m above sea level). The southwestern part of the cone has been destroyed. Several lava flows run down the still-preserved northern flank. Flows also stretch to the south, and one of these reaches the sea coast. The somewhat lower (991 m) and presently active Kudryavyi Cone closely adjoins Srednii Cone on the west.

To the north, between Srednii and Kudryavyi Cones, a ridge of distinctly older appearance sticks out at right angles to the slope; this is by-passed on both sides by the younger lava flows. This ridge is a remnant of some older structure, perhaps a remnant of the somma.

The summit of Kudryavyi Cone is somewhat elongate in an east-west direction; two shallow craters are located on it. The shallow, gentle eastern crater is almost completely enclosed. Its floor is uneven, with scattered depressions and barriers; these irregularities were produced in part by the work of extracting volcanic sulfur, which was carried on here on a large scale by the Japanese. The west crater has the form of a steep-walled amphitheater, open to the north, from which a thick, fresh-looking lava flow was emitted.

The flanks of Kudryavyi Cone are rather weakly dissected and are covered by numerous flows of lava (along with subordinate loose deposits) up to 2-3 km long that extend in all directions and reach the caldera walls.

On the west, near the base of Kudryavyi Cone, is the large extrusion dome of Men'shoi Brat, whose basal diameter exceeds 1 km and whose relative height reaches 200 m (563 m above sea level). On the northwest and southeast flanks of the dome, near its summit, are two small scoria cones. The first of these has a crater open to the west, from which a small stream of lava flows. The crater of the second cone is completely enclosed, but near its southern foot there is a lava bocca which has produced a large number of very fresh-looking lava flows that are up to 4-5 km long; some of them reach to the shore of Slavnoe Lake. These flows occupy an area of about 5 km$^2$.

A large, craterless lava pile of the effusion-dome type is located 2 km west of Men'shoi Brat Dome, on the floor of the caldera; lava tongues extend northward from it for up to 2.5 km. One of the thick flows bisects the valley of the Slavnaya River and forms a lake in the atrio. Slavnoe Lake is situated in the western part of the atrio near the caldera wall; its dimensions are 1 × 3 km.

Another small conelet lies in the southern part of the caldera, in line with the two old domes mentioned earlier. This cone produced a miniature lava flow that bends around the north side of one of the domes. Thus in the Medvezhii Volcano group of Holocene volcanic structures there are: three large cones, at least three scoria cones, three extrusion domes, and one effusion dome. Almost all these formations have large lava flows.

In the vicinity of the caldera are found deposits of dacite pumice; they are composed almost entirely of glass, sometimes with a fibrous structure. Very rare phenocrysts of plagioclase, augite, and hypersthene are scattered through the glass.

The lavas of Medvezhii Cone are normal two-pyroxene andesites with hyalopilitic groundmasses.

The lavas of the young flows in the Medvezhii massif are composed of basalt and andesite-basalt. Aphyric and porphyritic varieties are noted in the flows from Kudryavyi Cone; the latter type usually contains phenocrysts of labradorite, augite, and olivine. The olivine is often decomposed. In one of the flows there are very large crystals of hypersthene, which is almost the only mafic mineral present and which is surrounded by a thin rim of augite. Plagiophyric andesite-basalts are represented in other flows; almost the only phenocryst mineral in them is basic plagioclase, though augite is present in insignificant amounts. In the lavas from the summit cone there is a great deal of augite and hypersthene, often in glomeroporphyritic aggregates. The structure of the groundmass is hyalopilitic or microdoleritic, with abundant microlites of augite.

The lavas of the Men'shoi Brat scoria cones are typical basalts with phenocrysts of labradorite ($An_{65}$), abundant olivine, and subordinate augite. In another flow abundant hypersthene is found. The groundmass structure is hyalopilitic or pilotaxitic.

Information on eruptions in this group is very scanty. The first reference is to 1778 or 1779, when Antipin and Shabalin were on the island. A large-scale eruption apparently occurred in 1883, with the emission of large flows of lava. Just which cone erupted is not completely clear; at present very strong solfataric activity is shown by both craters of Kudryavyi Cone, on whose slopes are very fresh lava flows, and it was almost certainly this cone that erupted in 1883. In addition, the two cones of Men'shoi Brat and the third scoria cone also look very fresh, and eruptions must have occurred from them in the last 100-200 years.

Tsirk Caldera adjoins the somma of Medvezhii Volcano on the west. The rim of this caldera, strongly altered by later erosion, has a diameter of about 6 km. In all probability the caldera was formed at the end of the first glaciation or in interglacial times; there are cirques of the second glaciation on the inner walls of the caldera. The southern rim is breached by the valley of the Tsirk River, which drains the more or less flat, circular (about 5 km across) floor of the caldera. There is a hot lake on the caldera floor.

The strongly eroded (in part by glacial activity) remnants of the old volcanic massif of Sibetoro rise to the north of the caldera. Its highest point rises to 853 m.

Still farther west, near the boundary betweeen the lower Quaternary lavas and the Tertiary basement, the rather small Tornyi Cone (417 m) with a small lava flow is located in a broad glacial cirque of the first glaciation, at the source of the Porozhistaya River. This cone is late Pleistocene or early Holocene in age.

The western edge of this first volcanic massif is the fault that separates it from the graben of the Vetrovoi Isthmus. Here are located the remains of a Pleistocene volcano, Parusnaya Mountain, and of the late Pleistocene or early Holocene cone of Golets (442 m), which has produced lava flows that reach the sea coast.

Samples from the Mount Golets region are of andesite-dacite, with phenocrysts of plagioclase, pyroxene, and apparently brown basaltic hornblende that was described as biotite. The groundmass structure is hyaline (Popkova et al., 1961).

The developmental history of the volcanoes in this part of the island is depicted in the following summary. In the early (preglacial) Pleistocene several large shield-like volcanoes (Old Kamui, Old Medvezhii, Sibetoro, and probably others in the region of Tsirk, Dobrynin, and Parusnaya Calderas) arose on the Tertiary basement. Before the beginning of the first glaciation calderas were

formed on the first two of these.  Then during the first glaciation
all the volcanic structures were strongly eroded.  The glaciers
were dominantly of the mountain-valley type and ran down to the
present sea coast, where moraines can now be found (for example,
in the area of Medvezhii Gulf).  The preglacial calderas underwent
especially intense glaciation; as a result, their forms have been
badly altered.  In addition, the outlines of the calderas were disrupted by a series of faults, dominantly of northwest trend, along
which the eastern portions of the calderas were dropped a considerable distance.  Several valleys were also established along the lines
of tectonic movement and were later transformed into glacial troughs
(Slavnaya River, Skal'nyi Creek).

In interglacial times Tsirk Caldera was formed, and several
small cones arose in the ruined Kamui Caldera.  Tornyi and Golets
Cones were also formed at the end of the Pleistocene.  However, in
general interglacial volcanic activity was of rather moderate intensity.

The second glaciation affected only the highest peaks in the
area of Mount Sibetoro, Tsirk Caldera, and Kamui Volcano, where
some of the interglacial cones were partly destroyed.  Other cones
(in particular, the very high cone of Kamui) continued to be active
during the second glaciation.

In the Holocene the new, large cone of Demon was formed; its
lavas flooded U-shaped valleys of the second glaciation.  In the final
stage of development of Demon Volcano an eastward-directed explosion occurred that may have been initiated by tectonic movements, and subsequent erosion cut a large canyon through from the
caldera to the shore of Vries Strait.

The Medvezhii Volcano massif has been the arena of intense
Holocene activity.  Here, in the basin of the caldera and outside of
it, a series of varied cones and domes arose that were described
above.

Vetrovoi Isthmus.  The low-lying Vetrovoi Isthmus
separates the Medvezhii Peninsula from the rest of the island.  The
isthmus is a graben, bounded by faults with nearly north-south
trends.  The width of the graben averages 10 km, and it is younger
than the first glaciation.  This whole area is covered by thick pumice deposits; these in part form the cliffs of Belaya Mountain.  The

source of this pumice remains uncertain. Only during the working
out of the geology did we identify a wide caldera here that gave rise
to a pumice deposit.

The remains of the caldera, with a rim-to-rim diameter of
6-7 km, lie approximately in the center of the isthmus. The high
point on the caldera rim (264 m) is in the south. This caldera, like
all of the isthmus, is composed of pumice, and all the protruding
topographic highs have a peculiar finely dissected surface of the
"bad-land" type. The caldera is open to the southeast, and the eastern part of the isthmus is a flat pumice terrace lying 16 m above
sea level. A flat plain composed of pumice lies to the southwest of
the narrow rim of the caldera; its height is 20 m near the sea coast,
but it gradually rises toward the interior of the island. On the northwest the caldera structure is close to a fault surface and in part
covers it. A rather large amount of pumice also fell out onto the
flanks of the badly destroyed preglacial Shirokii Volcano that are
adjacent to the fault.

Judging from the topography, the caldera was formed as a result of a strong submarine eruption in interglacial times. The built-
up rim of the caldera stuck out of the sea as a ring-like peninsula
near the flanks of Shirokii Cone, and in all probability there was an
interglacial caldera bay within the basin of the caldera.

An explosion caldera about 1 km across has been blown out
of the caldera floor; in it lies Tainoe Lake, without an outlet. The
small Klumba Cone (163 m high), with the remains of a crater on
its top, rises on the edge of this crater. Still farther south, near
the wall of the caldera, are the remains of an older pumice cone
that is already badly ruined, but also with the remains of a crater
on its summit.

Groznyi Ridge Group. The next group of volcanoes
and volcanic land-forms extends along the Pacific coast of the island from the Vetrovoi Isthmus to Kosatko Bay, a distance of about
45 km. Several preglacial, strongly eroded massifs occur in this
group. In addition, six or seven Holocene cones are located here,
of which three are active at present (Fig. 61).

The first glaciation destroyed all the preglacial structures,
and the highest massifs apparently were also affected by the second
glaciation; as a result, the old centers of activity can be reconstruc-

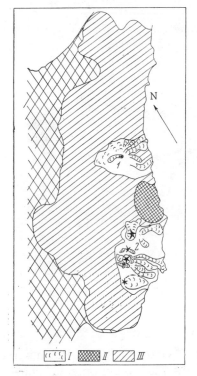

Fig. 61. Distribution map of the volcanoes of Groznyi Ridge: I) Lava flows;
II) interglacial volcanic edifices; III)
preglacial lavas; 1) Baranskii Volcano;
2) Groznyi group of volcanoes. (Other
symbols as in Fig. 9).

ted only with great difficulty. Three of the old volcanic massifs are preserved: Shirokaya Mountain (Shirokii Volcano, 373 m) close to the Vetrovoi Isthmus; Mount Verblyud (721 m) in the central part of the island; and the Gorelaya Mountain massif (971 m), which forms the northeast end of Groznyi Ridge. Moraines of the second glaciation lie in the broad valley of the upper course of the Kuibyshevka River, between Mount Verblyud and Groznyi Ridge. Similar moraines also occur in the headwaters of the Kurilko River in the area of Gorelaya Mountain.

Judging from the form of the preserved parts of the massifs, preglacial activity here was of the "linear-clustered" type.

Most of Groznyi Ridge is overlapped by Holocene volcanic deposits. No clear evidence of interglacial activity is preserved; it probably was rather limited in scale.

Perhaps the small isolated ridge extending to the sea coast between Machekh and Baranskii Volcanoes is the remnant of a chain of interglacial volcanoes. Its high point, Mount Kadilko, rises to 661 m. The flanks of the ridge clearly overlap a glacial topography, in addition to which they are strongly eroded, and although distinct glacial forms are not seen here, the intensity of erosion of this ridge is considerably greater than near the Holocene structures. This allows us provisionally to assign the part of Groznyi Ridge in question to interglacial times.

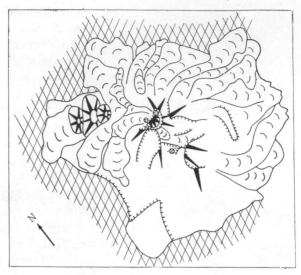

Fig. 62. Sketch of the structure of Baranskii Volcano
(symbols as in Fig. 9).

The extreme southwestern part of Groznyi Ridge, Rebunshiri Mountain, is also strongly eroded. This 782-m-high massif consists of two or three cones that have become fused together. They arose after the first glaciation and may belong to the late Pleistocene or early Holocene.

Baranskii Volcano, on the northeastern part of Groznyi Ridge, is a rather complex structure. This volcano rises as an isolated, strongly truncated cone 1126 m high.

Lava flows from the volcano descend southeastward for 4-5 km to the sea coast; they stretch for a similar distance in the opposite direction and occupy a generally oval area 6 × 9 km across, with its length perpendicular to the ridge (Fig. 62).

The flanks of the cone, except on the north, are composed of young, strongly eroded lavas, and seen from the south the volcano has a rather ancient appearance. The cone is "planted" on a glacial topography of the first glaciation; however, the rather wide and deep ravines near its summit lead to the suggestion that its crater may have served as a center of the second glaciation, i.e., that it is interglacial in age. The base of the cone is better preserved, and here on the south we can trace the old lava flows which have preserved their morphologic outlines very clearly. Some of the flows form capes that stick out into the sea, for example, near Vodopadnyi Creek.

The old summit crater is badly destroyed; of it there remain only isolated, badly worn down high points that, along with the adjacent parts of the flanks, are situated in the form of a star 700-800 m across. In the old crater there is a gently sloping, essentially pyroclastic inner cone that fills it completely and in part spills out beyond it, into the upper part of the ravines. Its crater is somewhat displaced to the north. The crater rim in the north is completely destroyed, and the crater is broadly open to the north-northwest. Numerous lava flows related to the inner crater cover a considerable sector of the northern and eastern parts of the cone. In the north they reach to the foot of Mount Volchok; in the east, to the bed of Dugovoi Creek. In the southeast one of the flows runs down as far as the sea coast.

On the floor of the crater a plug-dome rises as a gently sloping shield. Its low, steep flanks pass into a broad, weakly convex summit. The diameter of the dome is about 500 m, its height above the crater floor is about 40-50 m (in the northern, blown-away part the height is at least 100 m). Between the slopes of the dome and the walls of the crater a narrow, arcuate corridor remains that separates the dark-gray dome from the lighter-colored, variegated crater wall.

The northern part of the dome has been blown away; there is a chain of several small explosion craters here that extends to the northwest, and a large deep explosion pipe. A very young lava flow descends along the northwest slope from this blown-away part of the dome. It is 500 m wide at its source, but downstream it widens to 1 km; it is up to 2 km long and 50-60 m thick. The lower and middle parts of this flow are composed of normal two-pyroxene andesite with a microlitic groundmass (though at the source the groundmass is very glassy).

As noted earlier, the flanks of the volcano are furrowed by deep, wide ravines. Some of these are the result of normal fluvial or perhaps glacial erosion, others are lateral explosion craters that have been modified by erosion. The amphitheater on the southwest flank of the cone is undoubtedly of volcanic origin. There are strong

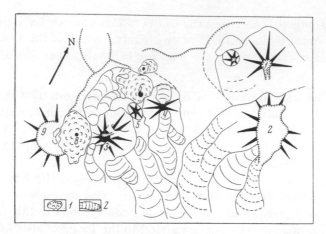

Fig. 63  Sketch of the structure of the volcanoes of the Groznyi
group: 1) Effusion domes; 2) pyroclastic and mud flows (explanation of other numbers given in text; (other symbols as in
Fig. 9).

solfataras and hot springs on the floor and walls of this amphitheater; farther down, near the base, there are mud cauldrons.

There are also subsidiary structures near the northwest base
of the volcano.  At least three closely spaced and strongly overgrown cones are located here, forming a northwest-trending chain.
Lava flows stretch to the northwest from them.  The lava flows of
the central crater surround this group of cones on two sides.

The volcano shows intense solfataric activity.  In addition to
the southwest parasitic crater, there are strong solfataras on the
summit, in the explosion pipe, and in some of the explosion craters.
There is no information on eruptions of the volcano in former times,
though according to oral information from the local inhabitants, weak
explosive activity occurred in 1951.

Baranskii Volcano arose in the last Pleistocene and displayed
mixed activity, with explosions and lava effusions.  As a result, a
rather large cone was thrown up.  In the final phase of the first stage
of activity a large explosion crater was formed in the southwestern
part of the cone, and the crater chain near the northwest foot was
probably formed at this same time.  Then a long hiatus in the activity of the volcano set in (second glaciation?), during which the flanks
of the cone were cut up by wide, deep ravines.  The destruction of

the north flank was particularly severe, and only scattered remnants of the crater rim were preserved.

Volcanic activity began again in the Holocene. Effusive activity was dominant now; pyroclastic deposits "healed" the summit portions of the old cone to a slight extent, but lava flows extended for 4-5 km. In the final stages the crater of the volcano was sealed by a plug of pumiceous andesite; then the northern part of the dome was destroyed (by an explosion?) and the last lava flow was emitted there.

The very last eruptions were moderate explosions that formed a chain of explosion craters. Fumarolic activity continues at the present time.

A chain of modern volcanoes begins 14 km southwest of Baranskii Volcano; it extends for 8-9 km, from Teben'kov Volcano to Motonopuri Cone (Fig. 63).

Teben'kov Volcano (1 in Fig. 63) is a rather regular cone about 500 m high (1212 m, absolute). This cone rises in the southeastern part of a broad caldera-like depression that is broadly open to the east; it is up to 2 km wide. This depression undoubtedly served as a source of supply for several large glaciers during the first glaciation.

The summit crater of Teben'kov Cone is surrounded by a crown of lava crags; it is about 200 m across and 50-70 m deep. The southern part of the rim has been destroyed, and a wide, deep ravine runs down the flank from here. There is a small depression in the northern rim from which a ravine also descends, but it is less clearly expressed. Near the northwest foot of the cone, on the lake shore, is located a small subsidiary cone with the remains of a crater; another subsidiary cone apparently occurs on the upper, northern part of the cone. All the flanks and even the crater of the volcano are covered with thick, difficultly penetrable, brushy stands of cedar.

The base of the cone on the north, west, and south is hemmed in by the slopes of the caldera-like depression, but on the east the slopes of the cone run down to the headwaters of Kedrovyi and Nagornyi Creeks; seen from this direction, the cone appears to be higher.

The lavas of Teben'kov Volcano are two-pyroxene andesites, with phenocrysts of labradorite ($An_{60-65}$), hypersthene, and augite.

The groundmass structures are microdoleritic, microlitic, and crystallitic.

Teben'kov Cone is closely adjoined on the south by a broad north-south depression, M a c h e k h   C r a t e r , 0.8 ×1.5 km across and up to 500 m deep (2 in Fig. 63). The inner walls of this crater are very steep and difficult to get down and the whole basin shows evidence of strong solfataric activity. The rocks have been converted into a white clayey mass and much sulfur, gypsum, and pyrite are formed. On the crater floor, in its central part, are solfatara vents and hot springs. A mineralized stream flows from the crater, cutting through its southern rim.

Rocks of various ages are exposed in the crater walls. The highest part in the north and the southeastern wall are the eroded remnants of Pleistocene structures. The remains of a Holocene cone lie in the central part of the crater. The thick pyroclastic flows that descend to the south and southeast as far as the sea coast are related to the activity of this young cone.

The lavas of Machekh Volcano are two-pyroxene andesites and andesite-basalts. They differ from the andesites of Teben'kov in their more acid plagioclase, $An_{43-60}$. Olivine is found in the andesite-basalts. The groundmass has an intersertal structure.

The present broad depression of Machekh Crater is a purely erosional landform that apparently was derived from a large explosion crater. In all probability the present form has been produced by glacial activity.

On the southwest Teben'kov Volcano is adjoined by the complex Groznyi Volcano group. Part of a small caldera-like depression has been preserved in the northwest part of this group. Its rim beheads U-shaped valleys of the first glaciation (tributaries of Blagodatnaya and Kuibyshevka Rivers) while on the north, decreasing in height, it joins the U-shaped valley of the Mnogooznernaya River.

In the rest of the area the caldera rim is either completely destroyed or overlapped by younger volcanic deposits. The rather steep cliff of the inner caldera wall rises on the northwest to a height of 200-240 m above its floor (about 800 m above sea level). The diameter of the caldera is estimated at about 3-3.5 km; its age is late Pleistocene or early Holocene.

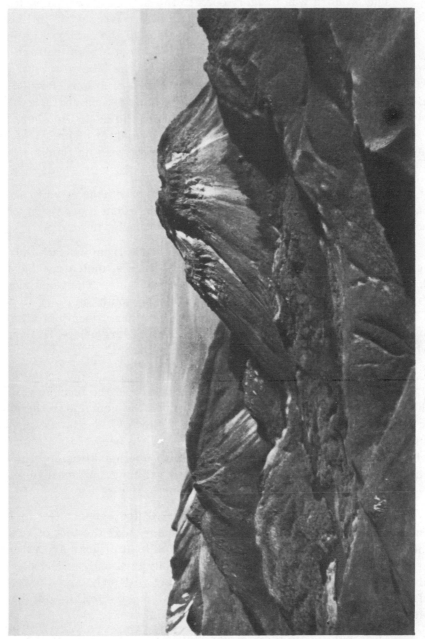

Fig. 64. Groznyi Volcano.

The caldera was broadly open to the south and here, near its southern edge, the large extrusion dome called Groznyi (3 in Fig. 63) rose to a height of 590 m above the floor of the caldera (1158 m, absolute). This dome consists of three large blocks that are separated by depressions. Most probably all the blocks were formed as a result of a single rather extended eruption. Two co-alesced explosion funnels occur on top of the western and largest block, forming a broad trench that is open to the east. Another, lower dome is located on the north flank. A cluster of intense sol-fataras lies at the boundary between this dome and the flanks of the main dome (Fig. 64).

The agglomerate mantle of the domes stretches down to their base almost everywhere so that an oval, somewhat elongate in a north-south direction and measuring $1.1 \times 1.7$ km, is formed.

Short, steep-sided lava tongues protrude from beneath the agglomerate mantle at many points on the flanks, indicating that the dome arose in the crater of an older central cone. The flows reach to the caldera walls in the north, while in the south the older flows stretch for several kilometers. One of them has a length of 6 km and reaches the sea coast, where it forms the large Cape Drakon which is about 1.5 km wide and 30-50 m high.

The large extrusion dome of Drakon (4 in Fig. 63) rises to the northeast of Groznyi Dome, approximately at the boundary of the caldera rim. Sharp lava teeth surround the remains of the sum-mit explosion crater, from which a thick flow of viscous lava poured out to the south. This lava flow reaches to the base of the dome by way of a large lava fall; going on for about another kilometer, it runs up against an old flow from the central cone. A small flow also reaches to the foot of the dome in the north.

A small extrusion dome (5 in Fig. 63) lies between Groznyi and Drakon Domes, and to the north of them, near the foot of Drakon, there is a large effusion dome (6 in Fig. 63). Its flanks are rather gentle and are composed of coarsely blocky lava without any in-dividual flows. Steep-sided, short offshoots extend from the foot of the dome. Another small effusion dome occurs in this vicinity, right in the valley of Mnogoozernyi Creek (7 in Fig. 63).

At an early date there was apparently a large lake in the northern part of the caldera, but at present flows from the central cone and the offshoots of the large effusion dome have considerably restricted the dimensions of this lake and have given its shore a marvelously sinuous outline.

The last member of this group, the Ermak effusion dome, adjoins Groznyi Dome on the west. This dome is also elongate in a north-south direction; on its summit are traces of an explosion crater (8 in Fig. 63).

All the rocks of the Groznyi volcanic group are very similar to one another; they are all two-pyroxene andesites. The plagioclase phenocrysts range from $An_{55}$ to $An_{67}$; augite and hypersthene phenocrysts are also present. The structure of the groundmass is hyalopilitic in the lava flows, microlitic or sometimes pilotaxitic in the dome lavas.

Ermak Dome is adjoined on the west by an older volcanic structure, Ermak Volcano (9 in Fig. 63). This is a rather gentle cone with a broad, gently sloping crater 600-700 m in diameter. The eastern part of the crater rim has been destroyed (Ermak Dome stands in its place). There is a small lake in the crater itself. The outer flanks of the cone have also been badly destroyed. This volcano is most probably of late Pleistocene age.

Ermak Cone is adjoined on the west by the small cone of Kedrovyi, which is ruined on the north, and a little farther on by Motonopuri Cone. The latter's crater and flanks show evidence of erosion, but the cone is undoubtedly Holocene. Its overgrown lava flows reach almost to the sea coast.

A small parasitic cone, Malysh Cone, rises in the pass between Motonopuri and Rebunshiri Cones.

Chirip Peninsula Volcano Group. About in the middle of the Okhotsk coast of the island, the Chirip Peninsula sticks out into the sea; its dimensions are approximately $10-12 \times 19$ km. Tertiary rocks occur in the basement on its southern (and western?) sides. Preglacial Quaternary volcanic rocks are found mainly along the west coast. A narrow, sharp ridge of preglacial rocks in the central part of the peninsula extends as far as the sea coast from a high point in the east. In all probability, preglacial rocks also make up the northern part of the peninsula.

The central part of the peninsula has been cut into on the west by a broad, deep depression up to 4 km long and 500 m deep. Rocks of a preglacial volcanic structure (lavas and pyroclastics of an-

desite-basaltic and basaltic composition) are exposed in the walls
of this depression.

The depressions, which is a unit in plan, is divided into two parts
by a cross piece and by the remnants of an old cone — divided, as it
were, into two broad valleys that resemble wide glacial troughs. According to investigations by Gushchenko, the barrier that splits the basin consists of two half-destroyed extrusion domes, while the floor
of the depression is covered by the deposits of incandescent avalanches. Fumaroles issue from near the base of one of these domes;
in addition, mineralized hot springs occur on the floor of the depression. The rocks in the walls of the depression are strongly
altered by postvolcanic processes, and there are sulfur deposits
here.

This depression cuts through postglacial as well as preglacial
lavas, so that its age is Holocene. However, it may be that in the
Holocene an earlier, essentially negative landform, possibly related
to glaciation, was simply rejuvenated. In addition, tectonic factors
certainly played a role in the development of this depression; the
whole western part of the Chirip massif is cut by north-south faults
(which in many cases have dropped the central part down in step-
wise fashion); one or more tectonic breaks with east-west trends
occur in the area of the depression. Explosive processes also
played a definite role. In general this depression is of complex
origin.

As noted above, fault tectonics have played a large role here.
The whole west coast is cut up by large-scale faults; most probably
the terrace of the east coast is also controlled by a fault. An older
tectonic break occurs near the base of the peninsula.

A whole series of rectilinear and arcuate step faults occur in
the northwestern part of the peninsula; the central part of the massif has been dropped down along these. These faults are very young;
they are partially expressed in the topography, and locally they affect the young lavas. In all probability tectonic movement has continued up to the present, as indicated from time to time by the occurrence in the Chirip area of shallow earthquakes that are regis-
tered by the nearby seismic station "Kuril'sk."

In the northern part of the massif, near its base, there are
four cone-shaped elevations, most probably the remnants of parasitic cones. There is also one crater on the southeast.

Fig. 65. Sketch of the structure of the Chirip Peninsula volcanoes: 1) Lines of tectonic dislocations (symbols as in Fig. 9).

Rocks of the preglacial volcano constitute the peninsula up to
a height of about 1000-1100 m. The lavas and pyroclastics of the
Holocene cones lie with marked angular unconformity on these older
rocks (Fig. 65).

The northernmost cone, Chirip, has an absolute height of 1561
m and rises approximately 400-500 m above the older structure. Its
flat, shallow crater, with a small lake on the floor, has a diameter
of about 200 × 250 m. Lava flows with strongly overgrown surfaces,
indicating their relative age, reach out in all directions.

On the southwest the flows are broken off at the edge of the
depression mentioned earlier. Several flows that moved northward
have passed over the scarps and reached the sea coast. Part of the
flows that moved northward also rest against the scarps of the east-

west faults. Finally, in the southeast the lavas also reach to the sea coast.

On the east flank of Chirip Cone, but a little lower down, there is a small double subterminal crater that apparently is related to the down-faulting of this part of the cone. This crater emitted voluminous flows of lava to the east; these extend as far as the sea coast. One of the flows forms Cape Chirip.

Bogdan Khmel'nitskii Cone rises 4 km to the south of Chirip Cone; they are separated by a saddle about 1100 m high. Lava flows run down and to the east from the saddle, but their source has been cut away by the depression. Ends of flows are found on the west flank of the massif, and judging from the position of all these flows, there were one or two more small cones in the area of the present headwaters of Yuzhnyi Chirip Creek.

The high point on Bogdan Khmel'nitskii (1589 m) is the eroded remnant of a volcanic cone whose near-crater part outcrops at the very head of Yuzhnyi Chirip Creek. A younger subterminal crater is exposed 300 m southeast of the summit and 40 m lower down. Perhaps this crater also occurs on a small tectonic break.

The lava eruptions of Bogdan Khmel'nitskii Cone and of its subterminal crater were very voluminous and have completely overlapped the whole southern part of the old massif. They reach the sea on both the east and the west sides of the peninsula. On the east the lavas from the subterminal crater form Cape Konservnyi.

The subterminal crater itself has a diameter of 250-270 m and a depth of 30-35 m. Within this crater lies the crater of the last eruption; it is about 100 m across and up to 25 m deep.

In the vicinity of the latter crater have been deposited scoria, lapilli, bombs, and blocks with "bread crust" surfaces.

The lavas of Bogdan Khmel'nitskii Cone are two-pyroxene andesites. The dominant phenocrysts are of labradorite and augite, along with subordinate hypersthene. The structure of the groundmass is hyalopilitic; microlites of plagioclase predominate, but are accompanied by less abundant augite and subordinate hypersthene. In the cliffs where the Chirip "caldera" cuts into this cone a variety of rocks are exposed, ranging from pyroxene basalts to welded tuffs of andesitic or dacitic composition.

According to information received from the local inhabitants by Gushchenko, the subterminal crater displayed fumarolic activity until 1956. Fumarolic activity in the Chirip massif is observed at present in three fields in the southern part of the large depression. In the literature eruptions are recorded in 1843 and 1860, but the sites are not closely identified. Eruptions might have occurred both in the areas of the modern fumarolic fields or from any of the sub-'terminal craters. The last eruption was probably an explosive one from the subterminal crater of Bogdan Khmel'nitskii.

The Chirip massif was formed in the Pleistocene on a Tertiary basement. In all probability there were several eruption centers here, and the manifestation was probably similar in character to the linear-clustered type. Perhaps explosions occurred in the central part of the massif and formed a rather large caldera-like depression. A rather thick layer of welded tuff testifies to explosions of such character; these tuffs are found in the valley walls of Severnyi Chirip Creek. The one or two depressions formed were widened and modified by normal fluvial and glacial erosion.

In postglacial times at least three cones arose along the axis of the massif. The northernmost and southernmost ones exceeded 1500 m in height; the central one or two were somewhat lower. The Pleistocene depression was considerably filled in by the products of new eruptions. However, renewed erosion later restored this depression; at the same time, the central cone was completely destroyed (of it, only the ends of lava flows remain). The southern cone, Bogdan Khmel'nitskii, was also partly destroyed.

The important role played by faulting is very characteristic of the Chirip massif. Disjunctive dislocations began in the Pleistocene and have continued down to the present day; some of the faults are expressed in the present topography. The migration of activity from the summit to the subterminal craters is apparently related to these dislocations, and to some degree they are responsible for the large caldera-like depression.

In the middle part of the island, southwest of Kosatka Bay, stretches the Bogatyr' Ridge. Here Quaternary volcanic rocks extend for 20-25 km and reach a width of up to 15 km (Fig. 66).

The whole region has undergone two stages of glaciation. The preglacial volcanic centers are not preserved. Moraines of the first

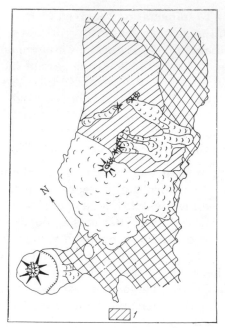

Fig. 66. Distribution map of the volcanoes
of Bogatyr' Ridge and Atsonupuri Volcano:
1) Pleistocene lavas (other symbols as in
Fig. 9).

glaciation extend a long way down; in particular, they cover the 200-
m marine terrace, near the east flank of the ridge, composed of the
Tertiary basement. In some places the interglacial centers have
been only partly destroyed, and locally remnants of interglacial
cones and lava flows can be seen, especially in the northeast part
of the ridge.

    Based on this distribution, we can infer that volcanic activity
here was of "linear-clustered" character. The second glaciation
"rejuvenated" the cirques and troughs of the first glaciation; it developed more strongly on the north flank of the ridge than on the
south.

    Some cones were active at the very end of the Pleistocene,
and all gradations can be found from strongly eroded to very fresh
cones.

    Holocene activity also has had a "linear-clustered" character.
Three clusters are recognized, extending along the trend of the

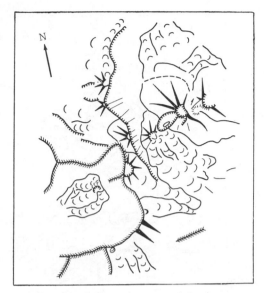

Fig. 67.  Sketch of the structure of the Holocene
cones in the northern Bogatyr' Ridge: 1) Burevestnik Cone (other symbols as in Fig. 9).

ridge.  All the cones, except for the complexly structured Stokap
Volcano, are more or less one-stage structures that are built on a
strongly dissected glacial topography.

The first "cluster" is located in the northeastern part of the
ridge (Fig. 67).  Here a chain of four or five small cones extends
at an angle (of about 45°) to the trend of the ridge.  The easternmost cone is Burevestnik Cone; its absolute height is 1427 m and
its relative height, about 500 m.  The crater and flanks of the cone
have been partly destroyed.  Strongly overgrown lava flows extend
to the northeast  from the base.  On the east a rather badly ruined
cone, apparently of early glacial age, sticks out from beneath Bogatyr' Cone.

To the west of Bogatyr' Cone is a small cone with a crater
open to the south from which were emitted broad lava flows up to
9 km long.  These flows fill the U-shaped valleys and extend almost to the Pacific shore.  Between these two cones, right on their
fused-together slopes, an elongate explosion crater has been blown
out.

Farther west, on the edge of a U-shaped valley, is a cone that
is destroyed on one side and that contains the remains of a rather

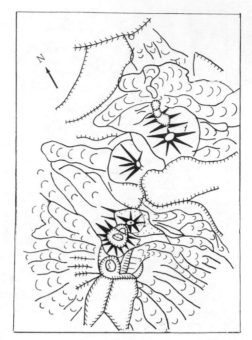

Fig. 68. Sketch of the structure of the Holocene
cones of the southern Bogatyr' Ridge (1) and
Stokap Volcano (2) (symbols as in Fig. 9).

wide but shallow crater. In all probability, this is an early Holocene
(or perhaps late glacial) cone.

To the north of this cone there are three more half-destroyed
cones on the axis of the ridge; these are tentatively referred to the
late Pleistocene.

Still farther west, at the head of a U-shaped valley, is a small
cone with a crater open to the south; this cone has produced a long
lava flow that stretches toward the Sea of Okhotsk. The last portions of the lava were rather viscous and formed short, steep-sided
tongues.

The second cluster is situated 5-6 km to the southwest, along
the ridge (Fig. 68). There are four fused-together cones that form
a small ridge (3 km long) on the axis of the main ridge. Long (up

to 10 km) tongues of lava extend from this group of cones to the
Pacific Ocean. On the lower slopes the lavas from this group of
cones join the lavas from the Burevestnik group.

There is one more older-looking cone 0.5 km west of this
group; it is probably early Holocene in age.

Bogatyr' Ridge comes to an end in the complex S t o k a p
V o l c a n o group. This volcano, like the other cones on the ridge,
is located on a rather high Pleistocene base. However, the scale
of activity of this volcano was vastly greater than that of all the
other cones on the ridge taken together. The Stokap lavas have completely leveled off all the irregularities in the glacial topography
and extend entirely across the island, from the Pacific Ocean to
the Sea of Okhotsk (up to 16 km). The area covered by lavas of this
volcano is estimated at 135 km$^2$.

The Stokap Volcano group is a complex cluster of eight to ten
cones and craters, part of which have been considerably destroyed
by glacial quarrying and erosion. On the north and south this group
is limited by the remnants of crater walls that form portions of a
single large crater about 1 km across. However, newer volcanic
structures and explosion craters rather complicate the present picture, so that it is difficult to speak with confidence about what are
parts of this single crater or what are remnants of individual lava
cones. The upper part of the massif is cut by two deep gorges, and
formerly there may have been a lava cone at the head of one of these.

A little to the north of the south wall are the remnants of another lava cone from which a large lava flow poured to the south.

All the rest of the volcanic forms in the Stokap group extend
in a line along the ridge axis.

In the western part of the large crater a large explosion
crater, about 0.5 km across and 50-80 m deep, has broken through.
Its floor is flat and is partly covered by a lake. There is a second
explosion crater, 300-400 m across from rim to rim, 500 m to the
east. Probably this crater was formed as a result of the explosion
of the summit dome that sealed up a small lava cone located at the
site of the eastern rim of the large crater. The easternmost member of this group is a small dome that outcrops just beyond the old
large crater. On the east the loose mantle of this dome overlaps

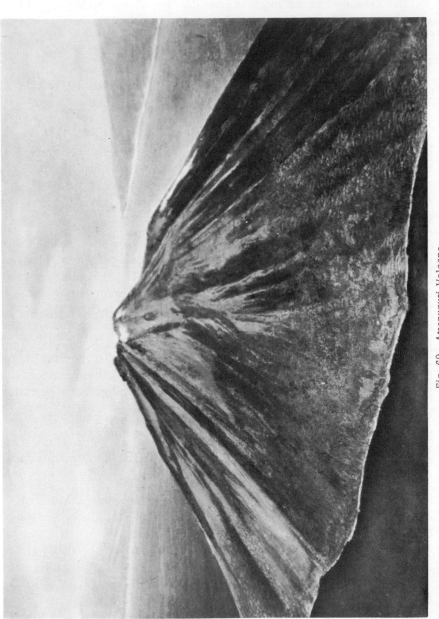

Fig. 69.  Atsonupuri Volcano.

the glacial topography, but to the north and south lava flows stick out from beneath it, indicating that the dome arose in the crater of a lava cone. A small parasitic crater has been cut into the western flank of this structure.

The lavas of Stokap Volcano are composed of basic andesites and andesite-basalts. Phenocrysts are of labradorite, hypersthene, and augite; olivine is rare. In some thin sections there are very large grains of plagioclase and hypersthene. The structure of the groundmass is hyalopilitic, hyaline, or sometimes microdoleritic.

The developmental history of the volcanoes of Bogatyr' Ridge is similar to the history of other "linear-clustered" groups. Beginning in preglacial times and continuing into the Holocene the eruptions were of uniform character, producing explosions and lava effusions from closely spaced, rather small cones. The only exception was the southwesternmost member of this group, Stokap Volcano, which very probably was active as a large stratovolcano from the very beginning. Later a group of smaller structures was developed on its summit.

At present none of the cones on Bogatyr' Ridge shows any evidence of activity.

Atsonupuri Volcano rises in the southern part of Iturup (on the Okhotsk shore), forming a peninsula that is joined to the island by a low isthmus scarcely 30 m high.

This volcano has a structure of the Somma-Vesuvius type (Fig. 69). The somma is preserved only on the southeast where, at a height of about 900 m above sea level, the sharp rim of the preserved portion of the caldera is well expressed. Its diameter is about 2 km. The eastern and western parts of the rim are hidden beneath the deposits of a younger cone, but its outlines "show through" the thin pyroclastic cover. The outer slopes of the somma are rather even and are broken only by the not very numerous, shallow valleys of intermittent streams. The upper flanks are covered by scoria from an eruption of the central cone, while the lower parts are overgrown by brush and woods. On the sea shore and near Lesozavodsk (6 km from the edge of the caldera) the ends of the basaltic flows from the somma are exposed.

The central cone rises to 1205 m above sea level, or about 300 m above the caldera rim. On the southeast, between the somma rim and the flank of the central cone, there is a small flat atrio, but on the northwest the flanks of the cone sweep down to and beyond the sea coast, without interruption and without a submarine terrace, to a depth of somewhat more than 1000 m. At the same time, the 140-m submarine terrace adjoins the remnants of the somma.

The crater of the central cone stretches from southwest to northeast and has the form of an oval funnel $400 \times 500$ m across and more than 100 m deep. On the northeast and southwest the crater rim has marked low points, "gateways"; a large ravine of the sciarra type extends to the northwest from the crater, and a similar ravine occurs on the northwest of the cone. The flanks are covered mainly by scoria and scoriaceous bombs. The black color of the pyroclastic deposits on the lower parts of the volcano changes upward to cherry-red, indicating a high temperature at the moment of eruption and later secondary oxidation.

In the cliff-like walls of the crater and at the heads of the ravines layers of lava are exposed. The ends of the flows, sometimes with wavy surfaces, crop out near the shore. The lava from the top of the volcano is a hypersthene basalt with phenocrysts of basic plagioclase, olivine, and clinopyroxene. The same minerals occur as microlites (Katsui, 1961).

The smooth, undissected character of the flanks of the somma indicate its youth. According to available poor data and on the basis of the coast and of the lava flows, we can surmise that the somma structure was superposed on the 140-m submarine terrace, i.e., is Holocene in age or in any case is no older than very late Pleistocene. The northwestern part of the somma is broken by a fault of large displacement; apparently this fault also affected part of the submarine terrace. In all probability the formation of the caldera and then of the central cone was related to this break.

The crater of the central cone and even its flanks show evidence of extremely recent movement along this break; apparently the fault is very young. The seismic station "Lesozavodsk" from time to time records a series of weak earthquakes beneath Atsonupuri Volcano, and apparently these are related to continued movement along the fault.

The volcano arose at the end of the Pleistocene or at the beginning of the Holocene as an isolated island up to 1.5 km high.

Later the cone was tied to the island by a low isthmus of detrital
material. In the middle Holocene a fault developed along the northwest coast of the island. Apparently, initial movement along this
fault brought on the development of an explosion caldera about 2 km
across. Then the northwestern part of the old cone was dropped
down and a new, inner cone arose in the basin. The composition
of the lava remained basaltic, as before. The eruptions have been
mainly Strombolian; more rarely, fluid lava has been emitted.

Information on the present activity of this volcano is rather
poor. At the beginning of September 1812 Captain Rikord observed
an eruption "with flames" in the southern part of Iturup, in all probability an eruption of Atsonupuri Volcano. According to information from the local inhabitants, a weak eruption occurred in 1932.
At present Atsonupuri shows no activity, but eruptions of the volcano are still quite possible.

Rokko massif is an ancient massif of volcano rocks to
the south of Atsonupuri Volcano. It is up to 20 km long and up to
10 km wide. Numerous valleys cut the massif, and it is impossible
in a brief study to determine the positions of the original eruptive
centers. In some places isolated buttes suggest the remains of
cones of the "linear-clustered" type. The high point in the southwestern part of the massif rises to 907 m. The deep valleys that
cut into the massif are modified glacial troughs and cirques of the
first glaciation.

The large Urbich Caldera is cut into the central part of this
massif. It is superposed onto the glacial topography, but its smooth
inner walls show no traces of glacial quarrying, which fixes its age
as later than the first glaciation. Tentatively its age is stated as
interglacial. The diameter of the caldera from rim to rim is 6 km;
its highest point is 622 m. On the floor of the caldera, at a height
at least 100 m above sea level, is the fresh-water caldera lake of
Krasivoe; the Urumpet River (1 in Fig. 70) flows from the lake to
the ocean.

Urbich Caldera has shown no activity during the Holocene, and the loose deposits inevitably related to the formation of
such a type of caldera have been removed from the flanks and from
the valleys and now occur only in the Mount Golubok area.

To the southwest of Rokko lies the Yuzhnyi [Southern] Isthmus,
50-60 m high, composed of dacitic pumice. This isthmus joins to-

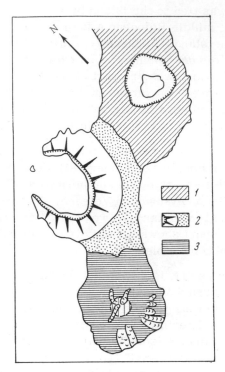

Fig. 70. Distribution map of the volcanoes
of southern Iturup: 1) Rokko massif and Urbich
Caldera; 2) L'vinaya Past' Caldera; 3) Berutarube Volcano (other symbols as in Fig. 9).

gether the three southern volcanoes of Iturup: Rokko Mountain,
L'vinaya Past', and Berutarube Volcano.

Genetically this pumice is related to the formation of L'vinaya Past' [Lion's Jaw] Caldera. This caldera has the form of a
ring, broken on the north, with gently sloping outer flanks and steep,
cliffy inner walls (2 in Fig. 70). The base of the cone has a diameter of 12-13 km; the caldera itself is elongate in a north-south direction and is 7 × 9 km in dimension. The caldera rim, as we
said, is broken on the north; a passage up to 5 km wide joins the
basin of the caldera with the waters of the Sea of Okhotsk. In the
middle of this passage a rock sticks out that, seen from the sea,
resembles the figure of a reclining lion.

The "sill" that separates the caldera bay from the sea is less
than 50 m deep. The caldera rim rises to a height of 400 m, and the

inner walls drop away steeply to depths of up to 550 m below sea level. Thus the depth of the caldera amounts to almost 1 km.

The structure of a stratovolcano is exposed in the walls of the caldera. The rocks are dominantly basic in composition; the pumice from the final explosion has a dacitic composition, with 65.6% $SiO_2$.

The absence of evidence of glacial scouring on the flanks allows us to infer that the original structure arose after the first glaciation and that the age of the caldera itself may even be Holocene. The northern part of the caldera structure is cut off by the same fault that runs through Atsonupuri Volcano, but it is not clear whether the development of the caldera is related to this fault; most probably the fault is younger than the caldera.

Berutarube Volcano forms the southwesternmost part of Iturup. It is a gently sloping cone with a basal diameter of 10–11 km. The flanks of the cone, especially in the upper part, are cut by wide valleys; apparently these are modified glacial troughs. Locally troughs can be recognized that are enclosed one within the other, apparently corresponding to the first and second glaciations (3 in Fig. 70).

The summit of the cone (1222 m) is rather broad; here the heads of the U-shaped valleys and cirques come together, and on them a small cone of Holocene age has been superposed. This cone is also badly destroyed; strongly altered rocks are exposed in the walls of the cone, and two or three lava flows extend from it. Strong solfataras that are depositing sulfur occur in the broad amphitheaters cut into the walls of the inner crater.

The preglacial volcano had the form of a gently sloping cone with a flat crater approximately 1.2 km across. In interglacial times continued volcanic activity partly "closed up" the glacial troughs of the first glaciation. The second glaciation "rejuvenated" several troughs of the first glaciation, but only on the very upper part of the cone. In the Holocene a small cone with a basal diameter of 1 km was formed. This cone was essentially a pyroclastic one: only two small lava flows can be observed. Eruptions ceased several hundred (or at most, a thousand) years ago here; the flanks of the cone have been partly destroyed, but fumarolic activity continues to the present day.

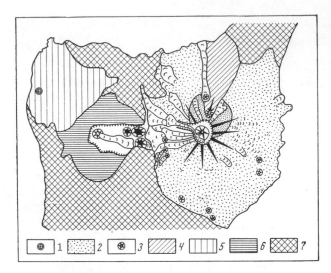

Fig. 71. Distribution map of the volcanoes of northern Kunashir: 1) Fumarole field of Rurui Volcano; 2) pyroclastic deposits of Tyatya Volcano; 3) subsidiary craters; 4) remnants
of the old volcanic structure of Tyatya; 5) Rurui Volcano
massif; 6) Smirnov Volcano massif; 7) basement (other symbols as in Fig. 9).

## Kunashir Island

Kunashir is the southwesternmost island of the Greater Kurile
chain. It is separated from Iturup by Yekaterina Strait, 22 km wide.
To the east of Kunashir, across the wide (60 km) South Kurile Strait,
lie the islands of the Lesser Kurile chain. Izmena Strait (17 km
wide) on the south and Kunashir Strait (25 km) on the west separate
Kunashir from Hokkaido.

Kunashir is the third largest of the islands, after Iturup and
Paramushir; it is 123 km long, 7-35 km wide, and 1490 km$^2$ in area.

There are four well-developed volcanoes on the island, and
all of them are active. The northern and widest part of the island
is occupied by the high Tyatya and Rurui Volcanoes (Fig. 71); the
other two volcanoes are situated in the southern part of the island.
The middle part of the island, between these two groups, is an area
of Tertiary volcanogenic rocks, with intrustions of granodiorite. There
are no clearly expressed volcanogenic landforms here.

Fig. 72. Tyatya Volcano.

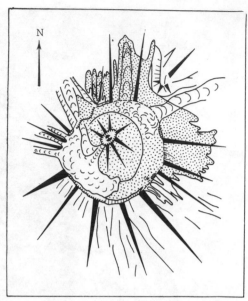

Fig. 73. Sketch of the structure of the summit
of Tyatya Volcano (symbols as in Fig. 9).

At the northeast end of Kunashir rises one of the most beautiful of the Kurile volcanoes, T y a t y a .   It is constructed like Somma-
Vesuvius, but the exceptional regularity of its form (like that of
Krenitsyn Peak on Onekotan) far surpasses the famous Vesuvius
with its Somma (Fig. 72).

The somma of Tyatya Volcano forms a large, regular, strongly
truncated cone 1485 m high, with a basal diameter of 15-18 km. On
the north and south its slopes drop off to the shores of the Sea of
Okhotsk and the Pacific Ocean; to the west Rurui Volcano rises beyond a high saddle; and on the east the Tertiary basement sticks
out from beneath the base of the cone.

On the summit of the cone there is a shallow, slightly oval
caldera, 2.1 × 2.4 km across.  The rather low (50-80 m) rim of the
caldera is well expressed only on the south; scattered parts of the
rim also rise above the caldera floor on the west and northwest.  A
rim is absent on the east and in part on the north; there the flat floor
of the caldera passes directly onto the outer slopes of the somma.

The slopes of the upper flank reach 30°, but they do not exceed 4-5° at the base.  On the whole, the form of the slopes has the

typical aspect of a logarithmic curve. The flanks of the somma are cut by numerous barrancas and erosional ravines; however, seen from afar the dissection of the flanks becomes less conspicuous and does not upset the regularity in form of the volcano to any great extent. More pronounced incision occurs only on the upper part of the northwest flank.

On the northeast flank of the volcano, 150 m below the rim of the caldera, rise the eroded remnants of a volcanic structure of some kind. In all probability these are remnants of an older volcanic structure, a narrow sector of which extends down to the Okhotsk shore.

On this same flank, at a height of 750 m, are the remnants of a subsidiary crater from which a lava flow stretches down to the sea coast. Remnants of three small parasitic cones, two of them with lava flows, are located near the west foot, in the headwaters of the Tyatinaya River; in addition, several small, badly ruined cones are located on the southern base.

Mainly lava flows are exposed in the gullies on the somma. In the caldera cliffs a layered structure is seen, but with lavas dominant over pyroclastics.

The rocks of the lower and middle parts of the somma are mainly olivine basalts; phenocrysts are of bytownite ($An_{70-90}$), augite, and olivine, though in some the augite is missing. The groundmass structure is intersertal, with microlites of labradorite ($An_{65}$) and augite.

The upper part of the somma and the rim of its caldera are composed of basic andesite and andesite-basalt. The phenocrysts are of labradorite ($An_{54-70}$), hypersthene, and augite. The structure of the groundmass is hyalopilitic, passing into pilotaxitic. The plagioclase of the microlites is andesine-labradorite, $An_{45-52}$.

The central cone is situated almost in the center of the caldera; its basal diameter is about 1.5 km. On the top are two explosion funnels separated by a wall; the rim of the crater containing them is strongly elongate in a northeast direction, and it is $400 \times 250$ m across. The southeastern part of the rim rises considerably higher than the rest. The high point on the volcano is located here at 1822 m above sea level or about 400 m above the floor of the caldera (Fig. 73).

Fig. 74. Vertical aerial photograph of the summit
of Tyatya Volcano.

The southwestern part of the rim is about 100 m lower. Lava flows descend everywhere, reach the base of the cone, and over-flowing, cover the whole southwest segment of the atrio right up to the scarp of the caldera wall. Two small branches stretch to the northwest as far as a ravine and then run down along it for at least 3-3.5 km; the older flows in this valley, which are probably also related to the central cone, extend to the very foot of the somma and have a total length of 8-9 km. In the west narrow flows from the central cone reach the outer slopes of the somma through the broken rim of the caldera and run down the barrancas as far as the base, into the headwaters of the Tyatinaya River. In addition, the flows of the central cone (which are half covered with scoria) over-flow the rim of the caldera and descend along the northern and north-eastern flanks to an altitude of about 700 m. The larger part of the central cone and of the caldera floor are covered with scoria. At some points on the caldera floor the outlines of the covered lava flows show through. Scoria reaches down the flanks of the somma to an altitude of about 700 m from the lower part of the caldera rim. Here the scoria smooths out the topographic irregularities, and only the highest ridges of the somma flows rise above the broad scoria plain* (Fig. 74).

The lava and pyroclastics of the central cone are olivine ba-salts. The phenocrysts are of bytownite ($An_{70-90}$), olivine, and augite. The structure of the groundmass is intersertal, microdoleritic, or hyaline. The plagioclase microlites are somewhat more acid ($An_{55-82}$).

Milne (1886) reported that the local Ainu inhabitants said that the outer crater of Tyatya Volcano was occupied by a lake. At pres-ent there is no lake, and the presence of a rather old gap in the northwestern part of the cone summit would not have allowed a lake to exist there in the recent past. Perhaps the tradition of a lake was preserved from more ancient times when a lake might have existed there, a lake which was, in the literal sense of the word, expelled with the continued growth of the central cone.

The absence of evidence of glacial activity on the flanks of the somma, in spite of its considerable height, testifies to the post-

---

*Markhinin (1959) recognized a "middle scoria series" on the somma of Tyatya Vol-cano. However, study of aerial photographs shows that this scoria is related to the central cone.

glacial age of the present surface of the somma. The remnants of the older structure that stick through the northeast flank indicate the existence of an older, now buried structure.

The somma of Tyatya Volcano arose no earlier than late Pleistocene (perhaps at the beginning of the Holocene) on the remains of a Pleistocene structure. The eruptions of the volcano were characterized by mixed Strombolian activity, with a dominance of fluid basaltic and andesite-basaltic flows. Only at the very end did some more acid two-pyroxene andesites appear, but even they were of rather basic composition. Then, apparently, some break in the eruptions occurred, ending with an explosion that cleared out the old conduit. Perhaps the several scoria cones and lava flows on the flanks and near the base were formed during this break in activity of the summit crater. The summit of the cone was partly blown away and partly collapsed into the conduit, and the crater was widened to 2-2.5 km. Renewed activity led to the development of the central cone. This activity had a dominantly Strombolian character, at times with the emission of basaltic lava. Scoria filled the atrio and locally spilled over onto the outer flanks of the somma.

The last eruption occurred in 1812, and the freshest flows in the atrio (on the southwest) may have been erupted at just that time. In the final stages of the eruption scoria and pyriform bombs were ejected. At present the volcano shows no signs of activity.

Rurui Volcano lies at the north end of Dokuchaev Ridge. On the northwest its base extends to the Sea of Okhotsk, on the northeast and southwest Tertiary basement is exposed, and on the southeast the volcano is bounded by the ruined volcano Smirnov.

The present cone (1486 m) is planted on a topography produced by the first glaciation. Three wide, deep canyons that reach to the top have completely destroyed the crater and have eroded a considerable part of the flanks. These canyons, which become narrower downslope, apparently are modified cirques of the second glaciation. The typical structure of a stratovolcano is exposed in the walls of these canyons. An analyzed rock from the west flank of the cone is a normal two-pyroxene andesite with 58% $SiO_2$ (Gumennyi and Neverov, 1961). On the summit light-colored rocks are exposed, altered by the action of fumaroles, but the latter now occur only in one area on the west slope of the volcano, at an altitude of 150-350 m above sea level and occupying an area of about 1 $km^2$.

The remnants of Smirnov Volcano adjoin Rurui Volcano on the southwest. This is a rather gently sloping cone 1182 m high. A portion of the north flank is more or less preserved, but the whole southern part has been badly destroyed by glacial activity. The Tertiary basement is exposed in the lower U-shaped valleys.

A particularly wide glacial trough stretches to the southeast of the source of the Tyatinaya River. At the very head of this trough is a small, half-ruined cone from which a long (about 4 km) lava flow descends. Near the end of this flow is a large cone (Mount Vil'yams, 675 m) that is about 1 km in diameter; a very short lava flow extends from it. The summit crater is not preserved.

A little farther to the north are two rather fresh-looking domes. One of them, Mount Gedroits (758 m), looks like an effusion dome about 700 m across; the other is an extrusion dome $1 \times 1.4$ km in diameter. The four volcanic structures mentioned are on the floor of a U-shaped valley of the first glaciation and may be considered to be parasitic formations on ancient Smirnov Volcano.

Two closely spaced volcanoes, Rurui and Smirnov, arose in the northern part of Dokuchaev Ridge in preglacial times. The first glaciation badly ruined both structures; the southern part of Smirnov Volcano, where the weak rocks of the Tertiary basement lay at rather shallow depth, was particularly strongly modified by glacial quarrying.

In interglacial times Rurui Volcano renewed its activity, and a new cone was formed here which may have been active up to the very end of the Pleistocene. At present only one fumarolic field continues to be active on its flank.

The southern volcano, Smirnov, showed activity after the first glaciation only in the form of eccentric eruptions. At first two cones were formed, then an effusion dome, and finally an extrusion dome. Judging from how well their forms are preserved, only the upper cone is early Pleistocene or perhaps even late Pliocene; the other structures are certainly Holocene.

Mendeleev Volcano is located in the southern third of the island of Kunashir, not far from Yuzhno-Kurilsk (Fig. 75). It is a complex volcano; the details of its structure are concealed by poor outcrops and by the thick, impenetrable undergrowth.

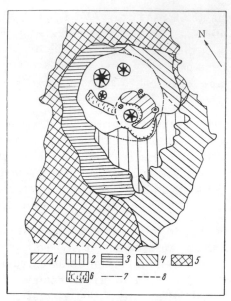

Fig. 75. Sketch of the structure of Mendeleev Volcano: 1) Central cone; 2) second somma; 3) first somma; 4) undissected series (in part Tertiary); 5) basement; 6) pyroclastic flow; 7) outline of the old caldera; 8) outline of the young caldera (other symbols as in Fig. 9).

The oldest unit of this massif is a large caldera, up to 6-7 km across. Its strongly smoothed rim is preserved only on the northwest; on the east the rim has been completely destroyed, and on the south it is overlapped by the younger somma of Mount Mechnikov. However, the lower part of the old cone sticks out from beneath the lavas of the second caldera on the south.

The preserved portions of the outer slopes show evidence of intense erosion in the form of numerous radial valleys. The Tertiary basement sticks out from beneath the lavas of the first somma in the valley floors and sides.

The first somma is composed of two-pyroxene andesites and andesite-basalts (sometimes with traces of olivine). The phenocrysts are of labradorite ($An_{57-58}$), augite, hypersthene, and rare olivine. The structure of the groundmass is pilotaxitic.

The old cone is covered almost completely by an unconsolidated pyroclastic series (Golovninsk suite, according to Markhinin).

We consider this series to be the product of eruptions related to the formation of a large caldera. Various andesites and dacites, including a thick sequence of dacite pumice, take part in the make-up of this series.

The second somma is located eccentrically to the first; it is offset to the south. The southern half of this structure is well pre-served, and the way in which its almost undissected flanks overlap the eroded topography of the first somma is clearly visible. Here the rim of the second caldera is preserved in the form of a semi-circle whose high point is called Mount Mechnikov (800 m). The northern half of the second somma partly fills the hollow of the first caldera. At present only strongly eroded remnants of the flanks are preserved here, and the position of the second caldera is fixed by scattered remnants and by the location of the fumarole fields. The dimensions of the second caldera, from rim to rim, are $3 \times 3.5$ km.

In the northern part of the first caldera are two strongly smoothed-down high points; apparently they are eccentric domes on the second somma. The formation of the second caldera prob-ably was related to a north-directed explosion.

Rather uniform pyroxene-olivine basalts and andesite-basalts take part in the construction of the second somma (Mount Mechnikov). Labradorite-bytownite ($An_{60-80}$), olivine, and augite (which is some-times developed around the olivine) occur as phenocrysts. The groundmass is microdoleritic or andesitic.

The two-pyroxene andesitic pyroclastic deposits in the basin of the first caldera apparently are related to the explosion of the second caldera.

In the basin of the second caldera arise the remains of an inner central cone, offset slightly to the north. On the south, between the second somma and the central cone, part of the narrow atrio is pre-served; to the east one can make out the juncture of the central cone and the second somma structure. The four solfatara fields that en-circle the cone on the east and north at a height of 300-400 m are confined just to this junction. The northeast field undoubtedly is re-lated to an eccentric explosion crater; perhaps the other fields are also related to lateral explosions. The lateral craters that have been widened by erosion and the channels cut by stream erosion have modified the contours of the eastern half of the central cone, while the western half is completely missing.

The high point on this part of the massif rises to approximately 850 m, or 300-500 m above the floor of the second caldera. The basal diameter of the cone was 3 km; the width of its crater, more than 1 km. In all probability the formation of the broad crater was caused by an westward-directed explosion that removed the western part of the structure. From the "breach" a strongly overgrown pyroclastic flow descends northwestward into the basin of the old caldera.

A rather well-preserved extrusion dome rises in the ruined crater of the central cone. Its summit rises to 890 m above sea level, or about 200 m above the floor of the crater.

The rocks of the central cone, judging from very scanty data (because of the poor exposures) are two-pyroxene andesites and andesite-basalts, with phenocrysts of basic plagioclase, hypersthene, and augite; olivine is sometimes present. The groundmass is microlitic. The central dome is composed of dacite with 65.5% $SiO_2$. Andesine ($An_{45}$), quartz, augite, hypersthene, and rare olivine are found as phenocrysts. The structure of the groundmass is hyalopilitic.

The complex form of the volcano and the repeated variation in the composition of its rocks point to a long and complex history of development. In its general outlines this history can be presented as follows.

A rather broad cone with a basal diameter of up to 12 km arose on a Tertiary basement. At first andesites and andesite-basalts were erupted, but gradually the rocks became more acid; this stage was concluded by a gigantic explosion with the ejection of acid pumice and the subsequent formation of a wide caldera. All these events certainly occurred in preglacial times, but a more accurate date than this is not yet possible.

Then on the southern edge of the first caldera a new cone was formed. Its southern slopes overlapped the erosional topography on the older structure, and its northern slopes partly filled the large old caldera. The dimensions of this cone were at least 6-8 km. As in the first cone, the eruptions began with basic effusions and the final directed blast that destroyed the northern part of the summit

produced acid andesites. The second caldera that resulted was
3-3.5 km across. The domes in the northern part of the first caldera apparently are related to its final phases. Tentatively this
structure can be assigned an interglacial age.*

The central cone is also apparently composed of rather basic
rocks. It almost completely filled the area of the second caldera.
At the end of its development several explosions occurred along the
edge of the cone; then the western part of the cone was blown away
and a dacite dome arose in the hollow that remained. The age of
the dome is Holocene.†

Thus, more or less similar phenomena were repeated three
times in the history of Mendeleev Volcano: formation of a cone of
basic composition, then explosion of acid material, followed by development of a new basic cone, and so on. The scale of the phenomena diminished each time: the first caldera is 6 km across,
the second caldera is 3 km across, and the crater of the central
cone is 1 km across.

Milne noted a single eruption of Mendeleev in 1880. The eruption occurred in the area of the northeast solfatara field and was
apparently a purely gas eruption, perhaps with weak explosions.

At present the volcano shows continuous solfataric activity
in the three lateral solfatara fields.

The last volcano of the Greater Kurile chain, G o l o v n i n
C a l d e r a , forms the southern end of Kunashir island. It forms a
very broad (more than 10 km across) but gently sloping cone with
a strongly truncated summit.

The northwest flank drops off rather steeply to the sea and
locally forms high coastal cliffs.

On the south the outer slopes of the cone are very gentle (they
do not exceed 7-8°) and they gradually pass into a wide coastal plain
here. The flanks are cut by a radial system of many creeks and

---

*Radiocarbon dating of the remains of vegetation from the base of this structure gives
  an age of 39,000 years B. P. (Sulerzhitskii) which corresponds to the Würm interstadial.
†Radiocarbon dating of the remains of vegetation from beneath the deposit of an incandescent avalanche related to the formation of the dome gives its age as 4200
  years B. P. (Sulerzhitskii).

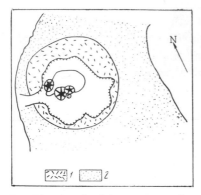

Fig. 76. Sketch of the structure of Golovnin Caldera: 1) Inner somma; 2) outer somma.

rivulets and are covered by woods and brush; outcrops are practically absent on them.

On the summit lies a caldera 4–5 km across; its outline has been severely altered by later erosion. The caldera rim has a height of 300–400 m; the inner walls are relatively gentle, no steeper than 25–28°.

A series of high points rises above the rather smooth outer slopes that encircle the whole caldera; among the highest of these are Mount Golovnin (547 m) on the south and Mount Vorob'ev in the north. All of these high points can be joined together into a circle approximately 7 km in diameter. The impression is created that the disposition of these high points indicates the existence of an old caldera which enclosed a second cone, now also crowned by a caldera. The fresher appearance of the flanks of the inner cone on the northwest and the pyroclastic flows on the south, which partly fill the old erosional gashes, apparently indicates this (Fig. 76). However, the suggestion made as to the existence on Golovnin Volcano of a second, outer caldera still must be proven.

In the not-too-numerous outcrops on the somma flanks and in the caldera dominantly pyroclastic rocks are exposed; these are tuffs of hypersthene- and two-pyroxene-andesitic composition. Large blocks of andesitic lava (56–58% $SiO_2$) are found. The phenocrysts in the andesites are of labradorite ($An_{54-65}$), augite, and hypersthene; in some, hypersthene is the only mafic phenocryst mineral. The groundmass is hyalopilitic, hyaline, or sometimes cryptocrystalline. The microlitic plagioclase is labradorite, $An_{50}$. Dacite blocks are also found; in them the plagioclase is andesine, $An_{47}$, accompanied by pyroxenes; quartz grains also occur. The structure of the dacite groundmass is hyalopilitic. The surface of the cone is covered by pumice from the caldera-forming explosions; its composition is dacitic (66–70% $SiO_2$).

The floor of the present caldera is about 3 km across. Its northern part is occupied by a caldera lake, 1 × 2.5 km in size. The lake surface lies at 130 m above sea level, and its depth reaches 62 m. Thus the caldera floor lies at an altitude of 63 m above sea level, and its depth below the rim is about 330 m. The lake waters flow through a narrow gap in the western part of the caldera into Kunashir Strait.

The southwestern part of the caldera floor ($1.5 \times 2$ km) rises 7-12 m above the lake level, or 70-80 m above the lake floor. A steeply dipping periclinal deposit of lacustrine conglomerates is seen along the edge of this higher area, suggesting the presence here of a concealed laccolithic intrusion similar to that formed on Usu Volcano (Hokkaido) in 1910 or 1944.

At the boundary between the lake and the higher area of the floor two extrusion domes are located along an east-west line; they are of two-pyroxene andesite-dacite (64.7% $SiO_2$). Their relative height above lake level is 130 m. The western extrusion is broken on the north by a small explosion crater in which solfataras are active. Near the base of the eastern extrusion, but on its south side, there is also an explosion crater that is in part cut into the flank of the dome and in part into the adjacent portion of the caldera floor. The crater is 350 m across, and its floor is covered by a hot lake (with a temperature of 36-65°C or, near the solfataras, up to 100°C). The solfataras located along the shore of the lake and on its floor give the illusion of boiling, as a result of which the lake has been named Kipyashchoe (Boiling Lake). It may be that these two extrusions are related to the lifting of the southern part of the crater.

They are composed of andesite-dacitic lavas. The phenocrysts are of andesine-labradorite ($An_{53}$), augite, hypersthene, and rarely quartz; the structure of the groundmass is hyaline, with some spherulites.

In addition to these two domes, Markhinin (1959) recognized two more within the caldera: Podushechnyi near the west shore of the lake and, less clearly expressed, Krutoi near the south edge of the caldera floor. The lavas of Podushechnyi Dome differ from those of the central domes of the caldera in their larger content of quartz and also in the fact that the hypersthene in them is altered to chlorite. Perhaps this is related to the subaqueous conditions of formation of the dome. Zelenov found active fumaroles on the subaqueous flanks of this dome.

An andesite dome (Vneshnii) is found on the outer slopes of the cone. Very probably Mount Golovnin and the other high points along the edge of the supposed first caldera are also domes.

In the Vneshnii Dome lavas, in addition to plagioclase ($An_{65}$) and two pyroxenes, olivine is also found; it is often surrounded by a rim of hypersthene or of opacite. The structure of the ground-mass is microdoleritic.

Golovnin Volcano is distinguished by the presence of an un-usually large amount of pyroclastic material; lava flows have not been recognized here. In all probability this volcano showed dom-inantly explosive activity. It may be that the process of caldera formation, with destruction of a considerable part of the volcanic structure and the outpouring of pumiceous pyroclastic flows, was repeated twice. In the final stage, apparently, several extrusion domes were squeezed out onto the floor of the caldera.

At present the volcano displays continuous solfataric activity at six points: in the two explosion craters on the central domes, at two sites on the north shore of the caldera lake, on the under-water slopes of Podushechnyi Dome, and on the shore of Kunashir Strait near Vneshnii Dome.

Kunashir is the only one of the large islands of the Greater Kurile chain where the linear-clustered type of volcanic activity occurred hardly at all. To this type (but very tentatively) can be assigned only the four small centers in the saddle between the rem-nants of Smirnov Volcano and Tyatya Volcano. A second peculiarity of the Kunashir volcanoes is the wide development of acid pumice. It covers an area of about 250 $km^2$, which constitutes more than a third of the area of the Quaternary volcanic rocks on this island.

## General Comments on the Volcanoes

As we said at the beginning of this chapter, there are 102–104 Holocene volcanoes on the Kurile Islands at the present time.

Of this number, 56 cones, i.e., somewhat more than half (54–55%) are of the linear-clustered type, whereas 46 cones (44–45%) are structures of the central type. If we consider the main zone and the western zone separately, then it appears that in the western zone all the present-day structures are of the central type, so that

the proportion of linear-clustered cones in the main zone is somewhat larger than the overall average.

However, the idea that linear-clustered eruptions had a dominant role in the Holocene would be wrong. In order to compare the
actual scales of volcanic activity we must proceed not from the number of cones, which may be very different in dimensions, but from
the volume of volcanic products. As a first approximation we can
use the areas occupied by the products of volcanic activity. We have
calculated these areas for all of the islands, and separately for the
Pleistocene and Holocene.

The total area of Holocene lavas in the main volcanic zone
amounts to 1255 $km^2$, of which 350 $km^2$ or 28% belong to the linear-
clustered eruptions. Thus the actual relationship of volcanism of
the central and linear-clustered types is substantially different than
would be indicated by counting the number of cones. If we go back
to the Pleistocene, of the total area of 5345 $km^2$ for the Quaternary
lavas of the main zone the share of the linear-clustered eruptions
is as much as 3150 $km^2$, or almost 60%. It follows from this that
the share of the linear-clustered eruptions in the Holocene has been
considerably reduced. This is particularly important on Urup, where
Pleistocene lavas of this type cover an area of more than 750 $km^2$,
whereas the area of the Holocene ones barely reaches 20 $km^2$.

Of the total number of 33 active volcanoes in the main zone,
29 belong to structure of the central type and only four (all on Paramushir) belong to linear-clustered groups. In completing our examination of volcanoes of the linear-clustered type, we can point
out that the extrusion domes are rarely formed in this type. They
are known on only five cones (i.e., they constitute less than 10%).
Calderas generally do not occur in massifs of this type. An isolated exception, perhaps, is Urbich Caldera on Iturup; however, the
association of the Rokko massif, in which the caldera has been blas-
ted, is not completely clear.

Holocene structures of the central type, of which there are
41 in the main zone, are rather varied. Volcanoes constructed on
the Somma-Vesuvius model (24 structures, or about 60%) predominate. Here all transitions occur from half-ruined, sometimes almost imperceptible remnants of sommas to such excellently expressed forms as Krenitsyn Peak on Onekotan or Tyatya Volcano

on Kunashir. The central structures in this type are usually Holocene, but the times of formation of the calderas are very different, from preglacial to Holocene.

Sometimes the central structure is a simple, solitary cone, but in other cases it is a complicated complex of individual cones and domes. There are up to ten cones and domes in Medvezh'ya Caldera on Iturup, for example.

Of the total number of 33 active volcanoes in the main zone, 21 (or two-thirds of the total) are of the Somma-Vesuvius type.

The number of single cones in the main zone is 15 (or 37% of the total; all of them are Holocene and only one of them (Baranskii Volcano) may be late Pleistocene.

We must note the rather common development of domes on the central structures. They are known in 17 cases: 16 are on volcanoes of the Somma-Vesuvius type (this amounts to more than 40% of all the volcanoes and to about 70% of those of the Somma-Vesuvius type) and only one is on a single cone.

In the western zone all the volcanoes, with perhaps the exception of the Pleistocene structure on Makanru, are single cones. Of the five Holocene cones, there are domes on the summits of three.

The character of the eruptions in the Kuriles is varied. In recent times, perhaps, moderate explosions of the Vulcanian and Strombolian types have predominated. Violent Plinian eruptions and directed explosions of the Bezymyannyi type are also known. Dome extrusions are frequent; on 14 of the active volcanoes the eruptions have concluded with the formation of domes. The effusion of lava flows and the formation of parasitic cones have occurred only rarely in historic times. Eruptions of the Hawaiian type are completely unknown.

In conclusion, here are some remarks on the intensity of volcanism in different parts of the arc. The opinion that exists on this score and that is often cited, that the maximum intensity of volcanism is in the central part of the arc (Goryachev, 1960), is based on a calculation of the number of eruptions. However, the number of eruptions cannot be used as an objective measure of intensity. First of all, eruptions have been recorded in the Kuriles for only the past

TABLE 1. Intensity of Volcanism

| Lava | Entire main zone | Northern part | Central part | Southern part | Western zone |
|---|---|---|---|---|---|
| Total area of Pleistocene lavas, km² | 5345 | 1350 | 495 | 3500 | 95 |
| Total area of Holocene lavas, km² | 1255 | 395 | 165 | 695 | 250 |
| Area of Pleistocene lavas per km along the arc, km²/km | 4.6 | 4.5 | 2.5 | 6.5 | 0.4 |
| Area of Holocene lavas per km along the arc, km²/km | 1.1 | 1.3 | 0.8 | 1.3 | 0.9 |
| Percentage of Holocene lavas | 23% | 30% | 33% | 20% | 260% |

250 years. Very many eruptions have remained unrecorded, while the known eruptions have been of such different strengths that an estimate only on the basis of the number of eruptions yields practically nothing.

An estimate of the intensity of volcanism can be derived from the summary quantities of volcanic products. As a first approximation we can use the calculated area of the volcanic rocks.

If we calculate the area of volcanic rocks that occur per kilometer along a given segment of the arc, the intensity of volcanism can be characterized rather objectively by this value. The ratio between the area of the Holocene and the area of Pleistocene lavas can characterize the intensity of present-day volcanism.

The relevant values are given in Table 1.

From this table it follows that in the Pleistocene volcanism developed most intensely in the southern part of the chain and most weakly in the central part. In postglacial times the intensity of volcanism in the south diminished considerably and became comparable with that in the northern part, while it continued to be somewhat weaker in the central part. Thus the objective data do not support the opinion offered on the greater intensity of volcanism in the central part of the chain; the actual relationships are probably just the opposite.

The material from the western zone is not very impressive, because a considerable part of the structure is covered

by the waters of the Sea of Okhotsk. However, all the available
data indicate a significant increase in the intensity of volcanism in
this zone in the Holocene. The new information supports the opinion expressed earlier by us (Gorshkov, 1960) concerning the "ad-
vance" of volcanism to the west.

If we examine the character and intensity of volcanism in relation to different types of crust, the following features appear:
the linear-clustered type of volcanism is unknown in the central part
of the arc, on crust of suboceanic type. In this same part of the arc, the
intensity of volcanism in the Quaternary lagged behind (and still lags
behind) that in the northern and southern parts. On the continental
(northern) portion of the arc, the intensity of volcanism remained approximately uniform during Quaternary. On the subcontinental crust
in the south of the arc, the intensity of volcanism declined in the
Holocene, mainly because of the almost complete cessation of effusions of the linear-clustered type.

It might be possible to relate directly a given feature of volcanic activity to the type of crustal structure. However, in our
opinion, the very development of the earth's crust in island arcs
to a considerable degree is a result of the phenomenon of volcanism (Gorshkov, 1963a). In this relation the suboceanic crust of the
Central Kurile arc is "rudimentary," which is explained by the
"flabbiness" of the subcrustal processes in this region.

*Chapter VI*

# Petrography and Petrochemistry of the Volcanic Rocks

## PETROGRAPHY OF THE VOLCANIC ROCKS

The lavas of the volcanoes of the Kurile Islands are rather varied in their petrographic relationships: all varieties from basic basalts to acid liparites can be found here. At the same time, we must note the uniformity in the petrochemical composition of the lavas; all of them belong to the calc-alkali family, specifically to its most calcic types.

The pyroxene andesites are the most widespread of the Kurile lavas. The basalts and andesite-basalts, to which a rather important group of rocks belongs, are also widespread. Dacites and acid andesites are found more rarely as massive lavas, but vast deposits of pumice are often composed of these rock types. Liparites are very rare and are found only as pumice deposits.

The lavas of the main and western zones are very similar in their petrographic relations; the only difference is the almost universal presence of hornblende in the volcanoes of the western zone. In the main zone hornblende occurs only as an exceptional mineral; even in the acid, pumiceous pyroclastics it is relatively rare. In the western zone hornblende is normal even in the massive lava flows, to say nothing of the domes and the more acid pyroclastics.

Basalts. The dominant phenocrysts in these are basic plagioclase, ranging from labradorite and labradorite-bytownite to anorthite; diopsidic augite with $2V = (+) 52-56°$ and with $Z\wedge c = 43-48°$ is an almost constant component. Olivine is commonly found; in some rocks it is dominant over augite. Grains of opaque minerals are rather numerous.* Because of the dominance of plagio-

---

* We have neglected to mention the opaque minerals in the petrographic characterization of the various volcanoes.

clase among the phenocrysts, the basalts of the Kurile Islands can be referred to as plagiobasalts. The groundmass of the basalts usually has an intersertal, microdoleritic, or hyaline structure. In the last case the rock takes the form of a hyalobasalt. Aphyric basalts with hyalopilitic groundmasses arc also found. The microlites are of plagioclase (somewhat more acid than the phenocrysts and ranging from labradorite to bytownite) and augite; olivine microlites are rare.

Basalts were rather widespread among the old central eruptions; they are often found in the somma lavas of complex volcanoes and in caldera volcanoes. There are contemporary or recent basalts on Alaid, Raikoke, Brat Chirpoev, Menshoi Brat Cone in the Medvezh'ya Caldera, Stokap, Atsonupuri, and Tyatya.

Andesite-basalts are a very widespread group of rocks. Part of the dated or recent effusion, as well as the scoria of the Strombolian eruptions are andesite-basalts (Ebeko, Karpinskii, Nemo, Matua, Kudryavyi Cone in Medvezh'ya Caldera, Teben'kov).

Plagioclase and augite are always present as phenocrysts in the andesite-basalts. The plagioclase ranges in composition from andesine-labradorite to labradorite-bytownite and sometimes to bytownite. The augite has an optic angle ranging from +51° to +54° or some times to +56°. Phenocrysts of olivine are often found and somewhat more rarely, those of hypersthene; the two minerals usually occur separately. Small amounts of opaque minerals can be seen rather often. The groundmass has hyalopilitic, pilotaxitic, and hyaline structures, with microlites of augite and plagioclase (from andesine to labradorite).

The andesite-basalts of Alaid Volcano and its crater Taketomi can be considered separately; these were studied in great detail by Kuno. These rocks were described by him as basalts, but their chemical composition corresponds more closely to the andesite-basalts. Plagioclase and olivine are ubiquitous as phenocrysts. The plagioclase is zoned, with cores of anorthite and rims of bytownite. The olivine is also zoned, with a large 2V in the cores (−81° to −89°) and a small 2V in the rims (−76° to −85°). Pyroxene is found in variable amounts, from vanishingly small to very large; it is diopsidic augite with 2V = (+) 54-57°. From the core toward the margin the color of the mineral intensifies and at the same time the 2V and Z∧c decrease (by 1-3°); in the same direction the mineral becomes visibly pleochroic. The groundmass is composed of microlites of these same minerals and of small amounts of opaque minerals, magnetite and ilmenite; in addition, potash feldspar occurs in the interstices. The large olivine and augite microlites are zoned, with the same type of change from core to rim as in the phenocrysts. The change in optical properties of the mafic minerals indicates that the role played by iron in their make-up increases during crystallization. This feature is characteristic of many other lavas in the Kurile volcanoes.

Andesites make up the most widespread and compositionally variable group of rocks. In the Kurile Islands the more basic members of the andesite family are dominant; there is a complete transition from them into the andesite-basalts.

Most widely distributed are the two-pyroxene andesites, which contain phenocrysts of both ortho- and clinopyroxene. They are found as lava flows, as lava domes, and also as pyroclastic flows. In these rocks the phenocrystic plagioclase is often zoned, with increasing acidity from core to rim; the composition ranges from labradorite (or more rarely, labradorite-bytownite) to andesine-labradorite. The hypersthene phenocrysts have marked pleochroism according to the usual scheme, from X = rose-colored to Z = almost colorless. The optic angle is negative and has an average value of about 60°, with a range of 52-66°. This corresponds to varieties with a rather high content of the ferrosilite molecule (up to 50%). The monoclinic pyroxene is diopsidic augite (sometimes close to pigeonite) with 2V = (+) 48-57°, but averaging 52-54°. The groundmass usually has a hyalopilitic or andesitic structure, sometimes with some orientation of the microlites, that often passes into a pilotaxitic structure. The plagioclase microlites range from andesine to andesine-labradorite. The pyroxene microlites usually are of augite or more rarely of hypersthene; both pyroxenes are rather commonly present.

Augite or hypersthene andesites, with phenocrysts of only the monoclinic or only the orthorhombic pyroxene, are found somewhat more rarely.

Sometimes olivine is present in the pyroxene andesites, along with the phenocrysts of hypersthene and augite. The 2V is close to 90°, ranging from −85° to +85°. In one case (on Trezubets Volcano) quartz is present in a rock along with olivine.

Hornblende is rarely found in the andesites of the main zone; as yet, it is known only from Kunashir and Trezubets Volcanoes. In the andesites of the western zone hornblende is a normal phenocryst mineral. Usually the hornblende is the brown, basaltic variety with a small extinction angle (0–10°). Along with the hornblende there are phenocrysts of zoned plagioclase, augite, and hypersthene.

There is a gradual passage from the acid andesites through intermediate rocks (andesite-dacites) to the dacites.

D a c i t e s are found only in the main zone. They occur mainly as pumice deposits around the large calderas on Iturup and Kunashir. Dacites also compose the domes on Ushishir and Mendeleev Volcanoes; lava flows of dacitic composition are not found. The pyroxene dacites have the greatest distribution; dacites containing phenocrysts of hornblende are much rarer. Usually even the pumiceous pyroclastics are pyroxene dacites. The phenocrysts present in them are of quartz (in small amounts), andesine or (more rarely) andesine-labradorite, diopsidic augite (2V = (+) 52–55°), and hypersthene (2V = (–) 60°). In some parts of the dacite dome of Mendeleev Volcano Markhinin (1959) found olivine along with the quartz. The structure of the dome dacites is hyalopilitic or hyaline, at times with some spherulitic structure. The plagioclase microlites are usually of acid andesine. The structure in the pumice is hyaline, at times with a perlitic fracture; phenocrysts are rather rare.

The pyroxene-hornblende dacites differ only in the presence of hornblende. Such rocks are found in some pumice deposits; generally the hornblende in them is of the normal, green variety. Lavas with the composition of pyroxene-hornblende dacite compose domes in the caldera of Ushishir Volcano. Here both normal, green hornblende and brown, basaltic hornblende with its small extinction angle are found.

In concluding these brief remarks on the petrographic features of the lavas of the Kurile volcanoes, we must emphasize that no important variations in the petrography of the volcanic products along the arc are known. At the same time, there is a very marked increase in the role of the acid rocks on the two southern islands, Iturup and especially Kunashir. On all the islands north of Iturup the area covered by dacites is very insignificant. On Iturup, out of the total area of 2060 km$^2$ for the Quaternary volcanic rocks, the acid rocks (dacite pumice) makes up 35 km$^2$, or 6.5%; on Kunashir the respective figures are 666 km$^2$, 250 km$^2$, and 37%.

## PETROCHEMISTRY OF THE VOLCANIC ROCKS

The first attempt at a petrochemical analysis of the volcanic rocks of the Kurile arc was made by the author (Gorshkov, 1960) some years ago on the basis of a rather small number of analyses. It was pointed out that, as in other island arcs, these rocks belong to an extreme calc-alkali type, similar to the Pelée type, and that toward the inner part of the arc the alkalic character noticeably increases.

The petrochemical analysis in this monograph is based on 209 selected analyses from all the islands of the Kurile chain. Of these, 127 of the analyses were made on material collected by the author in the Chemical-Analytical Laboratory of the Institute of Volcanology, Siberian Section, Academy of Sciences of the USSR, the majority of which were done specifically for the present work.

Almost all the analyses published by earlier authors were checked anew, with determination of the alkalies by flame photometry; all the analyses by one of the analysts (N. N. Postnikova) were excluded because the checks sometimes showed considerable deviations. Nineteen analyses are from published and in part unpublished material of E. K. Markhinin, and they were also done at the Institute of Volcanology; these analyses were not checked, but all the analyses made by one of the chemists were excluded on the basis of the criterion indicated earlier.

Forty-five analyses (39 of them from Paramushir) were taken from the published data of the volcanologists of the Sakhalin Complex Institute (SakhKNII). The number of analyses that they published is considerably larger than this, but the accuracy of many of the analyses raises considerable doubt (for example, by the equality in content of $K_2O$ and $Na_2O$ or even by the dominance of $K_2O$). Thus, only those analyses were chosen that raised no particular doubts; the rest were excluded.

Finally, 18 analyses were taken from various publications, mainly Japanese. A number of Japanese analyses made before 1930 were also excluded.

---

* Nevertheless, we must note that several analyses of one and the same rock done in the Institute of Volcanology and in the Sakhalin Complex Institute indicated a systematic lowering of the amount of alkalies in the analyses of the Sakhalin Complex Institute.

The author has considered the selection of more or less accurate chemical-analytical data carefully because not even the most complex and perfect system of checks can correct for the inaccuracy of analyses. Initial data of low quality lead to erroneous conclusions. An exceptionally important point in any system of checks is the accuracy of the $Na_2O$ and $K_2O$ determinations; particular attention must be paid to controlling the determination of these oxides.

### North Kurile Islands

For the North Kurile Islands 92 analyses are available; of these 49 were made in the Institute of Volcanology, 39 were made in the Sakhalin Complex Institute, and four were taken from the literature.

In Tables 2-4 are presented the analyses of the preglacial, interglacial, and postglacial lavas of Vernadskii Ridge. The attempt to discover some differences between lavas of different ages did not succeed; the scatter of the values lies within the same limits for all groups, and the curves of the average values practically coincide. Thus, no marked change occurred in the petrochemistry of the Vernadskii Ridge lavas during the Quaternary.

In Table 5 are given the analyses of the lavas of Karpinskii Ridge on Paramushir, and in Table 6 are given the analyses from all the remaining islands that belong to the northern group. In Fig. 77 it will be seen that the compositions of the lavas of Paramushir and of the other islands of this group completely coincide.* The points that are plotted are located in a band between the curves for the Pelée and Lassen Peak types.

### Western Zone

For the volcanoes on the western zone 22 analyses were used, of which 17 were made in the Institute of Volcanology and five were taken from literature sources.

All of the analyses are presented in Table 7; the points represented are plotted in Fig. 77 along with the points for the volcanoes

---

*In considering only the very general relationships we will use only the right-hand side of the Zavaritskii diagram (the *ASB* projection) and only the points of origin, without the vectors.

## TABLE 2. Preglacial Volcanoes of the Vernadskii Ridge

| Components | Vetrovoi Volcano | | | | | Ebeko group | | | | |
|---|---|---|---|---|---|---|---|---|---|---|
| | 1 | 2 | 3 | 4 | 5 | 6 | 7 | 8 | 9 | 10 |
| $SiO_2$ | 46.0 | 46.04 | 46.79 | 48.26 | 48.45 | 48.07 | 48.11 | 54.29 | 56.13 | 57.38 |
| $TiO_2$ | 0.67 | 0.94 | 1.06 | 0.83 | 0.74 | 1.10 | 0.83 | 0.97 | 0.74 | 0.61 |
| $Al_2O_3$ | 20.58 | 20.57 | 18.21 | 20.00 | 20.79 | 18.29 | 17.49 | 17.65 | 18.06 | 17.50 |
| $Fe_2O_3$ | 5.48 | 3.71 | 6.19 | 4.52 | 5.24 | 3.05 | 5.36 | 4.62 | 3.58 | 3.31 |
| $FeO$ | 5.86 | 7.20 | 5.97 | 5.95 | 4.26 | 6.04 | 4.35 | 4.20 | 4.23 | 2.44 |
| $MnO$ | 0.23 | 0.04 | 0.06 | 0.13 | 0.06 | 0.02 | 0.02 | 0.08 | 0.15 | 0.18 |
| $MgO$ | 5.01 | 5.26 | 6.26 | 4.94 | 5.59 | 7.35 | 8.34 | 4.14 | 3.73 | 3.24 |
| $CaO$ | 11.81 | 12.85 | 11.45 | 11.45 | 11.32 | 11.58 | 11.05 | 8.66 | 8.34 | 7.53 |
| $Na_2O$ | 2.50 | 1.76 | 2.08 | 2.32 | 2.99 | 2.14 | 2.41 | 2.84 | 2.65 | 2.66 |
| $K_2O$ | 0.75 | 0.82 | 0.73 | 0.83 | 1.01 | 1.09 | 1.02 | 1.64 | 1.75 | 1.82 |
| $P_2O_5$ | 0.13 | 0.10 | 0.11 | 0.13 | 0.34 | 0.25 | 0.23 | 0.35 | 0.14 | 0.18 |
| S | 0.30 | 0.49 | 0.59 | 0.23 | 0.07 | 0.01 | 0.14 | 0.16 | 0.11 | — |
| $H_2O^+$ | — | — | — | — | — | — | — | — | — | 0.93 |
| $H_2O^-$ | 0.11 | 0.25 | 0.55 | 0.39 | 0.33 | 0.38 | 0.68 | 0.46 | 0.41 | 1.04 |
| Ign. | — | — | 0.10 | 0.15 | — | 0.75 | 0.93 | — | 0.21 | — |
| Sum | 99.43 | 100.03 | 100.15 | 100.13 | 101.19 | 100.12 | 100.96 | 100.06 | 100.23 | 100.01 |
| $a$ | 7.0 | 5.7 | 6.5 | 6.8 | 8.3 | 6.5 | 6.9 | 8.9 | 8.7 | 9.0 |
| $c$ | 11.2 | 12.6 | 10.4 | 10.9 | 10.4 | 9.3 | 8.4 | 7.8 | 8.2 | 7.9 |
| $b$ | 24.3 | 21.9 | 22.0 | 23.0 | 22.9 | 26.7 | 28.4 | 18.6 | 16.4 | 13.3 |
| $s$ | 57.5 | 59.8 | 61.1 | 50.3 | 58.4 | 57.5 | 56.3 | 64.7 | 66.7 | 69.8 |
| $a'$ | — | — | — | — | — | — | — | — | — | 14 |
| $f'$ | 46 | 32 | 42 | 44 | 39 | 32 | 31 | 45 | 46 | 43 |
| $m'$ | 37 | 46 | 40 | 39 | 43 | 48 | 51 | 39 | 40 | 43 |
| $c'$ | 17 | 22 | 18 | 17 | 18 | 20 | 18 | 16 | 14 | — |
| $n'$ | 83 | 76 | 81 | 80 | 81 | 74 | 78 | 73 | 69 | 69 |

**1.** Basalt, Vetrovoi Volcano; coll. B. N. Shilov, sample No. 102v; anal. V. M. Bragina (SakhKNII). **2.** Basalt, Vetrovoi Volcano; coll. V. N. Shilov, sample No. 35v; anal. M. N. Zorin (SakhKNII). **3.** Basalt, Vetrovoi Volcano, Mount Zemleprokhodets; coll. V. N. Shilov, sample No. 36; anal. M. N. Zorin (SakhKNII). **4.** Basalt, Vetrovoi Volcano; coll. V. I. Fedorchenko, sample No. 1039; anal. Z. B. Ivleva (SakhKNII). **5.** Basalt, Vetrovoi Volcano; coll. V. N. Shilov, sample No. 101; anal. A. G. Zbrueva and A. G. Pinchuk (SakhKNII; analyses 1-5 from Gorkun et al., 1963). **6.** Basalt, Mount Smirnov; coll. V. I. Fedorchenko, sample No. 1098d; anal. E. N. Grigoryan (SakhKNII, Rodionova et al., 1963). **7.** Basalt, flank of Mount Smirnov; coll. V. I. Fedorchenko, sample No. 1098g; anal. E. N. Grigoryan (SakhKNII, Rodionova et al., 1963). **8.** Andesite-basalt, Zelenaya Mountain; coll. V. N. Shilov, sample No. 46; anal. M. N. Zorin (SakhKNII, Rodionova et al., 1963). **9.** Two-pyroxene andesite, Lagernyi Creek; coll. V. I. Fedorchenko, sample No. 1046; anal. L. K. Markova (SakhKNII, Rodionova et al., 1963. **10.** Andesite, divide between Yur'ev and Gorshkov Rivers; coll. K. K. Zelenov; anal. Kanakina (Zelenov, 1963).

## TABLE 2 (Continued)

| Components | Ebeko group | | | Lagernoe Plateau | | | | | | Vernadskii group | |
|---|---|---|---|---|---|---|---|---|---|---|---|
| | 11 | 12 | 13 | 14 | 15 | 16 | 17 | 18 | 19 | 20 | 21 |
| $SiO_2$ | 59.14 | 59.32 | 63.76 | 50.14 | 55.19 | 57.63 | 58.46 | 60.17 | 60.30 | 55.63 | 56.04 |
| $TiO_2$ | 0.63 | 0.58 | 0.44 | 0.59 | 0.63 | 0.57 | 0.58 | 0.68 | 0.56 | 0.55 | 0.63 |
| $Al_2O_3$ | 17.92 | 17.66 | 17.11 | 20.43 | 19.25 | 18.61 | 19.66 | 17.19 | 17.85 | 17.45 | 18.84 |
| $Fe_2O_3$ | 4.41 | 3.18 | 3.27 | 4.46 | 3.82 | 2.34 | 4.01 | 3.45 | 4.47 | 4.82 | 4.33 |
| FeO | 2.46 | 3.24 | 1.66 | 5.50 | 4.38 | 4.94 | 2.24 | 3.55 | 2.88 | 3.60 | 3.28 |
| MnO | 0.06 | 0.01 | 0.07 | 0.06 | 0.10 | 0.09 | 0.08 | 0.21 | 0.10 | 0.22 | 0.28 |
| MgO | 2.29 | 3.14 | 1.80 | 4.96 | 3.56 | 2.59 | 1.75 | 2.41 | 2.80 | 3.58 | 3.74 |
| CaO | 6.93 | 6.81 | 5.98 | 9.94 | 8.08 | 7.13 | 6.29 | 6.48 | 6.63 | 7.17 | 7.33 |
| $Na_2O$ | 2.67 | 3.37 | 3.26 | 2.58 | 2.44 | 2.73 | 2.52 | 2.46 | 3.13 | 2.71 | 2.40 |
| $K_2O$ | 1.85 | 1.41 | 1.34 | 1.02 | 1.58 | 1.90 | 1.65 | 1.93 | 1.08 | 1.76 | 1.41 |
| $P_2O_5$ | 0.19 | 0.22 | 0.15 | 0.06 | 0.07 | 0.10 | 0.11 | 0.34 | 0.22 | 0.59 | 0.15 |
| S | 0.26 | 0.30 | 0.27 | 0.08 | 0.18 | 0.05 | 0.04 | 0.23 | 0.11 | 0.86 | 0.36 |
| $H_2O^+$ | — | — | — | — | — | — | — | — | — | 0.26 | — |
| $H_2O^-$ | 0.65 | 0.21 | 0.39 | 0.58 | 0.14 | 0.19 | 1.27 | 0.40 | 0.89 | 0.28 | 1.42 |
| Ign. | 0.21 | 0.10 | 0.20 | 0.06 | 0.08 | 0.78 | 0.94 | 0.17 | 0.17 | — | — |
| Sum | 99.67 | 99.55 | 99.71 | 100.46 | 99.50 | 99.65 | 99.60 | 99.67 | 101.19 | 99.62 | 100.21 |
| $a$ | 9.1 | 11.1 | 9.4 | 7.7 | 8.1 | 9.2 | 8.4 | 8.5 | 8.7 | 9.1 | 7.8 |
| $c$ | 8.1 | 6.8 | 7.1 | 11.0 | 9.6 | 8.5 | 8.1 | 7.8 | 8.0 | 7.8 | 9.5 |
| $b$ | 11.3 | 13.0 | 8.2 | 20.4 | 15.2 | 12.4 | 12.4 | 11.5 | 12.0 | 16.0 | 14.3 |
| $s$ | 71.5 | 69.1 | 75.3 | 60.9 | 67.1 | 69.9 | 71.1 | 72.2 | 71.3 | 67.1 | 68.4 |
| $a'$ | — | — | — | — | — | — | 27 | — | — | — | — |
| $f'$ | 58 | 46 | 56 | 48 | 52 | 57 | 48 | 59 | 57 | 51 | 53 |
| $m'$ | 36 | 42 | 39 | 43 | 42 | 38 | 25 | 37 | 40 | 40 | 47 |
| $c'$ | 6 | 12 | 5 | 9 | 6 | 5 | — | 4 | 3 | 9 | 0 |
| $n'$ | 69 | 68 | 78 | 79 | 70 | 69 | 69 | 67 | 81 | 70 | 72 |

**11.** Two-pyroxene andesite, summit of Mount Zelena; coll. V. I. Fedorchenko, sample No. 1094a; anal. Z. I. Ivleva (SakhKNII, Rodionova et al., 1963). **12.** Andesite, Mount Smirnov; coll. V. I. Fedorchenko, sample No. 1098; anal. E. N. Grigoryan (SakhKNII, Rodionova et al., 1963). **13.** Dacite, head of Matroskaya River (average of three analyses; SakhKNII). **14.** Andesite-basalt; coll. V. N. Shilov, sample No. 2037d anal. M. N. Zorin and others (SakhKNII, Rodionova et al., 1964). **15.** Two-pyroxene andesite; coll. V. N. Shilov, sample No. 2036a; anal. M. N. Zorin and others (SakhKNII, Gorkun et al., 1963). **16.** Two-pyroxene andesite; coll. V. N. Shilov, sample No. 2040; anal. M. N. Zorin and others (SakhKNII, Gorkun et al., 1963). **17.** Two-pyroxene andesite; coll. V. N. Shilov, sample No. 2037; anal. L. K. Markova and others (SakhKNII, Rodionova et al., 1964). **18.** Two-pyroxene andesite; coll. V. I. Fedorchenko, sample No. 1053v; anal. Z. I. Ivleva (SakhKNII, Gorkun et al., 1963). **19.** Two-pyroxene andesite; coll. V. N. Shilov, sample No. 117; anal. A. Zbrueva and others (SakhKNII, Gorkun et al., 1963). **20.** Andesite, Skalistoe mine area; coll. G. M. Vlasov; anal. Il'inykh (KKGRÉ, Popkova et al., 1961, No. 2473). **21.** Andesite; data same as above (Popkova et al., 1961, No. 2466).

## TABLE 3. Interglacial Lavas of the Volcanoes of the Vernadskii Ridge

| Components | 22 | 23 | 24 | 25 | 26 | 27 | 28 | 29 | 30 |
|---|---|---|---|---|---|---|---|---|---|
| $SiO_2$ | 53.38 | 55.01 | 56.15 | 56.16 | 60.32 | 51.91 | 53.21 | 54.19 | 57.29 |
| $TiO_2$ | 0.70 | 0.72 | 0.61 | 0.95 | 0.47 | 0.70 | 0.63 | 0.53 | 0.52 |
| $Al_2O_3$ | 19.07 | 17.75 | 18.34 | 18.10 | 17.92 | 18.23 | 18.73 | 18.80 | 20.26 |
| $Fe_2O_3$ | 3.73 | 3.64 | 3.56 | 3.11 | 2.94 | 4.13 | 3.81 | 2.57 | 2.44 |
| FeO | 4.59 | 4.58 | 4.28 | 4.03 | 3.04 | 4.65 | 5.27 | 5.84 | 4.37 |
| MnO | 0.13 | 0.08 | 0.14 | 0.11 | 0.01 | 0.05 | 0.12 | 0.03 | 0.10 |
| MgO | 4.16 | 3.66 | 3.24 | 3.40 | 2.80 | 6.46 | 4.22 | 4.23 | 2.91 |
| CaO | 9.35 | 8.86 | 7.52 | 7.99 | 6.72 | 10.07 | 9.67 | 8.65 | 6.95 |
| $Na_2O$ | 2.45 | 2.70 | 3.10 | 2.40 | 2.89 | 1.95 | 2.63 | 2.25 | 2.58 |
| $K_2O$ | 1.54 | 1.51 | 1.66 | 1.63 | 2.05 | 1.09 | 1.32 | 1.50 | 1.90 |
| $P_2O_5$ | 0.10 | 0.17 | 0.11 | 0.12 | 0.19 | 0.19 | 0.06 | 0.03 | 0.05 |
| S | 0.03 | 0.24 | 0.11 | 0.17 | 0.40 | 0.30 | 0.15 | 0.08 | 0.10 |
| $H_2O^-$ | 0.17 | 0.43 | 0.24 | 0.67 | 0.07 | 0.22 | 0.11 | 0.29 | 0.22 |
| Ign. | 0.54 | 0.47 | 0.77 | 1.41 | 0.34 | 0.11 | — | 0.66 | 0.34 |
| Sum | 99.94 | 99.82 | 99.83 | 100.25 | 100.16 | 100.06 | 99.93 | 99.65 | 100.03 |
| $a$ | 8.0 | 8.5 | 9.6 | 8.1 | 9.7 | 6.1 | 8.2 | 7.5 | 8.9 |
| $c$ | 9.3 | 8.2 | 8.0 | 8.7 | 7.6 | 9.5 | 9.0 | 9.5 | 8.9 |
| $b$ | 18.1 | 17.5 | 15.0 | 14.5 | 11.3 | 22.6 | 19.9 | 17.3 | 13.6 |
| $s$ | 64.6 | 65.8 | 67.4 | 68.7 | 71.4 | 61.8 | 69.9 | 65.7 | 68.6 |
| $a'$ | — | — | — | — | — | — | — | — | 14 |
| $f'$ | 45 | 45 | 51 | 47 | 49 | 36 | 45 | 47 | 48 |
| $m'$ | 41 | 37 | 38 | 42 | 43 | 50 | 38 | 44 | 38 |
| $c'$ | 14 | 18 | 11 | 11 | 8 | 14 | 17 | 9 | — |
| $n'$ | 70 | 73 | 74 | 70 | 68 | 72 | 74 | 69 | 68 |

**22.** Andesite-basalt, Zelenaya Mountain; coll. V. N. Shilov, sample No. 269; anal. M. N. Zorin (SakhKNII, Rodionova et al., 1963). **23.** Two-pyroxene andesite, near the Revushchie fumaroles on Ebeko Volcano; coll. V. I. Fedorchenko, sample No. 1095; anal. Z. I. Ivleva (SakhKNII, Rodionova et al., 1963). **24.** Two-pyroxene andesite, same locality as 23; coll. V. N. Shilov, sample No. 271; anal. M. N. Zorin (SakhKNII, Rodionova et al., 1963). **25.** Two-pyroxene andesite, Lagernyi Creek; coll. V. I. Fedorchenko, sample No. 1048; anal. L. K. Markova (SakhKNII, Rodionova et al., 1963). **26.** Two-pyroxene andesite, headwaters of Savushkin Creek; coll. V. I. Fedorchenko, sample No. 1110a; anal. E. N. Grigoryan (SakhKNII, Rodionova et al., 1963). **27.** Basalt, Mount Nasedkin; coll. V. N. Shilov, sample No. 59; anal. M. N. Zorin (SakhKNII, Gorkun et al., 1963). **28.** Andesite-basalt, Bogdanovich Volcano; coll. V. N. Shilov, sample No. 2045; anal. M. N. Zorin and others (SakhKNII, Gorkun et al., 1963). **29.** Olivine-bearing two-pyroxene andesite, Krugozornyi Pass on Bogdanovich Volcano; coll. V. N. Shilov, sample No. 2010; anal. M. N. Zorin and others (SakhKNII, Gorkun et al., 1963). **30.** Two-pyroxene andesite, headwaters of Nasedkin River; coll. V. N. Shilov, sample No. 2004a; anal. L. K. Markova and others (SakhKNII, Gorkun et al., 1963).

TABLE 4. Modern Lavas of the Volcanoes
of Vernadskii Ridge (Paramushir Island)

| Components | Ebeko Volcano group | | | | |
|---|---|---|---|---|---|
| | 31 | 32 | 33 | 34 | 35 |
| $SiO_2$ | 52.95 | 54.35 | 58.07 | 58.14 | 56.91 |
| $TiO_2$ | 0.91 | 1.00 | 0.68 | 0.65 | 0.76 |
| $Al_2O_3$ | 18.01 | 17.61 | 17.21 | 18.69 | 17.74 |
| $Fe_2O_3$ | 3.26 | 2.69 | 3.61 | 2.09 | 3.31 |
| FeO | 5.85 | 5.83 | 4.23 | 4.60 | 5.03 |
| MnO | 0.10 | 0.26 | 0.16 | 0.05 | 0.08 |
| MgO | 4.79 | 4.33 | 3.04 | 3.13 | 4.03 |
| CaO | 9.10 | 8.66 | 7.41 | 7.49 | 7.60 |
| $Na_2O$ | 2.51 | 3.05 | 3.26 | 2.86 | 2.23 |
| $K_2O$ | 1.52 | 1.92 | 2.35 | 1.60 | 1.74 |
| $P_2O_5$ | 0.11 | 0.18 | 0.02 | 0.11 | 0.11 |
| $H_2O^+$ | — | 0.18 | 0.12 | — | — |
| $H_2O^-$ | 0.07 | 0.16 | 0.20 | 0.10 | 0.18 |
| S | 0.10 | 0.10 | 0.05 | 0.43 | 0.10 |
| Ign. | — | — | — | — | — |
| Sum | 99.28 | 100.32 | 100.14 | 99.94 | 99.82 |
| $a$ | 7.9 | 9.6 | 10.6 | 9.0 | 7.6 |
| $c$ | 8.6 | 7.2 | 6.4 | 8.5 | 8.5 |
| $b$ | 20.1 | 19.3 | 15.4 | 13.0 | 16.2 |
| $s$ | 63.4 | 63.9 | 67.6 | 69.5 | 67.7 |
| $f'$ | 44 | 43 | 48 | 49 | 49 |
| $m'$ | 42 | 39 | 34 | 43 | 44 |
| $c'$ | 14 | 18 | 18 | 8 | 7 |
| $n'$ | 71 | 71 | 68 | 73 | 67 |

**31.** Andesite-basalt, northwest flow of the lower crater of Ebeko Volcano; coll. R. I. Rodionova, sample No. 348b; anal. M. N. Zorin (SakhKNII, Rodionova et al., 1963). **32.** Andesite-basalt, flow from the lower crater of Ebeko Volcano; coll. author, sample No. 46124; anal. V. P. Énman. **33.** Andesite, inner part of a breadcrust bomb, 1934-1935 eruption of Ebeko Volcano; coll. author, sample No. 53657b; anal. I. I. Tovarova. **34.** Andesite, bomb from the 1934-1935 eruption of Ebeko Volcano; coll. V. N. Shilov, sample No. 25; anal. M. N. Zorin (SakhKNII, Rodionova et al., 1963). **35.** Two-pyroxene andesite, flow from Neozhidannyi Volcano; coll. V. I. Fedorchenko, sample No. 1032a; anal. M. N. Zorin (SakhKNII, Rodionova et al., 1963).

## TABLE 4 (Continued)

| Components | Ebeko | Bogdanovich Volcano group | | | | | | | | Fersman Volcano |
|---|---|---|---|---|---|---|---|---|---|---|
| | 36 | 37 | 38 | 39 | 40 | 41 | 42 | 43 | 44 | 45 |
| $SiO_2$ | 60.69 | 56.14 | 55.32 | 57.37 | 58.39 | 60.68 | 56.53 | 58.58 | 59.74 | 59.11 |
| $TiO_2$ | 0.59 | 0.74 | 0.65 | 0.61 | 0.64 | 0.73 | 0.70 | 0.81 | 0.55 | 0.56 |
| $Al_2O_3$ | 17.91 | 17.82 | 18.20 | 19.04 | 18.01 | 16.97 | 18.18 | 17.32 | 18.32 | 18.59 |
| $Fe_2O_3$ | 1.53 | 3.37 | 3.85 | 3.09 | 2.51 | 2.22 | 2.82 | 1.95 | 1.80 | 6.45 |
| $FeO$ | 4.44 | 4.37 | 4.53 | 4.31 | 4.53 | 4.05 | 4.87 | 5.17 | 4.20 | 0.62 |
| $MnO$ | 0.10 | 0.21 | 0.12 | 0.10 | 0.04 | 0.25 | 0.07 | 0.21 | 0.06 | 0.16 |
| $MgO$ | 3.05 | 3.63 | 3.83 | 2.83 | 3.37 | 2.78 | 3.68 | 3.30 | 2.81 | 2.53 |
| $CaO$ | 7.34 | 8.77 | 8.91 | 7.62 | 7.30 | 6.28 | 7.67 | 6.00 | 6.76 | 7.63 |
| $Na_2O$ | 2.55 | 2.60 | 3.06 | 2.60 | 2.56 | 3.42 | 2.38 | 3.06 | 2.80 | 3.48 |
| $K_2O$ | 1.92 | 1.61 | 1.79 | 1.86 | 1.74 | 2.63 | 1.80 | 2.31 | 2.16 | 1.00 |
| $P_2O_5$ | 0.07 | 0.21 | 0.08 | 0.13 | 0.06 | 0.10 | 0.21 | 0.17 | 0.05 | — |
| $H_2O^+$ | — | — | — | — | — | 0.26 | — | 1.16 | — | 0.09 |
| $H_2O^-$ | 0.03 | - - | 0.06 | 0.41 | 0.11 | 0.04 | 0.02 | 0.15 | 0.26 | 0.05 |
| S | 0.12 | 0.16 | 0.11 | 0.14 | 0.31 | 0.02 | 0.27 | — | 0.16 | — |
| Ign. | 0.15 | 0.10 | 0.20 | 0.25 | 0.04 | — | 0.30 | — | 0.05 | — |
| Sum | 100.49 | 99.73 | 100.71 | 100.36 | 99.61 | 100.43 | 99.50 | 100.19 | 99.72 | 100.27 |
| $a$ | 8.5 | 8.3 | 9.7 | 8.9 | 8.4 | 11.4 | 8.3 | 10.3 | 9.7 | 9.5 |
| $c$ | 8.0 | 8.2 | 7.7 | 8.9 | 8.7 | 5.8 | 8.6 | 6.8 | 7.8 | 8.1 |
| $b$ | 12.1 | 16.8 | 17.9 | 13.0 | 13.5 | 12.8 | 15.1 | 13.5 | 11.4 | 12.4 |
| $s$ | 71.4 | 66.7 | 64.7 | 69.2 | 69.4 | 70.0 | 68.0 | 69.4 | 71.1 | 70.0 |
| $f'$ | 47 | 45 | 44 | 55 | 50 | 48 | 49 | 52 | 51 | 52 |
| $m'$ | 44 | 38 | 37 | 39 | 44 | 37 | 43 | 43 | 43 | 36 |
| $c'$ | 9 | 17 | 19 | 6 | 6 | 15 | 8 | 5 | 6 | 12 |
| $n'$ | 67 | 71 | 72 | 68 | 70 | 66 | 67 | 67 | 66 | 84 |

**36.** Two-pyroxene andesite, extrusion on Lagernyi Creek; coll. V. I. Fedorchenko, sample No. 1047; anal. L. K. Markova (SakhKNII, Rodionova et al., 1963). **37.** Two-pyroxene andesite with olivine, crater of Bogdanovich Volcano; coll. R. I. Rodionova, sample No. 318; anal. Z. I. Ivleva (SakhKNII, Gorkun et al., 1963). **38.** Two-pyroxene andesite, Krasheninnikov Volcano; coll. V. N. Shilov; anal. M. N. Zorin and others (SakhKNII, Gorkun et al., 1963). **39.** Two-pyroxene andesite, Krasheninnikov Volcano; coll. V. N. Shilov, sample No. 63; anal. M. N. Zorin (SakhKNII, Gorkun et al., 1963). **40.** Two-pyroxene andesite, Krasheninnikov Volcano; coll. V. N. Shilov, sample No. 2034; anal. L. K. Markova and others (SakhKNII, Gorkun et al., 1963). **41.** Pyroxene andesite, Krasheninnikov Volcano; coll. author, sample No. 53706; anal. V. P. Énman. **42.** Two-pyroxene andesite, Kozyrevskii Volcano; coll. V. N. Shilov, sample No. 72; anal. M. N. Zorin (SakhKNII, Gorkun et al., 1963). **43.** Two-pyroxene andesite, Kozyrevskii Crater; coll. author, sample No. 53702; anal. V. P. Énman. **44.** Two-pyroxene andesite, east flow from Kozyrevskii Crater; coll. V. N. Shilov, sample No. 76; anal. M. N. Zorin (SakhKNII, Gorkun et al., 1963). **45.** Two-pyroxene andesite, dome of Fersman Volcano; coll. for author by K. I. Shmulovich, sample No. 63504; anal. Z. I. Beletskaya.

## TABLE 5.   Lavas of the Karpinskii Ridge Volcanoes

| Components | Chikurachki Volcano | | | Karpinskii Volcano | | | |
|---|---|---|---|---|---|---|---|
| | 46 | 47 | 48 | 49 | 50 | 51 | 52 |
| $SiO_2$ | 52.27 | 53.06 | 53.86 | 54.62 | 70.46 | 53.52 | 55,02 |
| $TiO_2$ | 0.66 | 0.72 | 0.84 | 0.86 | 0.40 | 0.86 | 0.98 |
| $Al_2O_3$ | 20.90 | 21.04 | 19.50 | 17.38 | 14.34 | 18.45 | 18.40 |
| $Fe_2O_3$ | 3.84 | 2.80 | 4.15 | 3.49 | 0.74 | 4.41 | 3.19 |
| FeO | 4.66 | 5.17 | 4.67 | 5.93 | 1.27 | 5.16 | 5.07 |
| MnO | 0.22 | 0.17 | 0.16 | 0.21 | 0.09 | 0.16 | 0.17 |
| MgO | 3.29 | 3.40 | 3.01 | 4.41 | 0.62 | 4.50 | 4.20 |
| CaO | 10.53 | 10.61 | 9.50 | 8.14 | 2.70 | 9.24 | 8.74 |
| $Na_2O$ | 1.90 | 2.00 | 3.21 | 2.82 | 4.73 | 2.78 | 2.98 |
| $K_2O$ | 0.92 | 0.85 | 1.08 | 1.10 | 2.09 | 1.09 | 1.24 |
| $P_2O_5$ | 0.06 | 0.06 | — | 0.08 | — | — | — |
| $H_2O^+$ | 0.47 | — | 0.19 | 0.92 | 2.83 | 0.14 | 0.18 |
| $H_2O^-$ | 0.44 | — | 0.14 | 0.11 | 0.08 | 0.15 | 0.24 |
| S | 0.33 | 0.10 | — | 0.03 | 0.04 | — | — |
| Sum | 100.49 | 99.98 | 100.31 | 100.10 | 99.75 | 100.46 | 100.41 |
| $a$ | 6.1 | 6.1 | 9.2 | 8.1 | 13.4 | 8.0 | 8.6 |
| $c$ | 12.1 | 12.4 | 9.1 | 8.1 | 2.8 | 8.6 | 8.4 |
| $b$ | 16.4 | 15.8 | 16.8 | 18.6 | 3.4 | 19.9 | 17.7 |
| $s$ | 65.4 | 65.7 | 64.9 | 65.2 | 80.4 | 63.5 | 65.3 |
| $f'$ | 52 | 50 | 51 | 49 | 55 | 46 | 45 |
| $m'$ | 37 | 39 | 31 | 39 | 31 | 39 | 41 |
| $c'$ | 11 | 11 | 18 | 12 | 14 | 15 | 14 |
| $n'$ | 76 | 77 | 81 | 79 | 78 | 79 | 79 |

**46.** Andesitic ash, May 1958 eruption of Chikurachki Volcano; coll. V. N. Shilov, sample No. 57; anal. R. I. Evseev (SakhKNII, Shilov and Voronova, 1962). **47.** Andesitic bomb, May 1958 eruption of Chikurachki Volcano; coll. V. N. Shilov, sample No. 58; anal. R. I. Evseeva (SakhKNII, Shilov and Voronova, 1962). **48.** Andesite, 1854 flow from Chikurachki Volcano; coll. author, sample No. 53688d; anal. I. I. Tovarova. **49.** Andesite-basalt, preglacial flow from Karpinskii Volcano; coll. author, sample No. 53676; anal. V. P. Énman. **50.** Rhyolitic pumice from moraine on Karpinskii Volcano; coll. author, sample No. 53684; anal. V. P. Énman. **51.** Andesite-basalt, young flow from Karpinskii Volcano (latest part); coll. author, sample No. 53673; anal. I. I. Tovarova. **52.** Andesite-basalt, earliest part of the same flow; coll. author, sample No. 53671; anal. N. S. Klassova.

## TABLE 6. Lavas of the Volcanoes on the Islands of Onekotan, Kharimkotan, and Shiashkotan

| Components | Onekotan Isnald, Nemo Peak | | | | | | | | |
|---|---|---|---|---|---|---|---|---|---|
| | 53 | 54 | 55 | 56 | 57 | 58 | 59 | 60 | 61 |
| $SiO_2$ | 58.14 | 64.31 | 64.31 | 53.76 | 55.66 | 56.12 | 56.88 | 57.24 | 59.32 |
| $TiO_2$ | 1.14 | 0.79 | 0.81 | 1.17 | 1.07 | 1.12 | 0.86 | 1.02 | 0.91 |
| $Al_2O_3$ | 16.78 | 15.80 | 16.57 | 16.46 | 17.51 | 17.75 | 19.56 | 16.93 | 17.73 |
| $Fe_2O_3$ | 2.52 | 1.77 | 2.98 | 4.50 | 2.98 | 3.01 | 3.22 | 5.88 | 2.78 |
| $FeO$ | 6.26 | 3.97 | 3.28 | 5.37 | 5.40 | 4.91 | 4.37 | 2.47 | 3.76 |
| $MnO$ | 0.19 | 0.16 | 0.17 | 0.24 | 0.22 | 0.19 | 0.22 | 0.16 | 0.13 |
| $MgO$ | 2.89 | 2.48 | 1.80 | 4.76 | 4.72 | 3.77 | 1.89 | 3.70 | 2.71 |
| $CaO$ | 6.60 | 5.75 | 4.70 | 7.96 | 8.45 | 8.46 | 8.31 | 8.00 | 7.82 |
| $Na_2O$ | 3.72 | 3.47 | 4.13 | 2.70 | 2.82 | 3.13 | 3.25 | 3.18 | 3.51 |
| $K_2O$ | 1.02 | 1.28 | 0.96 | 0.76 | 0.70 | 0.87 | 0.65 | 0.94 | 0.98 |
| $P_2O_5$ | — | 0.08 | — | — | 0.07 | — | — | — | — |
| $H_2O^+$ | 0.78 | 0.35 | 0.48 | 1.24 | 0.66 | 0.69 | 0.77 | 0.69 | 0.46 |
| $H_2O^-$ | 0.31 | 0.08 | 0.04 | 1.09 | 0.19 | 0.15 | 0.36 | 0.08 | 0.04 |
| $S$ | — | 0.04 | — | — | 0.04 | — | — | — | — |
| Sum | 100.35 | 100.33 | 100.23 | 100.01 | 100.49 | 100.17 | 100.34 | 100.29 | 100.15 |
| $a$ | 10.0 | 9.4 | 10.6 | 7.4 | 7.4 | 8.4 | 8.6 | 8.5 | 9.4 |
| $c$ | 6.6 | 5.8 | 5.8 | 7.7 | 8.3 | 8.0 | 9.6 | 7.2 | 7.5 |
| $b$ | 15.2 | 10.8 | 9.2 | 20.3 | 18.5 | 16.8 | 12.1 | 16.7 | 13.1 |
| $s$ | 68.2 | 74.0 | 74.4 | 64.6 | 65.8 | 66.8 | 69.7 | 67.3 | 70.0 |
| $a'$ | — | — | 2 | — | — | — | — | — | — |
| $b'$ | 50 | 50 | 64 | 47 | 44 | 46 | 62 | 46 | 47 |
| $m'$ | 48 | 39 | 34 | 41 | 44 | 39 | 28 | 38 | 36 |
| $c'$ | 12 | 11 | — | 12 | 12 | 15 | 10 | 16 | 17 |
| $n'$ | 84 | 80 | 86 | 85 | 87 | 84 | 89 | 84 | 84 |

**53.** Andesite, somma of Nemo Volcano; coll. author, sample No. 61578; anal. I. M. Bender. **54.** Andesite-dacitic ignimbrite, Nemo Bay; coll. author, sample No. 46275 (average of two analyses). **55.** Andesite-dacitic ignimbrite, Nemo Bay; coll. for author by G. E. Bogoyavlenskaya, sample No. 62231; anal. L. S. Mazalova. **56.** Andesite-basalt, northwest flank of Nemo Peak, at height of 600 m; coll. for author by G. E. Bogoyavlenskaya, sample No. 61573; anal. L. S. Mazalova. **57.** Andesite-basalt, scoria bomb from Nemo Peak; coll. author, sample No. 46285; anal. V. P. Énman. **58.** Andesite, Nemo Peak at a height of 600 m; coll. for author by G. E. Bogoyavlenskaya, sample No. 61569; anal. L. S. Mazalova. **59.** Andesite, flow near base of Nemo Peak; coll. for author by G. E. Bogoyavlenskaya, sample No. 61626; anal. L. S. Mazalova. **60.** Andesite, Nemo Peak; coll. for author by G. E. Bogoyavlenskaya, sample No. 61575; anal. I. M. Bender. **61.** Andesite, summit extrusion of Nemo Peak; coll. for author by G. E. Bogoyavlenskaya, sample No. 61562; anal. L. S. Mazalova.

## TABLE 6 (Continued)

| Components | Onekotan Island, Tao-Rusyr Caldera | | | | | | | | | |
|---|---|---|---|---|---|---|---|---|---|---|
| | 62 | 63 | 64 | 65 | 66 | 67 | 68 | 69 | 70 | 71 |
| $SiO_2$ | 49.21 | 50.02 | 50.60 | 52.66 | 53.11 | 53.35 | 55.31 | 65.10 | 65.51 | 58.68 |
| $TiO_2$ | 1.09 | 1.20 | 0.98 | 1.10 | 1.10 | 1.10 | 0.95 | 0.60 | 0.65 | 1.36 |
| $Al_2O_3$ | 17.72 | 17.93 | 18.89 | 18.16 | 17.49 | 17.53 | 15.79 | 15.58 | 14.87 | 16.07 |
| $Fe_2O_3$ | 5.02 | 3.11 | 3.31 | 2.55 | 2.90 | 4.06 | 2.69 | 2.62 | 2.41 | 3.84 |
| $FeO$ | 8.11 | 8.32 | 6.78 | 8.20 | 7.32 | 6.90 | 8.08 | 2.15 | 2.58 | 5.26 |
| $MnO$ | 0.22 | 0.29 | 0.28 | 0.26 | 0.26 | 0.19 | 0.18 | 0.20 | 0.19 | 0.21 |
| $MgO$ | 4.82 | 5.26 | 5.85 | 4.57 | 4.62 | 4.38 | 3.86 | 1.67 | 1.29 | 2.79 |
| $CaO$ | 11.10 | 10.22 | 10.43 | 9.10 | 9.88 | 9.22 | 8.12 | 4.18 | 3.86 | 6.60 |
| $Na_2O$ | 2.57 | 2.49 | 3.21 | 2.76 | 2.84 | 2.96 | 3.44 | 4.20 | 4.62 | 3.72 |
| $K_2O$ | 0.36 | 0.57 | 0.48 | 0.84 | 0.69 | 0.76 | 1.12 | 1.52 | 1.65 | 1.30 |
| $P_2O_5$ | 0.07 | — | — | — | — | — | — | — | — | — |
| $H_2O^+$ | 0.14 | 0.24 | 0.02 | 0.11 | 0.12 | — | 0.31 | 2.07 | 1.99 | 0.62 |
| $H_2O^-$ | 0.09 | 0.20 | 0.20 | 0.06 | 0.10 | 0.06 | 0.18 | 0.17 | 0.32 | 0.15 |
| $S$ | 0.02 | — | — | — | — | — | — | — | — | — |
| Sum | 100.54 | 99.85 | 101.03 | 100.37 | 100.43 | 100.51 | 100.03 | 100.06 | 99.94 | 100.60 |
| $a$ | 6.4 | 6.5 | 7.9 | 7.6 | 7.4 | 7.8 | 9.2 | 11.6 | 12.7 | 10.2 |
| $c$ | 9.1 | 9.2 | 8.9 | 8.7 | 8.3 | 8.1 | 6.1 | 4.9 | 3.7 | 5.7 |
| $b$ | 25.9 | 24.1 | 23.8 | 21.2 | 21.8 | 21.3 | 20.7 | 7.6 | 8.0 | 15.7 |
| $s$ | 58.6 | 60.2 | 59.4 | 62.5 | 62.5 | 62.8 | 64.0 | 75.9 | 75.6 | 68.4 |
| $a'$ | — | — | — | — | — | — | — | — | — | — |
| $f'$ | 49 | 46 | 41 | 50 | 45 | 49 | 50 | 59 | 59 | 55 |
| $m'$ | 32 | 39 | 42 | 38 | 37 | 35 | 32 | 38 | 28 | 30 |
| $c'$ | 19 | 15 | 17 | 12 | 18 | 16 | 18 | 3 | 13 | 15 |
| $n'$ | 91 | 87 | 91 | 83 | 87 | 86 | 82 | 81 | 80 | 81 |

**62.** Basalt, preglacial structure in the base of Tao-Rusyr Caldera; coll. author, sample No. 46305; anal. V. G. Sil'nichenko. **63.** Basalt, foot of Tao-Rusyr Caldera; coll. for author by G. E. Bogoyavlenskaya, sample No. 61529a; anal. L. A. Basharina. **64.** Basalt, Tao-Rusyr Caldera; coll. for author by G. E. Bogoyavlenskaya, sample No. 61555; anal. L. A. Basharina. **65.** Andesite-basalt, Tao-Rusyr Caldera; coll. for author by G. E. Bogoyavlenskaya, sample No. 61551; anal. L. A. Basharina. **66.** Andesite-basalt, upper part of Tao-Rusyr Caldera; coll. for author by G. E. Bogoyavlenskaya, sample No. 61556; anal. L. A. Basharina. **67.** Andesite-basalt, upper part of Tao-Rusyr Caldera; coll. for author by G. E. Bogoyavlenskaya, sample No. 61550; anal. L. A. Basharina. **68.** Andesite-basalt, Tao-Rusyr Caldera; coll. for author by G. E. Bogoyavlenskaya, sample No. 61548; anal. L. A. Basharina. **69.** Dacite pumice, northwest foot of Tao-Rusyr Caldera; coll. author, sample No. 46325 (average of two analyses). **70.** Dacite pumice, northeast foot of Tao-Rusyr Caldera; coll. for author by G. E. Bogoyavlenskaya, sample No. 61548a; anal. L. A. Basharina. **71.** Andesite from pyroclastic flow, age 7040 years. B.P., northeast foot of Tao-Rusyr Caldera; coll. for author by G. E.Bogoyavlenskaya, sample No. 61615; anal. G. F. Nekrasova.

## TABLE 6 (Continued)

| Onekotan Island, Krenitsyn Peak | | | | Kharimkotan Island | | | | | |
|---|---|---|---|---|---|---|---|---|---|
| 72 | 73 | 74 | 75 | 76 | 77 | 78 | 79 | 80 | 81 |
| 65.16 | 61.91 | 62.64 | 61.90 | 59.40 | 62.02 | 60.67 | 67.44 | 59.86 | 58.78 |
| 0.86 | 0.75 | 0.86 | 0.97 | 0.83 | 0.81 | 0.47 | 0.87 | 0 76 | 0.85 |
| 15.39 | 16.89 | 15.84 | 16.78 | 17.14 | 17.28 | 16.25 | 15.01 | 16.41 | 18.54 |
| 4.09 | 3.14 | 2.10 | 2.29 | 1.96 | 2.41 | 3.06 | 1.52 | 2.86 | 2.52 |
| 1.78 | 3.89 | 4.41 | 5.05 | 5.11 | 3.59 | 5.04 | 3.05 | 5.20 | 3.71 |
| 0.18 | 0.21 | 0.17 | 0.20 | 0.13 | 0.14 | 0.05 | 0.12 | 0.18 | 0.15 |
| 1.86 | 2.20 | 2.26 | 2.70 | 3.29 | 2.07 | 3 33 | 1.31 | 3.08 | 2.30 |
| 4.78 | 5.96 | 5.96 | 5.56 | 6 74 | 6.60 | 7.50 | 5.42 | 7.06 | 8.00 |
| 4.05 | 3.21 | 4.08 | 3 66 | 3.14 | 3.42 | 3.05 | 3.60 | 3.16 | 3.42 |
| 1.39 | 1.01 | 1.91 | 1.29 | 1.03 | 1.06 | 0.60 | 1.38 | 0.86 | 1.08 |
| — | — | — | — | 0.07 | — | — | — | — | — |
| 0.35 | 0.32 | 0.51 | 0.11 | 1.22 | 0.09 | 0.15 | 0.21 | 0.56 | 0.02 |
| 0.10 | 0.26 | 0.05 | — | 0.14 | 0.06 | 0.19 | 0.07 | 0.02 | 0.13 |
| — | — | — | — | — | — | — | — | — | — |
| 99.99 | 99.75 | 100.79 | 100.51 | 100.20 | 99.55 | 100.36 | 100.00 | 100.01 | 99.50 |
| 11.0 | 8.8 | 11.6 | 9.9 | 8.8 | 9.4 | 7.6 | 9.9 | 8.4 | 9.6 |
| 4.8 | 7.2 | 4.7 | 6.0 | 7.4 | 7.2 | 7.6 | 5.1 | 6.9 | 8.1 |
| 9.7 | 10.8 | 12.3 | 12.3 | 13.5 | 10.3 | 15.0 | 7.9 | 14.8 | 12.1 |
| 74.5 | 73.2 | 71.4 | 71.8 | 70.3 | 73.1 | 69.8 | 77.1 | 69.9 | 70.2 |
| — | — | — | — | — | — | — | — | — | — |
| 56 | 62 | 49 | 56 | 50 | 55 | 50 | 52 | 52 | 50 |
| 33 | 36 | 31 | 37 | 43 | 35 | 38 | 28 | 36 | 33 |
| 11 | 2 | 20 | 7 | 7 | 10 | 12 | 20 | 12 | 17 |
| 81 | 83 | 77 | 81 | 82 | 83 | 90 | 79 | 84 | 82 |

**72.** Andesite-dacite, summit extrusion of Krenitsyn Peak; coll. for author by G. E. Bogoyavlenskaya, sample No. 61439; anal. L. A. Basharina. **73.** Andesitic ash, 1952 eruption of Krenitsyn Peak; coll. for author by B. I. Piip, sample No. 1346a; anal. I. I. Tovarova. **74.** Andesitic pumice, 1952 eruption of Krenitsyn Peak; coll. for author by G. E. Bogoyavlenskaya, sample No. 61444; anal. L. A. Basharina. **75.** Andesite, 1952 dome on Krenitsyn Peak; coll. for author by G. E. Bogoyavlenskaya, sample No. 61453; anal. L. A. Basharina. **76.** Andesite bomb, ancient portion of Kharimkotan; coll. author, sample 46250; anal. V. P. Énman. **77.** Andesite-dacite, ancient portion of Kharimkotan; coll. author, sample No. 63005; anal. T. V. Dolgova. **78.** Andesite pumice, 1933 eruption of Kharimkotan; coll. T. Nemoto; anal. Geol. Survey of Japan (Nemoto, 1934). **79.** Dacite pumice, 1933 pyroclastic flow, Kharimkotan; coll. author, sample No. 63003; anal. T. V. Dolgova. **80.** Andesite, 1933 summit dome, Kharimkotan; coll. E. K. Markhinin; anal. M. I. Bel'skaya. **81.** Andesite, 1933 dome lava flow, Kharimkotan; coll. author, sample No. 63004; anal. T. V. Dolgova.

## TABLE 6  (Continued)

| Shiashkotan Island, Sinarko Volcano | | | | Shiashkotan Island, Kuntomintar Volcano | | | | | | |
|---|---|---|---|---|---|---|---|---|---|---|
| 82 | 83 | 84 | 85 | 86 | 87 | 88 | 89 | 90 | 91 | 92 |
| 57.09 | 59.14 | 56.18 | 58.77 | 58.56 | 56.25 | 59.28 | 55.98 | 57.40 | 59.53 | 60.14 |
| 0.66 | 0.76 | 0.73 | 0.61 | 1.50 | 0.83 | 0.64 | 0.89 | 0.70 | 0.63 | 1.07 |
| 16.48 | 15.63 | 16.60 | 16.87 | 17.22 | 17.46 | 15.98 | 18.00 | 16.02 | 17.82 | 16.84 |
| 5.69 | 7.00 | 4.62 | 4.12 | 2.87 | 3.05 | 4.82 | 4.89 | 6.06 | 2.86 | 2.39 |
| 3.48 | 1.58 | 4.68 | 2.96 | 4.86 | 5.74 | 3.88 | 3.62 | 3.62 | 4.04 | 4.21 |
| 0.19 | 0.18 | 0.17 | 0.15 | 0.23 | 0.14 | 0.16 | 0.14 | 0.18 | 0.18 | 0.18 |
| 3.97 | 3.17 | 3.90 | 3.59 | 3.52 | 3.82 | 3.33 | 4.08 | 3.82 | 2.71 | 3.01 |
| 8.00 | 6.95 | 8.20 | 6.62 | 7.10 | 7.70 | 7.45 | 8.35 | 7.15 | 6.49 | 6.67 |
| 3.90 | 4.00 | 3.80 | 2.92 | 3.15 | 3.34 | 3.26 | 2.59 | 3.44 | 3.17 | 3.68 |
| 1.02 | 1.20 | 0.95 | 1.16 | 1.05 | 0.88 | 1.13 | 1.00 | 0.80 | 1.21 | 0.57 |
| — | — | — | — | — | 0.27 | — | — | — | — | — |
| 0.05 | 0.48 | 0.44 | 1.36 | 0.11 | 0.13 | 0.15 | 0.11 | 0.39 | 1.05 | 0.63 |
| 0.08 | 0.38 | 0.32 | 0.36 | 0.10 | 0.14 | 0.13 | — | 0.39 | 0.23 | 0.54 |
| — | — | — | — | — | — | — | — | — | — | — |
| 100.61 | 100.47 | 100.59 | 99.49 | 100.37 | 99.75 | 100.21 | 99.65 | 99.97 | 99.92 | 99.93 |
| 10.1 | 10.6 | 9.8 | 8.5 | 8.7 | 8.9 | 8.9 | 7.5 | 8.9 | 9.2 | 9.1 |
| 6.1 | 5.1 | 6.2 | 7.5 | 7.4 | 7.5 | 6.3 | 8.8 | 6.4 | 7.8 | 7.0 |
| 18.1 | 16.4 | 19.1 | 13.9 | 14.9 | 16.9 | 16.5 | 16.8 | 17.9 | 11.9 | 12.9 |
| 65.7 | 67.9 | 64.9 | 70.1 | 69.0 | 66.7 | 68.3 | 66.9 | 66.8 | 71.1 | 71.0 |
| — | — | — | — | — | — | — | — | — | — | — |
| 46 | 47 | 45 | 49 | 49 | 49 | 48 | 47 | 50 | 56 | 50 |
| 33 | 33 | 35 | 45 | 41 | 39 | 35 | 42 | 36 | 40 | 40 |
| 21 | 20 | 20 | 6 | 10 | 12 | 17 | 11 | 14 | 4 | 10 |
| 85 | 83 | 85 | 79 | 82 | 84 | 81 | 79 | 86 | 80 | 92 |

**82.** Block from tuff, Cape Krasnyi, Shiashkotan; coll. for author by G. E. Bogoyavlenskaya, sample No. 63015; anal. N. V. Voronkova. **83.** Andesitic ignimbrite, Sinarko Volcano; coll. for author by G. E. Bogoyavlenskaya sample No. 63034; anal. N. V.Voronkova. **84.** Andesite-basalt, crater rim of Sinarko Volcano; coll. for author by G. E. Bogoyavlenskaya, sample No. 63029; anal. N. V. Voronkova. **85.** Andesite, crater rim of Sinarko Volcano; coll. for author by G. E. Bogoyavlenskaya, sample No. 63028. **86.** Pumiceous andesite, most recent eruption of Sinarko Volcano; coll. for author by G. E. Bogoyavlenskaya, sample No. 63025. **87** . Andesite, dome of Sinarko Volcano; coll. for author by G. E. Bogoyavlenskaya, sample No. 63019; anal. I. M. Bender. **88.** Andesite, dome lava flow of Sinarko Volcano; coll. for author by G. E.Bogoyavlenskaya, sample No. 63070; anal. N. V. Voronkova. **89.** Andesite block from tuff, on flank of Kuntomintar Volcano; coll. for author by G. E. Bogoyavlenskaya, sample No. 63049; anal. I. M. Bender. **90.** Andesite, crater rim of Kuntomintar Volcano; coll. for author by G. E. Bogoyavlenskaya, sample No. 63052; anal. I. M. Bender. **91.** Andesitic tuff, eastern crater rim of Kuntomintar Volcano; coll. for author by G. E. Bogoyavlenskaya, sample No. 63053; anal. M. I. Bel'skaya. **92.** Pumiceous andesite, bomb from Kuntomintar Volcano; coll. author, sample No. 46189; anal. N. S. Klassova.

## TABLE 7. Lavas of the Volcanoes of the Western Subzone

| Components | Alaid Volcano | | | | | | Fuss Peak | | | Shirinki Volcano | |
|---|---|---|---|---|---|---|---|---|---|---|---|
| | 93 | 94 | 95 | 96 | 97 | 98 | 99 | 100 | 101 | 102 | 103 |
| $SiO_2$ | 50.56 | 50.74 | 50,83 | 50.27 | 50,29 | 50.35 | 54.27 | 56.21 | 58.88 | 57.34 | 57.36 |
| $TiO_2$ | 1.03 | 0.74 | 0,35 | 1.05 | 1.28 | 1.28 | 0.75 | 0.89 | 0.67 | 1,19 | 0,67 |
| $Al_2O_3$ | 17.73 | 20.41 | 21.48 | 19.24 | 18.96 | 19.29 | 18.94 | 17,44 | 17.66 | 15.41 | 16.92 |
| $Fe_2O_3$ | 3.45 | 2.96 | 3.70 | 3.88 | 3.44 | 4.07 | 4.13 | 4.06 | 2.91 | 5.17 | 3.55 |
| $FeO$ | 5.88 | 6.15 | 5.89 | 5.88 | 6.75 | 5.81 | 3.33 | 3.22 | 2.83 | 3.38 | 3.41 |
| $MnO$ | 0.20 | 0.17 | 0.18 | 0.18 | 0.33 | 0.45 | 0.17 | 0.21 | 0.23 | 0.11 | 0.23 |
| $MgO$ | 6.17 | 3.55 | 3.03 | 4.11 | 4.14 | 4.11 | 3.77 | 3.91 | 2.54 | 1.92 | 3.09 |
| $CaO$ | 9.79 | 10.25 | 10.64 | 10.13 | 10.25 | 10.22 | 8.46 | 7.86 | 7.18 | 9.32 | 6.90 |
| $Na_2O$ | 2.46 | 2.32 | 3.11 | 3.15 | 2.85 | 2.69 | 3.40 | 3.38 | 3.82 | 3.64 | 4.22 |
| $K_2O$ | 1.76 | 1.97 | 1.06 | 1.99 | 1.25 | 1.35 | 2.67 | 2.55 | 2.26 | 1.76 | 1.96 |
| $P_2O_5$ | 0.35 | 0.37 | 0.08 | — | 0.40 | 0.39 | — | 0.01 | 0.03 | — | 0.17 |
| $H_2O^+$ | | — | | 0.15 | 0.20 | 0.13 | 0.13 | 0.07 | 1.18 | 0.36 | 1.04 |
| $H_2O^-$ | 0.06 | 0.08 | — | 0.08 | 0.09 | 0.09 | 0.26 | 0.46 | 0.08 | 0.11 | 0.20 |
| $S$ | — | — | — | — | 0.02 | 0.01 | — | — | 0.03 | — | — |
| $SO_3$ | 0.03 | 0.02 | — | — | — | — | — | — | — | — | — |
| Sum | 99.54 | 99.73 | 100.35 | 100.11 | 100.25 | 100.24 | 100.28 | 100.27 | 100.30 | 99.71 | 99.72 |
| $a$ | 8.2 | 8.5 | 9.1 | 10.0 | 8.5 | 8.5 | 11.8 | 11.3 | 11.9 | 10.6 | 12.4 |
| $c$ | 8.0 | 10.4 | 10.9 | 8.3 | 9.1 | 9.3 | 7.1 | 6.1 | 6.3 | 5,1 | 5.4 |
| $b$ | 23.8 | 18.6 | 18.0 | 21.4 | 21.5 | 21.0 | 17.1 | 17.1 | 12.6 | 17,4 | 15.1 |
| $s$ | 60.0 | 62.5 | 62.0 | 60.3 | 60.9 | 61.2 | 64.0 | 65.5 | 69.2 | 66,9 | 67.1 |
| $a'$ | — | — | — | — | — | — | — | — | — | — | — |
| $f'$ | 37 | 49 | 53 | 44 | 47 | 47 | 41 | 40 | 43 | 45 | 43 |
| $m'$ | 45 | 35 | 20 | 34 | 34 | 35 | 38 | 40 | 35 | 19 | 36 |
| $c'$ | 18 | 16 | 17 | 22 | 19 | 18 | 21 | 20 | 22 | 36 | 21 |
| $n'$ | 68 | 64 | 81 | 71 | 78 | 75 | 66 | 67 | 72 | 70 | 76 |

**93.** Basalt, Alaid Volcano; coll. G. M. Vlasov; anal. K. G. Slovetskaya (DVGU; Popkova et al., 1963, No. 2554). **94.** Andesite-basalt, Alaid Volcano; coll. G. M. Vlasov, sample No. 2551; anal. K. G. Slovetskaya (DVGU; Popkova et al., 1961). **95.** Andesite-basalt, summit of somma of Alaid Volcano; coll. Sasa; anal. Kannari. **96.** Andesite-basalt, Taketomi Crater; coll. author, sample No. 46155; anal. I. I. Tovarova. **97.** Andesite-basalt, Taketomi Crater; coll. Kuno; anal. Japanese Geol. Survey (Kuno, 1935). **98.** Andesite-basalt, scoria from Taketomi Crater; coll. H. Kuno; anal. Japanese Geol. Survey (Kuno, 1935). **99.** Hornblende andesite bomb, Fuss Peak; coll. author, sample No. 53680; anal. V. P. Énman. **100.** Hornblende-bearing two-pyroxene andesite, Fuss Peak; coll. author, sample No. 53666; anal. V. P. Énman. **101.** Pumiceous pyroxene-hornblende andesite bomb, Fuss Peak; coll. author, sample No. 53685; anal. V. P. Énman. **102.** Hornblende andesite, young cone of Shirinki Volcano; coll. E. K. Markhinin, sample No. 607; anal. I. M. Bender. **103.** Pumiceous hornblende andesite, from submarine flank of Shirinki Volcano; coll. author, sample No. 51858; anal. V. P. Énman.

## TABLE 7 (Continued)

| Components | Makanru Volcano 104 | Ekarma Volcano 105 | 106 | 107 | Chirinkotan Volcano 108 | 109 | 110 | 111 | 112 | 113 | Brouton Island 114 |
|---|---|---|---|---|---|---|---|---|---|---|---|
| $SiO_2$ | 47.92 | 57.68 | 58.84 | 60.26 | 54.16 | 54.58 | 57.02 | 57.58 | 57.00 | 58.40 | 53.62 |
| $TiO_2$ | 1.59 | 1.40 | 0.61 | 0.76 | 1.65 | 0.92 | 1.02 | 1.14 | 1.70 | 0.43 | 1.17 |
| $Al_2O_3$ | 13.54 | 16.82 | 16.46 | 14.04 | 17.41 | 19.30 | 18.93 | 18.42 | 20.11 | 16.17 | 18.45 |
| $Fe_2O_3$ | 5.80 | 2.96 | 3.71 | 5.32 | 4.55 | 3.66 | 3.34 | 3.94 | 4.83 | 3.32 | 3.25 |
| $FeO$ | 5.07 | 4.02 | 3.97 | 3.88 | 3.39 | 3.88 | 3.16 | 2.50 | 1.62 | 3.11 | 4.43 |
| $MnO$ | 0.17 | 0.19 | 0.14 | 0.07 | 0.19 | 0.17 | 0.18 | 0.19 | 0.14 | 0.14 | 0.19 |
| $MgO$ | 5.41 | 3.60 | 3.55 | 3.68 | 4.16 | 4.45 | 3.98 | 3.95 | 3.07 | 3.01 | 5.47 |
| $CaO$ | 9.39 | 8.25 | 7.48 | 7.20 | 8.95 | 8.00 | 7.50 | 6.90 | 5.89 | 6.88 | 7.72 |
| $Na_2O$ | 2.84 | 3.78 | 3.80 | 3.22 | 3.24 | 3.19 | 3.26 | 3.69 | 3.40 | 3.66 | 2.94 |
| $K_2O$ | 0.91 | 1.35 | 1.09 | 1.60 | 1.58 | 1.74 | 1.82 | 1.90 | 1.95 | 1.75 | 1.69 |
| $P_2O_5$ | — | — | — | — | — | — | — | — | — | — | — |
| $H_2O^+$ | 4.05 | 0.27 | 0.58 | 0.37 | 1.31 | 0.01 | 0.16 | 0.33 | 0.87 | 2.71 | 0.64 |
| $H_2O^-$ | 1.04 | 0.05 | 0.20 | 0.09 | 0.10 | 0.30 | 0.06 | 0.08 | 0.20 | 0.40 | 0.40 |
| $S$ | — | — | — | — | — | — | — | — | — | 0.33 | — |
| $SO_3$ | 2.81 | — | — | — | — | — | — | — | — | — | — |
| | | | | | | | | | | | |
| Sum | 100.54 | 100.37 | 100.43 | 100.49 | 100.69 | 100.20 | 100.43 | 100.62 | 100.78 | 100.31 | 90.97 |
| | | | | | | | | | | | |
| $a$ | 8.1 | 10.3 | 10.0 | 9.3 | 9.7 | 10.0 | 10.0 | 11.1 | 10.6 | 10.9 | 9.3 |
| $c$ | 5.5 | 6.1 | 6.2 | 4.6 | 7.1 | 4.8 | 8.0 | 6.9 | 7.4 | 5.7 | 8.2 |
| $b$ | 26.9 | 16.6 | 16.2 | 18.2 | 18.8 | 20.2 | 14.6 | 14.4 | 13.6 | 14.4 | 18.4 |
| $s$ | 59.5 | 67.0 | 67.6 | 67.9 | 64.4 | 65.0 | 67.4 | 67.6 | 68.4 | 69.0 | 64.1 |
| $a'$ | — | — | — | — | — | — | — | — | 17 | — | — |
| $f'$ | 39 | 39 | 43 | 45 | 40 | 36 | 43 | 42 | 44 | 43 | 40 |
| $m'$ | 36 | 37 | 37 | 33 | 38 | 38 | 48 | 47 | 39 | 36 | 52 |
| $c'$ | 25 | 24 | 20 | 22 | 22 | 26 | 9 | 11 | — | 21 | 8 |
| $n'$ | 82 | 81 | 84 | 75 | 76 | 74 | 73 | 75 | 73 | 76 | 73 |

**104.** Basalt, Makanru Volcano; coll. E. K. Markhinin, sample No. 611; anal. I. M. Bender. **105.** Pyroxene andesite, foot of Ekarma Volcano; coll. E. K. Markhinin, sample No. 672; anal. I. M. Bender. **106.** Andesite bomb, Ekarma Volcano; coll. author, sample No. 46246; anal. V. P. Énman. **107.** Hornblende-pyroxene andesite, summit extrusion of Ekarma Volcano; coll. for author by G. E. Bogoyavlenskaya, sample No. 63057; anal. I. M. Bender. **108.** Andesite-basalt, east coast of Chirinkotan Volcano; coll. for author by G. E.Bogoyavlenskaya, sample No. 61425; anal. I. M. Bender. **109.** Andesite-basalt, southeast coast of Chirinkotan Volcano; coll. E. K. Markhinin, sample No. 681; anal. M. I. Bel'skaya. **110.** Hornblende andesite, east flank of Chirinkotan Volcano; coll. for author by G. E. Bogovlenskaya, sample No. 61427; anal. I. M. Bender. **111.** Hornblende andesite, east flank of Chirinkotan Volcano; coll. for author by G. E. Bogoyavlenskaya, sample No. 61428; anal. I. M. Bender. **112.** Hornblende andesite, summit of Chirinkotan Volcano; coll. for author by G. E.Bogoyavlenskaya, sample No. 63074. **113.** Pumiceous hornblende andesite, submarine flank of Chirinkotan Volcano; coll. author, sample No. 51890; anal. V. P. Énman. **114.** Andesite-basalt, Brouton Island; coll. E. K. Markhinin, sample No. 752; anal. M. I. Bel'skaya

## TABLE 8.  Lavas of the Volcanoes of the Central Kurile Islands

| Components | Raikoke Island | | Matua Island | | | | | | Rasshua Island | | Ushishir Island | | |
|---|---|---|---|---|---|---|---|---|---|---|---|---|---|
| | 115 | 116 | 116a | 116б | 117 | 118 | 119 | 120 | 121 | 122 | 123 | 124 | 125 |
| $SiO_2$ | 51.00 | 52.00 | 55.82 | 55.10 | 53.26 | 53.40 | 50.85 | 53.84 | 55.36 | 55.10 | 56.72 | 58.02 | 60.90 |
| $TiO_2$ | 0.97 | 1.17 | 1.00 | 1.00 | 1.15 | 1.09 | 1.07 | 0.96 | 0.91 | 1.00 | 1.14 | 1.14 | 0.72 |
| $Al_2O_3$ | 16.82 | 17.73 | 19.89 | 20.63 | 19.81 | 19.14 | 18.88 | 18.58 | 16.97 | 16.92 | 19.09 | 17.92 | 16.80 |
| $Fe_2O_3$ | 3.61 | 4.05 | 2.89 | 2.91 | 2.52 | 2.64 | 4.83 | 4.43 | 2.52 | 2.65 | 3.63 | 3.77 | 3.35 |
| FeO | 5.63 | 5.23 | 3.80 | 3.45 | 5.17 | 5.76 | 5.06 | 4.26 | 6.01 | 6.27 | 3.99 | 3.42 | 3.58 |
| MnO | 0.14 | 0.17 | 0.23 | 0.23 | 0.15 | 0.28 | 0.42 | 0.22 | 0.17 | 0.26 | 0.17 | 0.19 | 0.30 |
| MgO | 6.85 | 5.54 | 2.85 | 3.07 | 3.67 | 4.42 | 4.38 | 3.96 | 4.48 | 3.89 | 3.99 | 2.88 | 2.83 |
| CaO | 11.74 | 10.66 | 8.08 | 9.56 | 9.26 | 8.70 | 9.30 | 8.91 | 9.12 | 7.97 | 5.75 | 7.35 | 7.04 |
| $Na_2O$ | 2.40 | 2.38 | 3.54 | 2.45 | 3.14 | 3.22 | 2.88 | 3.24 | 2.40 | 2.97 | 3.72 | 3.52 | 3.58 |
| $K_2O$ | 0.79 | 1.10 | 1.03 | 0.66 | 1.03 | 1.08 | 0.99 | 1.06 | 0.55 | 0.82 | 0.69 | 0.60 | 0.64 |
| $P_2O_5$ | — | — | 0.28 | 0.26 | 0.10 | 0.21 | 0.10 | 0.08 | — | 1.16 | — | — | 0.19 |
| $H_2O^+$ | 0.19 | 0.10 | 0.09 | 0.12 | 0.11 | 0.11 | 0.27 | 0.21 | 1.31 | 1.11 | 0.59 | 0.44 | 0.12 |
| $H_2O^-$ | 0.05 | 0.08 | 0.19 | 0 10 | 0.16 | 0.06 | 0.18 | 0.06 | 0.54 | 0.43 | 0.82 | 0.25 | 0.13 |
| S | — | — | — | — | 0.31 | — | 1.02 | — | — | — | — | — | — |
| Ign. | — | — | — | — | — | — | — | — | — | — | 0.13 | 0.39 | — |
| Sum | 100.19 | 100.21 | 99.69 | 99.54 | 99.84 | 100.11 | 100.35 | 99.61 | 100.34 | 100.55 | 100.43 | 99.89 | 100.18 |
| $a$ | 6.5 | 7 0 | 9.9 | 8.8 | 9.0 | 9.0 | 8.3 | 9.1 | 6.3 | 8.0 | 9.4 | 8.8 | 8.9 |
| $c$ | 8 0 | 8.6 | 9.1 | 11.5 | 9.4 | 8.8 | 9.1 | 8.4 | 8.7 | 7 8 | 7.1 | 8.1 | 7.0 |
| $b$ | 26.5 | 23.0 | 13.0 | 13.1 | 16.7 | 18.3 | 20.5 | 18.2 | 19.0 | 18.0 | 16.7 | 13.3 | 13.3 |
| $s$ | 59.0 | 61.4 | 68.0 | 68.6 | 64.9 | 63.9 | 62.1 | 64.3 | 66.0 | 66.2 | 66.8 | 69.8 | 70.8 |
| $a'$ | — | — | — | — | — | — | — | — | — | — | — | — | — |
| $f'$ | 32 | 38 | 51 | 49 | 46 | 45 | 48 | 46 | 44 | 49 | 44 | 52 | 50 |
| $m'$ | 44 | 42 | 39 | 43 | 39 | 43 | 38 | 38 | 41 | 38 | 41 | 39 | 37 |
| $c'$ | 24 | 20 | 10 | 8 | 15 | 12 | 14 | 16 | 15 | 13 | 15 | 9 | 13 |
| $n'$ | 81 | 76 | 83 | 86 | 82 | 81 | 81 | 82 | 88 | 84 | 90 | 90 | 90 |

115. Basalt, Raikoke Island; coll. E. K. Markhinin, sample No. 688; anal. L. S. Mazalova. 116. Basalt, Raikoke Island; coll. for author by G. E. Bogoyavlenskaya, sample No. 64005. 116 a. Lava, somma of Matua Volcano; coll. author, sample No. 54730a; anal. T. V. Dolgova, 116b. Lava, somma of Matua Volcano; coll. author, sample No. 54730b; anal. T. V. Dolgova. 117. Andesite, Sarychev Peak (Matua); coll. author, sample No. 54726; anal. N. S. Klassova. 118. Andesite-basalt bomb, 1930 eruption of Matua; coll. author, sample No. 46332; anal. V. P. Énman. 119. Andesite-basalt ash, from incandescent avalanche of 1946 eruption; coll. author, sample No. 54724; anal. N. S. Klassova. 120. Andesite-basalt bomb, 1946 eruption; coll. author, sample No. 54728 (average of two analyses). 121. Andesite-basalt, caldera structure of Rasshua Volcano; coll. E. K. Markhinin, sample No. 692; anal. L. S. Mazalova. 122. Andesite-basalt bomb, central cone of Rasshua Volcano; coll. author, sample No. 46368; anal. V. P. Énman. 123. Pyroxene andesite, somma of Ushishir Volcano; coll. E. K. Markhinin, sample No. 741; anal. I. M. Bender. 124. Pyroxene andesite, Ryponkich Island, somma of Ushishir Volcano; coll. E. K. Markhinin, sample No. 756; anal. I. M. Bender. 125. Hornblende andesite, somma of Ushishir Volcano; coll. author, sample No. 59207; anal. V. P. Énman.

## TABLE 8 (Continued)

| Compo-nents | Ushishir Island | | Ketoi Island | | Simushir Island, Zavaritskii Caldera | | | | | | |
|---|---|---|---|---|---|---|---|---|---|---|---|
| | 126 | 127 | 128 | 129 | 130 | 131 | 132 | 133 | 134 | 135 | 136 |
| $SiO_2$ | 62.28 | 63.10 | 49.26 | 58.10 | 53.68 | 51.69 | 53.19 | 54.43 | 67.10 | 64.60 | 64.40 |
| $TiO_2$ | 0.76 | 0.67 | 0.86 | 1,07 | 1.00 | 1.03 | 0.49 | 0.89 | 0.72 | 1.67 | 0.96 |
| $Al_2O_3$ | 16.09 | 16.37 | 19.74 | 16.82 | 18.70 | 17.00 | 19.24 | 18.03 | 13.74 | 13.96 | 14.62 |
| $Fe_2O_3$ | 3.25 | 3.54 | 3.76 | 2.31 | 4.85 | 4.05 | 2.29 | 4.54 | 1.88 | 6.91 | 2.59 |
| $FeO$ | 3.56 | 2.15 | 5.92 | 5.83 | 5.27 | 8.27 | 6.80 | 5.63 | 3.68 | 0.68 | 4.56 |
| $MnO$ | 0.20 | 0.19 | 0.15 | 0.29 | 0.16 | 0.23 | 0.17 | 0.16 | 0.15 | 0.49 | 0.49 |
| $MgO$ | 2.42 | 2.31 | 6.10 | 3.71 | 3.50 | 4.88 | 4.92 | 2.56 | 1.20 | 1.74 | 1.88 |
| $CaO$ | 6.66 | 5.74 | 11.14 | 7.42 | 9.60 | 9.15 | 9.67 | 9.47 | 3.80 | 5.26 | 5.21 |
| $Na_2O$ | 3.72 | 3.50 | 1.41 | 2.44 | 2.64 | 2.66 | 2.94 | 2.83 | 4.00 | 4.16 | 4.20 |
| $K_2O$ | 0.61 | 0.83 | 0.38 | 1.15 | 0.52 | 0.44 | 0.40 | 0.60 | 0.92 | 0.63 | 0.75 |
| $P_2O_5$ | — | 0.25 | — | 0.01 | — | — | 1.06 | — | — | 0.10 | 0.10 |
| $H_2O^+$ | 0.90 | 0.93 | 0.82 | 0.23 | 0.14 | 0.13 | 0.10 | 0.23 | 2.49 | 0.15 | 0.14 |
| $H_2O^-$ | 0.10 | 0.96 | 0.33 | 0.15 | | 0.16 | 0.07 | 0.27 | | 0.14 | 0.05 |
| S | — | — | — | — | — | — | — | — | — | — | — |
| Ign. | — | — | — | — | — | — | — | — | — | — | — |
| Sum | 100.55 | 100.54 | 99.87 | 99.53 | 100.56 | 99.69 | 100.34 | 100.63 | 99.68 | 100.49 | 99.95 |
| $a$ | 9.1 | 9.2 | 3.9 | 7.3 | 6.9 | 6.6 | 7.2 | 7.5 | 10.3 | 9.9 | 10.3 |
| $c$ | 6.3 | 6.6 | 12.2 | 7.9 | 9.6 | 8.4 | 9.7 | 8.8 | 4.0 | 4.3 | 4.6 |
| $b$ | 12.4 | 10.0 | 23.0 | 15.8 | 18.5 | 23.3 | 19.9 | 17.6 | 8.6 | 11.6 | 11.6 |
| $s$ | 72.2 | 74.2 | 60.9 | 69.0 | 65.0 | 61.7 | 63.2 | 66.1 | 77.1 | 74.2 | 73.5 |
| $a'$ | — | — | — | — | — | — | — | — | — | — | — |
| $f'$ | 52 | 55 | 41 | 50 | 52 | 51 | 45 | 56 | 61 | 57 | 58 |
| $m'$ | 33 | 40 | 48 | 41 | 34 | 36 | 43 | 26 | 32 | 25 | 27 |
| $c'$ | 15 | 5 | 11 | 9 | 14 | 13 | 12 | 18 | 7 | 18 | 15 |
| $n'$ | 91 | 86 | 85 | 74 | 89 | 92 | 92 | 89 | 87 | 92 | 90 |

**126.** Hornblende andesite, somma of Ushishir Volcano; coll. E. K. Markhinin, sample No. 710; anal. L. S. Mazalova. **127.** Dacite, intracrater dome of Ushishir Volcano; coll. author, sample No. 59208; anal. V. P. Énman. **128.** Basalt, Ketoi; coll. E. K. Markhinin, sample No. 744; anal. M. I. Bel'skaya. **129.** Andesite, lava flow from Pallas Peak, Ketoi; coll. author, sample No. 59952; anal. V. P. Énman. **130.** Andesite-basalt, Zavaritskii Caldera, lava flow of the first somma; coll. author, sample No. 58867; anal. I. I. Tovarova. **131.** Basalt, dike in the first somma of Zavaritskii Caldera; coll. author, sample No. 58883; anal. V. P. Énman. **132.** Andesite-basalt, Zavaritskii Caldera; coll. author, sample No. 58858; anal. I. I. Tovarova. **133.** Pyroxene andesite, flow from the second somma of Zavaritskii Caldera; coll. author, sample No. 58880; anal. V. P. Énman. **134.** Dacite pumice, second somma of Zavaritskii Caldera; coll. author, sample No. 58870A; anal. I. I. Tovarova. **135.** Andesite ignimbrite, red ground mass, Zavaritskii Caldera; coll. author, sample No. 59900A; anal. V. P. Énman. **136.** Andesite ignimbrite, black glass, Zavaritskii Caldera; coll. author, sample No. 59900B; anal. V. P. Énman.

## TABLE 8 (Continued)

| Compo-nents | Simushir Island | | | | | | | | | |
| --- | --- | --- | --- | --- | --- | --- | --- | --- | --- | --- |
| | Zavaritskii Caldera | | | | | | | | Milne Volcano | |
| | 137 | 138 | 139 | 140 | 141 | 142 | 143 | 144 | 145 | 146 |
| SiO$_2$ | 53.98 | 58.99 | 57.86 | 66.10 | 61.67 | 58.78 | 58.50 | 56.51 | 53.30 | 53 34 |
| TiO$_2$ | 0.79 | 0.70 | 0.96 | 0.70 | 0.79 | 1.20 | 0.90 | 0.80 | 1 02 | 1.68 |
| Al$_2$O$_3$ | 17.62 | 16.79 | 17.31 | 13.85 | 16.51 | 16.93 | 17.08 | 18.48 | 18 71 | 18.64 |
| Fe$_2$O$_3$ | 3.93 | 3.38 | 2.78 | 2.99 | 4.48 | 3.36 | 2.88 | 3.01 | 2.20 | 2.20 |
| FeO | 6.00 | 5.88 | 5.95 | 3.55 | 3.39 | 5.06 | 5.31 | 5.41 | 5.37 | 5.71 |
| MnO | 0.28 | 0.18 | 0.19 | 0.18 | 0.23 | 0.21 | 0.18 | 0.17 | 0.17 | 0.16 |
| MgO | 4.13 | 2.83 | 3.34 | 1.50 | 2.16 | 2.95 | 2.99 | 3.50 | 4.66 | 4.71 |
| CaO | 9.44 | 7.00 | 7.32 | 4.29 | 5.93 | 7.12 | 7.46 | 8.02 | 9 75 | 9.34 |
| Na$_2$O | 2.27 | 3.86 | 3.85 | 4.36 | 3.17 | 3.32 | 3.28 | 2.90 | 2.90 | 3.16 |
| K$_2$O | 0.65 | 0.71 | 0.60 | 0.85 | 1.02 | 0.97 | 0.99 | 0.97 | 1.35 | 1.25 |
| P$_2$O$_5$ | — | — | — | — | — | — | — | — | — | — |
| H$_2$O$^+$ | 0.14 | 0.20 | 0.05 | 1.98 | 0.24 | 0.20 | 0.10 | 0.24 | 0.34 | 0.05 |
| H$_2$O$^-$ | 0.40 | 0.14 | 0.08 | 0.31 | 0.15 | 0.03 | 0.10 | 0.07 | 0.08 | 0.10 |
| S | — | — | — | — | — | — | — | — | — | — |
| Ign. | — | — | — | — | — | — | — | — | — | — |
| Sum | 99.63 | 100.66 | 100.29 | 100.66 | 99.74 | 100.13 | 99.77 | 100.08 | 99.85 | 100.34 |
| $a$ | 6.0 | 9.5 | 9.4 | 10.8 | 8.7 | 8.9 | 9.0 | 8.2 | 8.2 | 9.1 |
| $c$ | 9.4 | 6.6 | 7.1 | 3.8 | 7.1 | 7.1 | 7.3 | 8.7 | 7.3 | 7.5 |
| $b$ | 19.6 | 15.6 | 16.0 | 10.1 | 11.6 | 14.9 | 15.0 | 15.8 | 18.7 | 19.9 |
| $s$ | 65.0 | 68.3 | 67.5 | 75.3 | 72.6 | 69.1 | 68.7 | 67.3 | 65.8 | 63.5. |
| $a'$ | — | — | — | — | — | — | — | — | — | — |
| $f'$ | 49 | 56 | 52 | 61 | 64 | 54 | 53 | 52 | 37 | 39 |
| $m'$ | 37 | 31 | 36 | 25 | 32 | 34 | 34 | 39 | 41 | 41 |
| $c'$ | 14 | 13 | 12 | 14 | 4 | 12 | 14 | 9 | 22 | 20 |
| $n'$ | 86 | 90 | 91 | 88 | 82 | 83 | 83 | 81 | 76 | 78 |

137. Andesite-basalt, central cone of Zavaritskii Caldera; coll. author, sample No. 58877A; anal. V. P. Énman. 138. Andesite, central cone of Zavaritskii Caldera; coll. author, sample No. 58857; anal. I. I. Tovarova. 139. Andesite, central cone of Zavaritskii Caldera; coll. author, sample No. 58856; anal. I. I. Tovarova. 140. Dacite pumice, central cone of Zavaritskii Caldera; coll. author, sample No. 58864V; anal. V. P. Énman. 141. Acid andesite, eastern dome in Zavaritskii Caldera; coll. author, sample No. 58854; anal. V. P. Énman. 142. Andesite bomb, Zavaritskii Caldera, 1957 eruption; coll. author, sample No. 58856; anal. I. I. Tovarova. 143. Andesite, 1957 dome in Zavaritskii Caldera; coll. author, sample No. 58853; anal. I. I. Tovarova. 144. Andesite, Zavaritskii Caldera, dome flow of 1957; coll. author, sample No. 58854; anal. I. I. Tovarova. 145. Andesite-basalt, somma of Milne Volcano; coll. for author by G. E. Bogoyavlenskaya and K. I. Shmulovich, sample No. 61675; anal. I. M. Bender. 146. Andesite-basalt, somma of Milne Volcano; coll. for author by K. I. Shmulovich, sample No. 61672; anal. I. S. Mazalova.

TABLE 8 (Continued)

| Compo-nents | Milne Volcano | | | | | | Goryashchaya Sopka | | |
|---|---|---|---|---|---|---|---|---|---|
| | 147 | 148 | 149 | 150 | 151 | 152 | 153 | 154 | 155 |
| $SiO_2$ | 55.90 | 57.52 | 53.00 | 55.46 | 59.74 | 58.90 | 58.12 | 61.34 | 63.24 |
| $TiO_2$ | 1.02 | 1.02 | 1.25 | 1.27 | 1.06 | 0.92 | 0.92 | 1.27 | 1.46 |
| $Al_2O_3$ | 18.30 | 18.73 | 16.38 | 15.67 | 17.02 | 17.10 | 17.48 | 16.26 | 14.91 |
| $Fe_2O_3$ | 2.58 | 2.49 | 3.00 | 5.99 | 2.90 | 3.69 | 3.58 | 3.59 | 4.97 |
| $FeO$ | 4.25 | 4.11 | 5.92 | 4.02 | 3.79 | 2.67 | 4.10 | 3.59 | 2.01 |
| $MnO$ | 0.19 | 0.13 | 0.19 | 0.16 | 0.16 | 0.16 | 0.12 | 0.16 | 0.18 |
| $MgO$ | 4.52 | 3.60 | 5.52 | 3.64 | 3.39 | 3.15 | 2.79 | 3.20 | 3.22 |
| $CaO$ | 7.80 | 6.45 | 9.32 | 9.24 | 6.80 | 7.24 | 7.64 | 6.30 | 5.20 |
| $Na_2O$ | 3.34 | 3.43 | 2.75 | 3.06 | 3.48 | 3.61 | 2.82 | 3.34 | 3.24 |
| $K_2O$ | 1.44 | 1.61 | 0.72 | 1.15 | 1.40 | 1.53 | 1.50 | 1.66 | 1.53 |
| $P_2O_5$ | — | — | — | — | — | — | — | — | — |
| $H_2O^+$ | 0.81 | 0.71 | 0.12 | 0.50 | 0.07 | 0.15 | 0.06 | 0.15 | 0.37 |
| $H_2O^-$ | 0.21 | 0.21 | 1.08 | 0.42 | 0.69 | 0.81 | 0.19 | 0.07 | 0.28 |
| S | — | — | — | — | — | — | 0.17 | — | — |
| Ign. | — | — | — | — | — | — | — | — | — |
| Sum | 100.36 | 100.01 | 99.25 | 100.58 | 100.50 | 99.93 | 99.49 | 100.02 | 100.61 |
| $a$ | 9.7 | 10.3 | 7.2 | 8.5 | 9.8 | 10.3 | 8.8 | 9.8 | 9.3 |
| $c$ | 7.6 | 7.8 | 7.6 | 6.2 | 6.6 | 6.6 | 8.0 | 6.0 | 5.2 |
| $b$ | 16.6 | 13.1 | 22.2 | 20.5 | 13.8 | 13.8 | 14.2 | 13.0 | 12.9 |
| $s$ | 66.1 | 68.8 | 63.0 | 64.8 | 69.8 | 69.3 | 69.0 | 71.2 | 72.6 |
| $a'$ | — | — | — | — | — | — | — | — | — |
| $f'$ | 40 | 49 | 39 | 45 | 45 | 39 | 53 | 45 | 48 |
| $m'$ | 47 | 48 | 43 | 30 | 42 | 43 | 35 | 42 | 42 |
| $c'$ | 13 | 3 | 18 | 25 | 13 | 18 | 12 | 13 | 10 |
| $n'$ | 78 | 77 | 86 | 80 | 79 | 78 | 74 | 75 | 76 |

147. Andesite, somma of Milne Volcano; coll. for author by K. I. Shmulovich, sample No. 61678; anal. I. M. Bender. 148. Andesite, somma of Milne Volcano; coll. for author by K. I. Shmulovich, sample No. 61677; anal. I. M. Bender. 149. Andesite-basalt, central cone of Milne Volcano; coll. for author by G. E. Bogoyavlenskaya and K. I. Shmulovich, sample No. 61661; anal. M. I. Bel'skaya. 150. Andesite-basalt, Milne Volcano; coll. for author by K. I. Shmulovich, sample No. 62665; anal. I. M. Bender. 151. Andesite, central cone of Milne Volcano; coll. for author by G. E. Bogoyavlenskaya and K. I. Shmulovich, sample No. 61664; anal. M. I. Bel'skaya. 152. Andesite, summit dome of Milne Volcano; coll. for author by G.E . Bogoyavlenskaya and K. I. Shmulovich, sample No. 61660; anal. M. I. Bel'skaya. 153. Andesite, lava flow from Goryashchaya Sopka; coll. author, sample No. 58893; anal. I. I. Tovarova. 154. Andesite, small dome on Goryashchaya Sopka; coll. for author by K. I. Shmulovich, sample No. 62099; anal. I. M. Bender. 155. Andesite, main dome of Goryashchaya Sopka; coll. for author by G. E. Bogoyavlenskaya, sample No. 61680A; anal. I. M. Bender.

TABLE 9.  Lavas of the Volcanoes of the South Kurile Islands

| Components | Chirpoi | | | Brat Chirpoev | | Urup | | | |
|---|---|---|---|---|---|---|---|---|---|
| | 156 | 157 | 158 | 159 | 160 | 161 | 162 | 163 | 164 |
| $SiO_2$ | 67.34 | 60.00 | 59.14 | 50.04 | 50.06 | 57.86 | 61.68 | 55.52 | 60.43 |
| $TiO_2$ | 0.68 | 0.90 | 0.93 | 1.18 | 1.18 | 0.50 | 0.57 | 0.66 | 0.53 |
| $Al_2O_3$ | 14.78 | 18.43 | 17.66 | 18.03 | 16.95 | 17.92 | 17.61 | 17.29 | 17.94 |
| $Fe_2O_3$ | 0.84 | 0.38 | 1.79 | 4.39 | 5.76 | 3.32 | 3.75 | 4.56 | 4.82 |
| FeO | 2.73 | 6.18 | 5.58 | 6.06 | 5.21 | 4.17 | 3.30 | 4.60 | 2.59 |
| MnO | 0.19 | 0.19 | 0.19 | 0.22 | 0.26 | 0.09 | 0.19 | 0.23 | 0.14 |
| MgO | 0.82 | 3.23 | 2.91 | 6.07 | 6.39 | 3.86 | 2.95 | 4.53 | 2.34 |
| CaO | 3.00 | 5.86 | 6.90 | 10.80 | 10.84 | 7.98 | 6.99 | 9.00 | 5.97 |
| $Na_2O$ | 4.46 | 3.48 | 3.19 | 2.64 | 2.50 | 2.65 | 2.12 | 2.35 | 3.70 |
| $K_2O$ | 2.13 | 1.33 | 1.31 | 0.67 | 0.62 | 1.06 | 0.98 | 0.40 | 0.84 |
| $P_2O_5$ | 0.18 | — | — | 0.15 | 0.14 | — | — | — | — |
| $H_2O^+$ | 0.58 | 0.23 | 0.17 | 0.25 | 0.30 | 0.54 | 0.37 | 1.19 | 0.58 |
| $H_2O^-$ | 0.31 | 0.25 | 0.20 | 0.07 | 0.07 | 0.30 | 0.10 | | |
| S | 0.55 | — | — | — | — | — | — | — | — |
| Ign. | — | — | — | — | — | — | — | — | — |
| Sum | 99.71 | 100.46 | 99.97 | 100.57 | 100.28 | 100.25 | 100.61 | 100.33 | 99.88 |
| $a$ | 13.0 | 9.8 | 9.2 | 6.8 | 6.5 | 7.8 | 6.3 | 5.8 | 9.7 |
| $c$ | 3.5 | 7.2 | 7.5 | 9.1 | 8.3 | 8.5 | 8.8 | 9.0 | 7.7 |
| $b$ | 5.5 | 13.0 | 13.6 | 24.9 | 26.3 | 15.5 | 12.4 | 19.0 | 11.0 |
| $s$ | 78.0 | 70.0 | 69.7 | 59.2 | 58.9 | 68.2 | 72.5 | 66.2 | 71.6 |
| $a'$ | — | 6 | — | — | — | — | 4 | — | — |
| $f'$ | 68 | 51 | 55 | 40 | 39 | 46 | 55 | 46 | 63 |
| $m'$ | 26 | 43 | 37 | 42 | 42 | 44 | 41 | 42 | 36 |
| $c'$ | 6 | — | 8 | 18 | 19 | 10 | — | 12 | 1 |
| $n'$ | 76 | 79 | 79 | 88 | 87 | 78 | 75 | 90 | 87 |

156. Rhyolite pumice, Lapka peninsula, Chirpoi; coll. for author by G. E. Bogoyavlenskaya, sample No. 62211; anal. T. V. Dolgova. 157. Andesite, Chernyi Volcano, Chirpoi; coll. for author by G. E. Bogoyavlenskaya, sample No. 62098; anal. T. V. Dolgova. 158. Andesite, Snow Volcano, Chirpoi; coll. for author by G. E. Bogoyavlenskaya, sample No. 62070; anal. T. V. Dolgova. 159. Basalt, Brat Chirpoev; coll. author, sample No. 46377; anal. V. P. Énman. 160. Basaltic lapilli, Brat Chirpoev Volcano; coll. author, sample No. 46372; anal. V. P. Énman. 161. Andesite, somma of Trezubets Volcano, Urup (Nemoto, 1933). 162. Hornblende andesite, central cone of Trezubets Volcano, Urup (Nemoto, 1933). 163. Andesite-basalt, Rudakov Volcano, Urup (Nemoto, 1933). 164. Andesite bomb, Cape Neproidesh', Urup (Nemoto, 1933).

## TABLE 9 (Continued)

| Compo-nents | Iturup Island | | | | | | | | | |
|---|---|---|---|---|---|---|---|---|---|---|
| | Medvezh'ya Caldera | | | | | | Vetrovoi Isthmus | | | Mt. Golets |
| | 165 | 166 | 167 | 168 | 169 | 170 | 171 | 172 | 173 | 174 |
| $SiO_2$ | 58.60 | 64.20 | 66.40 | 67.30 | 51.02 | 55.13 | 65.99 | 66.91 | 67.81 | 66.26 |
| $TiO_2$ | 0.91 | 0.92 | 0.87 | 0.96 | 0.90 | 1.49 | 0.48 | 0.56 | 0.54 | 0.66 |
| $Al_2O_3$ | 15.61 | 14.58 | 14.17 | 14.90 | 18.29 | 15.71 | 14.61 | 14.66 | 14.30 | 16.69 |
| $Fe_2O_3$ | 0.10 | 3.06 | 1.38 | 1.84 | 0.89 | 3.01 | 2.13 | 2.61 | 1.60 | 2.18 |
| $FeO$ | 4.14 | 2.18 | 2.88 | 2.84 | 9.34 | 7.12 | 2.25 | 1.98 | 2.34 | 3.62 |
| $MnO$ | 0.23 | 0.21 | 0.21 | 0.19 | 0.23 | 0.26 | 0.08 | 0.11 | 0.12 | 0.17 |
| $MgO$ | 1.29 | 1.18 | 1.27 | 1.56 | 6.73 | 4.87 | 1.33 | 1.26 | 1.31 | 1.20 |
| $CaO$ | 3.64 | 3.50 | 3.54 | 4.26 | 10.06 | 8.62 | 5.30 | 4.38 | 4.49 | 4.87 |
| $Na_2O$ | 2.99 | 3.90 | 4.20 | 4.39 | 2.26 | 2.17 | 3.01 | 2.96 | 3.43 | 3.52 |
| $K_2O$ | 0.80 | 1.42 | 1.08 | 1.10 | 0.35 | 0.68 | 1.04 | 1.46 | 1.33 | 0.37 |
| $P_2O_5$ | — | 0.09 | 0.09 | 0.07 | — | — | 0.10 | 0.10 | 0.40 | 0.22 |
| $H_2O^+$ | 4.66 | 3.94 | 3.03 | 0.34 | 0.06 | 0.31 | — | — | — | — |
| $H_2O^-$ | 2.36 | 0.81 | 0.35 | 0.16 | 0.02 | 0.13 | 0.91 | — | — | — |
| $S$ | 0.04 | 0.10 | 0.05 | — | — | — | — | — | — | — |
| Ign. | 4.17 | — | — | — | — | — | 2.12 | 2.79 | 2.58 | — |
| Sum | 99.54 | 100.09 | 99.52 | 99.92 | 100.15 | 99.50 | 99.35 | 99.78 | 100.25 | 99.76 |
| $a$ | 8.6 | 11.1 | 11.2 | 11.2 | 5.5 | 5.8 | 8.3 | 9.0 | 9.7 | 8.2 |
| $c$ | 4.9 | 4.5 | 4.0 | 4.4 | 9.7 | 7.8 | 6.0 | 5.6 | 4.9 | 6.0 |
| $b$ | 11.7 | 7.5 | 6.7 | 7.8 | 24.8 | 21.1 | 7.2 | 6.6 | 6.7 | 9.8 |
| $s$ | 74.8 | 76.9 | 78.1 | 76.6 | 60.0 | 65.3 | 78.5 | 78.8 | 78.7 | 76.0 |
| $a'$ | 39.7 | 5 | — | — | — | — | — | — | — | 24 |
| $f'$ | 39.7 | 68 | 63 | 56 | 41 | 47 | 61 | 65 | 57 | 55 |
| $m'$ | 20.6 | 27 | 32 | 33 | 47 | 40 | 29 | 33 | 33 | 21 |
| $c'$ | — | — | 5 | 11 | 12 | 13 | 10 | 2 | 10 | — |
| $n'$ | 84.3 | 81 | 85 | 86 | 91 | 83 | 81 | 75 | 79 | 93 |

**165.**Andesitic pumice, Medvezh'ya Caldera, Iturup; sample No. 59512A; anal. V. P. Énman.**166.** Dacite pumice, Medvezh'ya Caldera, Iturup; coll. for author by I. I. Gushchenko, sample No. 59516; anal. V. P. Énman. **167.** Dacitic pumice, Medvezh'ya Caldera, Iturup; coll. for author by I. I. Gushchenko, sample No. 59517; anal. V. P. Énman. **168.** Dacite pumice, Medvezh'ya Caldera, Iturup; coll. for author by I. I. Gushchenko, sample No. 59513; anal. V. P. Énman. **169.** Basaltic flow, Men'shoi Brat Volcano, Iturup; coll. for author by I. I. Gushchenko, sample No. 59528; anal. T. V. Dolgova. **170.** Andesite-basalt, Kudryavyi Cone, Iturup; coll. for author by I. I. Gushchenko, sample No. 59547; anal. V. P. Énman. **171.** Dacitic pumice, Vetrovoi Isthmus, Iturup; coll. O. M. Bent; anal. SakhKNII Laboratory (Bent, 1962). **172.** Dacitic pumice, Vetrovoi Isthmus, Iturup; coll. N. P. Savrasov, sample No. 2577; anal. Laboratory of DVGU (Popkova et al., 1961). **173.** Dacitic pumice, Vetrovoi Isthmus, Iturup; coll. Yu. S. Zhelubovskii. **174.** Andesite-dacite, Mount Golets area, Iturup; coll. N. P. Savrasov, sample No. 2380; anal. Laboratory of DVGU (Popkova et al., 1961).

## TABLE 9  (Continued)

| Compo-nents | Teben'-kov | Ivan Groznyi Volcano | | | | | | | Bogdan Khmel'-nitskii Cone | |
|---|---|---|---|---|---|---|---|---|---|---|
| | 175 | 176 | 177 | 178 | 179 | 180 | 181 | 182 | 183 | 184 |
| $SiO_2$ | 54.71 | 57.28 | 60.46 | 56.16 | 58.70 | 59.15 | 60.68 | 58.34 | 59.16 | 55.19 |
| $TiO_2$ | 0.74 | 0.70 | 0.74 | 0.88 | 1.11 | 1.34 | 1.11 | 0.87 | 1.25 | 1.12 |
| $Al_2O_3$ | 18.01 | 17.23 | 16.19 | 17.65 | 16.14 | 15.31 | 15.30 | 15.95 | 17.02 | 18.38 |
| $Fe_2O_3$ | 2.95 | 3.60 | 2.81 | 1.93 | 3.48 | 3.30 | 4.19 | 2.90 | 3.54 | 4.59 |
| FeO | 5.74 | 4.88 | 5.24 | 4.61 | 4.45 | 5.36 | 4.18 | 5.34 | 3.86 | 3.78 |
| MnO | 0.17 | 0.69 | 0.28 | 0.21 | 0.13 | 0.24 | 0.21 | 0.14 | 0.23 | 0.29 |
| MgO | 4.52 | 4.47 | 3.47 | 4.87 | 3.93 | 3.93 | 3.53 | 4.03 | 3.07 | 3.74 |
| CaO | 9.10 | 7.58 | 6.46 | 9.40 | 7.71 | 7.67 | 7.09 | 7.66 | 6.64 | 6.44 |
| $Na_2O$ | 3.21 | 2.00 | 2.69 | 2.76 | 2.98 | 2.60 | 2.63 | 2.69 | 3.17 | 2.52 |
| $K_2O$ | 1.10 | 1 24 | 1 29 | 0.82 | 1.11 | 1.03 | 1.29 | 1.24 | 1.85 | 1.43 |
| $P_2O_5$ | — | 0.02 | 0.02 | 0.17 | 0.06 | 0.05 | 0.05 | 0.16 | 0.08 | 0.11 |
| $H_2O^+$ | 0.10 | 0.60 | 0.16 | 0.45 | 0.28 | 0.17 | 0.16 | 0.31 | 0.08 | 1.63 |
| $H_2O^-$ | 0.11 | 0.18 | 0.06 | 0.07 | 0.12 | 0.07 | 0.06 | 0.20 | 0.09 | 0.54 |
| S | — | — | — | — | 0.10 | — | — | — | 0.02 | 0.06 |
| Ign. | — | — | — | — | — | — | — | — | — | — |
| Sum | 100.46 | 100.47 | 99.87 | 99.98 | 100.30 | 100.25 | 100.48 | 99.83 | 100.19 | 99.82 |
| $a$ | 8.8 | 6.4 | 8.1 | 7.5 | 8.2 | 7.3 | 7.7 | 8.0 | 9.9 | 7.9 |
| $c$ | 7.9 | 8.7 | 7.0 | 8.0 | 6.7 | 6.6 | 6.3 | 6.9 | 6.7 | 8.2 |
| $b$ | 19.4 | 17.3 | 14.6 | 18.1 | 16.6 | 17.7 | 16.0 | 17.3 | 13.8 | 16.4 |
| $s$ | 63.9 | 67.6 | 70.3 | 66.0 | 68.5 | 68.4 | 70.0 | 67.8 | 69.6 | 67.5 |
| $a'$ | — | — | — | — | — | — | — | — | — | 10 |
| $f'$ | 43 | 50 | 52 | 35 | 44 | 46 | 48 | 45 | 50 | 50 |
| $m'$ | 40 | 45 | 41 | 46 | 40 | 38 | 37 | 40 | 39 | 40 |
| $c'$ | 17 | 5 | 7 | 19 | 16 | 16 | 15 | 15 | 11 | — |
| $n'$ | 81 | 70 | 76 | 83 | 80 | 78 | 75 | 77 | 72 | 73 |

**175.** Andesite-basalt, summit of Teben'kov Volcano; coll. author, sample No. 54710; anal. N. S. Klassova. **176.** Pyroxene andesite, effusive dome of Ivan Groznyi Volcano, Iturup; coll. for author by G. E. Bogoyavlenskaya, sample No. 59200; anal. V. P. Énman. **177.** Andesite, Ermak Dome, Ivan Groznyi Volcano, Iturup; coll. for author by G. E. Bogoyavlenskaya, sample No. 59223; anal. V. P. Énman. **178.** Andesite-basalt, flow near Cape Drakon from Ivan Groznyi Volcano, Iturup; coll. author, sample No. 54720; anal. T. V. Dolgova. **179.** Andesite, southern lava flow from Ivan Groznyi Volcano, Iturup; coll. for author by G. E. Bogoyavlenskaya, sample No. 59220; anal. V. P. Énman. **180.** Andesite, southeastern lava flow, Ivan Groznyi Volcano, Iturup; coll. for author by G. E. Bogoyavlenskaya, sample No. 59226; anal. V. P. Énman. **181.** Andesite, southwestern lava flow, Ivan Groznyi Volcano, Iturup; coll. G.E. Bogoyavlenskaya, sample No. 59224A; anal. V. P. Énman. **182.** Andesite, summit dome of Ivan Groznyi Volcano, Iturup; coll. author, sample No. 54715; anal. N. S. Klassova. **183.** Andesite, Bogdan Khmel'nitskii Cone, Chirip massif, Iturup; coll. for author by I. I. Gushchenko, sample No. 59479; anal. V. P. Énman. **184.** Andesitic scoria, Bogdan Khmel'nitskii Cone, Chirip massif, Iturup; coll. for author by I. I. Gushchenko, sample No. 59480; anal. V. P. Énman.

## TABLE 9 (Continued)

| Compo-nents | Iturup Island | | | | Kunashir Island | | | | | |
|---|---|---|---|---|---|---|---|---|---|---|
| | Bogdan Khmel'-nitskii Cone | Stokap | Atsonu-puri | Yuzhnyi Isthmus | Tyatya Volcano | | | | Rurui | Mende-leev Volcano |
| | 185 | 186 | 187 | 188 | 189 | 190 | 191 | 192 | 193 | 194 |
| $SiO_2$ | 56.99 | 52.60 | 50.45 | 65.60 | 52.25 | 56.07 | 51.75 | 50.84 | 58.06 | 60.04 |
| $TiO_2$ | 0.99 | 0.68 | 1.64 | 0.42 | 1.11 | 0.96 | 1.20 | 0.94 | 1.07 | 0.91 |
| $Al_2O_3$ | 17.30 | 18.38 | 17.99 | 16.12 | 16.42 | 17.32 | 17.20 | 17.81 | 16.04 | 15.76 |
| $Fe_2O_3$ | 4.05 | 5.20 | 3.08 | 2.84 | 3.70 | 2.64 | 3.95 | 3.40 | 3.23 | 7.14 |
| FeO | 3.59 | 5.04 | 8.87 | 2.12 | 9.04 | 7.18 | 7.58 | 7.86 | 5.95 | 1.31 |
| MnO | 0.21 | — | 0.17 | 0.15 | 0.18 | 0.16 | 0.22 | 0.15 | 0.16 | 0.23 |
| MgO | 3.96 | 4.74 | 4.02 | 1.65 | 4.67 | 3.74 | 4.02 | 4.89 | 3.41 | 2.35 |
| CaO | 7.45 | 9.52 | 10.59 | 5.25 | 8.34 | 8.09 | 10.32 | 10.21 | 6.27 | 6.50 |
| $Na_2O$ | 3.01 | 2.50 | 2.97 | 3.49 | 3.27 | 2.60 | 2.12 | 2.48 | 2.01 | 3.30 |
| $K_2O$ | 1.81 | 0.39 | 0.58 | 1.85 | 1.24 | 0.66 | 0.96 | 1.07 | 1.30 | 0.83 |
| $P_2O_5$ | 0.03 | — | 0.16 | 0.22 | 0.19 | 0.03 | 0.50 | 0.18 | 0.13 | 0.15 |
| $H_2O^+$ | 0.13 | 0.71 | 0.07 | | 0.25 | 0.37 | — | 0.23 | | 0.40 |
| $H_2O^-$ | 0.05 | 0.23 | 0.08 | 0.46 | 0.07 | — | 0.51 | 0.02 | 1.47 | 0.38 |
| S | — | — | — | — | — | 0.03 | — | — | — | — |
| Ign. | — | — | — | 0.39 | — | | — | — | — | — |
| Sum | 99.57 | 99.99 | 100.67 | 100.56 | 100.65 | 99.85 | 100.33 | 100.08 | 99.10 | 99.30 |
| $a$ | 9.4 | 6.3 | 7.6 | 10.4 | 9.1 | 6.9 | 6.3 | 7.3 | 6.5 | 8.7 |
| $c$ | 7.2 | 9.7 | 8.7 | 5.6 | 6.5 | 8.6 | 8.8 | 8.7 | 8.0 | 6.4 |
| $b$ | 16.2 | 20.6 | 23.0 | 8.2 | 23.6 | 17.7 | 22.5 | 23.5 | 15.3 | 13.4 |
| $s$ | 67.2 | 63.4 | 60.7 | 75.8 | 60.8 | 66.8 | 62.4 | 60.5 | 70.2 | 71.5 |
| $a'$ | — | — | — | — | — | — | — | — | — | — |
| $f'$ | 45 | 47 | 50 | 55 | 50 | 54 | 50 | 46 | 59 | 57 |
| $m'$ | 42 | 41 | 30 | 35 | 34 | 37 | 31 | 36 | 40 | 30 |
| $c'$ | 13 | 12 | 20 | 10 | 13 | 9 | 19 | 18 | 1 | 13 |
| $n'$ | 72 | 91 | 89 | 74 | 80 | 87 | 76 | 77 | 70 | 85 |

**185.** Andesite, Bogdan Khmel'nitskii Cone, Iturup; coll. for author by I. I. Gushchenko, sample No. 59481; anal. V. P. Énman. **186.** Basalt, summit of Stokap Volcano, Iturup; coll. for author by I. I. Gushchenko, and G. E. Bogoyavlenskaya, sample No. 60255; anal. T. V. Dolgova. **187.** Basalt, Atsonupuri Volcano, Iturup (Katsui, 1961). **188.** Dacitic pumice, Yuzhnyi Isthmus, Iturup; coll. O. M. Bent; anal. Laboratory of SakhKNII (Bent, 1962). **189.** Basalt, foot of Tyatya Volcano, Kunashir; coll. E. K. Markhinin, sample No. 251; anal. V. G. Sil'nichenko (Markhinin, 1959). **190.** Pyroxene andesite, rim of somma of Tyatya Volcano; coll. E. K. Markhinin, sample No. 171; anal. V. G. Sil'nichenko. **191.** Basaltic bomb, central cone of Tyatya Volcano, Kunashir Island; coll. Yu. S. Zhelubovskii; anal. Tikhonenko. **192.** Basalt, central cone of Tyatya Volcano, Kunashir; coll. E. K. Marhinin, sample No. 243; anal. V. G. Sil'nichenko (Markhinin, 1959). **193.** Andesite, Rurui Volcano, Kunashir (Gumennyi and Neverov, 1961). **194.** Andesite, base of first somma of Mendeleev Volcano, Kunashir; coll. author, sample No. 52650; anal. I. I. Tovarova.

## TABLE 9  (Continued)

| Compo- nents | Kunashir Island | | | | | | | | |
|---|---|---|---|---|---|---|---|---|---|
| | Mendeleev Volcano | | | | Golovnin Volcano | | | | |
| | 195 | 196 | 197 | 198 | 199 | 200 | 201 | 202 | 203 |
| $SiO_2$ | 67.18 | 52.27 | 58.43 | 65.30 | 57.16 | 58.60 | 59.08 | 66.32 | 69.58 |
| $TiO_2$ | 0.73 | 0.80 | 0 77 | 1.22 | 0.58 | 0.62 | 0.62 | 0.52 | 0.47 |
| $Al_2O_3$ | 14.18 | 16.91 | 15 80 | 15.84 | 16.44 | 16.88 | 17.57 | 16.73 | 13.73 |
| $Fe_2O_3$ | 1.63 | 3.84 | 6.92 | 3.08 | 7.15 | 4.60 | 3.14 | 2.25 | 1.80 |
| $FeO$ | 2.16 | 6.58 | 2.00 | 2.40 | 2.60 | 4.00 | 2.72 | 2.28 | 1.68 |
| $MnO$ | 0.10 | 0.21 | 0.18 | 0.14 | 0.13 | 0.15 | 0.12 | 0.11 | 0.07 |
| $MgO$ | 1.21 | 4.94 | 3 44 | 1.70 | 2.86 | 3.40 | 1.86 | 1.24 | 0.75 |
| $CaO$ | 4.03 | 10.24 | 7.43 | 5.12 | 8.66 | 7.00 | 6.50 | 5.29 | 3.77 |
| $Na_2O$ | 3.28 | 2.10 | 2.90 | 3.70 | 2.56 | 2.14 | 2.95 | 3.40 | 3.50 |
| $K_2O$ | 1.71 | 0.69 | 0.55 | 1.65 | 0.51 | 0.46 | 0.55 | 0.97 | 0.91 |
| $P_2O_5$ | — | 0.06 | 0.13 | 0.09 | 0.10 | 0.06 | 0.27 | 0.02 | 0.03 |
| $H_2O^+$ | 2.41 | 0.82 | 0.92 | 0.13 | 0.72 | 0.98 | 1.13 | 0.65 | 0.98 |
| $H_2O^-$ | 0.79 | 0.35 | 0.04 | 0 11 | 0.42 | 0.72 | | | |
| $S$ | — | 0.02 | — | — | — | — | — | — | — |
| Ign. | — | — | — | — | — | — | 3.16 | 0.31 | 3.07 |
| Sum | 99.41 | 99.83 | 99.51 | 100.48 | 99.89 | 99.61 | 99.67 | 100.09 | 100.34 |
| $a$ | 11.3 | 5.7 | 7.4 | 10.6 | 6 6 | 5.6 | 7.9 | 9.2 | 9.2 |
| $c$ | 5.4 | 8.9 | 7.2 | 5.4 | 8.2 | 9.0 | 8.8 | 6.6 | 4 7 |
| $b$ | 6.8 | 22.7 | 16 4 | 8.0 | 16.7 | 14.7 | 9.8 | 6.8 | 4.6 |
| $s$ | 76.5 | 62.7 | 69.0 | 76.0 | 68.5 | 70.7 | 73.5 | 77.4 | 81.5 |
| $a'$ | — | — | — | — | — | 2 | 6 | 6 | 3 |
| $f'$ | 60 | 44 | 50 | 52 | 53 | 57 | 58 | 63 | 69 |
| $m'$ | 35 | 38 | 36 | 36 | 30 | 41 | 36 | 31 | 28 |
| $c'$ | 5 | 18 | 14 | 11 | 17 | — | — | — | — |
| $n'$ | 75 | 80 | 90 | 78 | 82 | 88 | 90 | 83 | 85 |

**195.** Dacitic pumice, southeast base of Mendeleev Volcano, Kunashir; coll. author, sample No. 52644; anal. I. I. Tovarova. **196.** Basalt, head of Shkol'naya Creek, Mendeleev Volcano, Kunashir; coll. E. K. Markhinin, sample No. 305; anal. V. P. Énman. **197.** Andesite, inner cone of Mendeleev Volcano; Kunashir; coll. author, sample No. 52652; anal. I. I. Tovarova. **198.** Dacite, extrusion dome on Mendeleev Volcano, Kunashir; coll. author, sample No. 52651A; anal. I. I. Tovarova. **199.** Andesitic tuff, Golovnin Volcano, Kunashir; coll. E. K. Markhinin, sample No. 313; anal. V. P. Énman (Markhinin, 1959). **200.** Andesite, Golovnin Volcano, Kunashir; coll. author, sample No. 52638; anal. I. I. Tovarova. **201.** Dacitic pumice, Lagunnoe Lake, Kunashir (Bent, 1962). **202.** Dacitic pumice, Sernobodsk, Kunashir (Bent, 1962). **203.** Liparitic pumice, Golovinino, Kunashir (Bent, 1962).

## TABLE 9 (Continued)

| Compo-<br>nents | Kunashir Island | | | |
|---|---|---|---|---|
| | Golovnin Volcano | | | |
| | 204 | 205 | 206 | 207 |
| $SiO_2$ | 56.56 | 62.37 | 64.35 | 64.68 |
| $TiO_2$ | 0.77 | 0.53 | 0.71 | 0.50 |
| $Al_2O_3$ | 17.79 | 14.41 | 15.73 | 16.00 |
| $Fe_2O_3$ | 4.58 | 7.94 | 2.83 | 4.24 |
| FeO | 4.26 | 0.68 | 3.56 | 2.22 |
| MnO | 0.19 | 0.08 | 0.16 | 0.08 |
| MgO | 3.86 | 1.68 | 1.87 | 1.54 |
| CaO | 8.10 | 5.64 | 5.56 | 6.12 |
| $Na_2O$ | 2.72 | 2.79 | 2.43 | 2.26 |
| $K_2O$ | 0.43 | 0.80 | 0.79 | 0.49 |
| $P_2O_5$ | 0.06 | 0.06 | — | — |
| $H_2O^+$ | 0.63 | 1.22 | 0.43 | 1.33 |
| $H_2O^-$ | 0.29 | 1.59 | 1.37 | 0.26 |
| S | 0.01 | 0.01 | — | 0.04 |
| Ign. | — | — | — | — |
| Sum | 100.25 | 99.80 | 99.79 | 99.87 |
| $a$ | 6.8 | 7.7 | 6.7 | 5.8 |
| $c$ | 9.0 | 6.1 | 7.0 | 7.8 |
| $b$ | 16.5 | 11.7 | 10.4 | 9.6 |
| $s$ | 67.7 | 74.5 | 75.9 | 76.8 |
| $a'$ | — | — | 10 | 11 |
| $f'$ | 52 | 65 | 59 | 62 |
| $m'$ | 41 | 25 | 31 | 27 |
| $c'$ | 7 | 10 | — | — |
| $n'$ | 92 | 83 | 82 | 88 |

**204.** Andesite, Vneshnii Dome, Golovnin Volcano, Kunashir; coll. E. K. Markhinin, sample No. 326; anal. V. P. Énman (Markhinin, 1959). **205.** Andesite, Podushechnyi Dome, Golovnin Volcano, Kunashir; coll. E. K. Markhinin, sample No. 308; anal. V. P. Énman (Markhinin, 1959). **206.** Andesite-dacite, eastern intracaldera dome, Golovnin Volcano, Kunashir; coll. author, sample No. 52634; anal. I. I. Tovarova. **207.** Andesite-dacite, central dome of Golovnin Volcano, Kunashir; coll. E. K. Markhinin, sample No. 174; anal. V. G. Sil'nichenko.

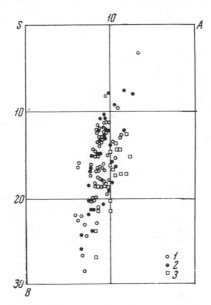

Fig. 77. Petrochemical diagram for the lavas of the volcanoes of the northern Kurile Islands: 1) Lavas of Paramushir; 2) lavas of the other islands of the northern group; 3) lavas of the volcanoes of the western zone.

of the northern group. From this figure and from Table 7 it can be seen that the points for the volcanoes of the northern and western groups in part overlap one another, but on the whole the rocks of the western zone prove to be significantly more alkalic. The points for the analyses fall between the Lassen Peak and Yellowstone curves.

Thus, the conclusion that we came to earlier regarding an increase in the alkalinity in the volcanoes of the western zone is also confirmed by additional information.

*Central Kurile Islands*

There are 43 analyses for the volcanoes of the Central Kurile Islands; all were made at the Institute of Volcanology. The analytical data are presented in Table 8 and are represented in Fig. 78. The points on Fig. 78 representing analyses from the largest island, Simushir, and from the other, smaller islands practically co-

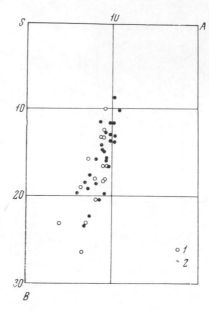

Fig. 78. Petrochemical diagram for
the lavas of the volcanoes of the Central Kurile Islands:  1) lavas of Simushir; 2) lavas of the other islands.

incide.  As in the North Kurile Islands, all the points fall into a band
between the variation curves for the Pelée and Lassen Peak types.

*South Kurile Islands*

There are 52 analyses available to characterize the South
Kurile Islands; of these 37 were made in the Institute of Volcanology,
six (mainly of pumice) in the Sakhalin Complex Institute, and nine
were taken from literature sources.  The analyses are presented
in Table 9, and the results of recalculation are indicated in Fig. 79.
The analysis points for the different islands of the group are indicated by different symbols.  Moreover, the volcanoes that form
peninsulas on the west coast of Iturup (Chirip Peninsula, Atsonupuri Volcano, and L'vinaya Past' Caldera) are indicated separately.

The points plotted show a rather broad scatter on both sides
of the Pelée-type variation curve and in general are somewhat displaced toward the *SB* axis relative to the more northern islands.
The volcanoes of the islands of Chernye Brat'ya and of the western

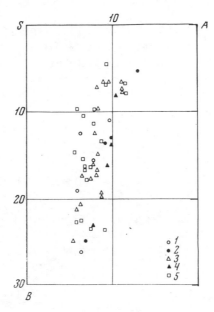

Fig. 79. Petrochemical diagram for the lavas of the volcanoes of the South Kurile Islands: 1) lavas of the islands of Chernye Brat'ya; 2) lavas of Urup; 3) lavas of Iturup; 4) lavas of the western peninsulas of Iturup; 5) lavas of Kunashir.

peninsulas of Iturup occupy a position farthest to the right, i.e., possess a somewhat higher than usual alkalinity. In this we see the same tendency of the alkalinity to increase toward the inner side of the arc that we noted earlier, although in this case the shift is not so great that these volcanoes can be assigned to the western zone.

## General Observations

It follows from the summary just presented that the characteristic petrochemical features of the Kurile lavas change somewhat along the arc (to which É. N. Érlikh called attention) and across its trend. In order to see the character of these changes in more detail, it is convenient to examine the corresponding variation curves, and in order to avoid subjectivity in the construction of these curves, we will construct them on the basis of average values of analyses.

TABLE 10. Average Chemical
Compositions of the Lavas
of the North Kurile Islands

| Compo-nents | 1 | 2 | 3 | 4 | 5 | 6 |
|---|---|---|---|---|---|---|
| $SiO_2$ | 48.9 | 53.9 | 56.5 | 59.7 | 65.9 | 72.5 |
| $TiO_2$ | 0.9 | 0.9 | 0.8 | 0.7 | 0.7 | 0.4 |
| $Al_2O_3$ | 19.1 | 18.0 | 17.9 | 18.0 | 15.8 | 14.8 |
| $Fe_2O_3$ | 4.5 | 3.6 | 4.2 | 3.0 | 2.6 | 0.7 |
| FeO | 6.1 | 5.9 | 4.3 | 4.0 | 2.8 | 1.3 |
| MnO | 0.1 | 0.2 | 0.2 | 0.1 | 0.2 | 0.1 |
| MgO | 5.9 | 4.5 | 3.7 | 2.9 | 1.8 | 0.6 |
| CaO | 11.3 | 9.0 | 8.2 | 7.1 | 4.8 | 2.7 |
| $Na_2O$ | 2.4 | 2.8 | 2.9 | 3.0 | 4.0 | 4.8 |
| $K_2O$ | 0.8 | 1.2 | 1.3 | 1.5 | 1.4 | 2.1 |
| $a$ | 6.8 | 8.1 | 8.5 | 9.0 | 10.9 | 13.4 |
| $c$ | 9.8 | 8.3 | 8.0 | 8.0 | 5.1 | 2.8 |
| $b$ | 24.9 | 20.0 | 16.6 | 12.8 | 8.8 | 3.4 |
| $s$ | 58.5 | 63.6 | 66.9 | 70.2 | 75.2 | 80.4 |
| $f'$ | 40 | 45 | 48 | 53 | 57 | 55 |
| $m'$ | 42 | 40 | 39 | 39 | 35 | 31 |
| $c'$ | 18 | 15 | 13 | 8 | 8 | 14 |
| $n'$ | 81 | 78 | 77 | 75 | 81 | 78 |

1. Basalt, average of 11 analyses; 2. An-
desite-basalt, average of 14 analyses; 3.
Basic andesite, average of 25 analyses; 4.
Acid andesite, average of 34 analyses; 5.
Dacite, average of 5 analyses; 6   Rhyolite,

We must note immediately that average values taken simply
from the sums of the individual analyses (without considering their
relative abundances) can be used for the construction of variation
curves only because the character of these curves does not depend
on the quantitative relationships of the different rock types.  However, it would be a mistake to consider the values arrived at as real
average values for the rock types, as is done in some works, and
on this basis to draw conclusions as to the composition of the parent
magma, as to the quantitative changes in acidity or alkalinity, and
so on.  In all these cases the number of analyses has a random character and is not related to the relative volumes occupied (or even
areas unerlain) by the different rock types.

TABLE 11. Average Chemical
Compositions of the Lavas
of the Western Zone of the
Kurile Islands

| Components | 1 | 2 | 3 | 4 | 5 |
|---|---|---|---|---|---|
| $SiO_2$ | 51.4 | 52.1 | 58.1 | 56.0 | 59.0 |
| $TiO_2$ | 1.4 | 1.0 | 1.0 | 1.0 | 0.7 |
| $Al_2O_3$ | 16.3 | 19.5 | 17.9 | 17.3 | 17.5 |
| $Fe_2O_3$ | 4.8 | 3.7 | 3.7 | 4.5 | 3.3 |
| FeO | 5.7 | 5.3 | 3.1 | 3.3 | 3.1 |
| MnO | 0.2 | 0.2 | 0.2 | 0.2 | 0.2 |
| MgO | 6.0 | 4.1 | 3.5 | 3.2 | 2.8 |
| CaO | 10.0 | 9.5 | 7.2 | 8.6 | 7.2 |
| $Na_2O$ | 2.8 | 3.0 | 3.6 | 3.5 | 4.0 |
| $K_2O$ | 1.4 | 1.6 | 1.7 | 2.4 | 2.2 |
| | | | | | |
| $a$ | 8.3 | 9.3 | 10.6 | 11.2 | 12.2 |
| $c$ | 6.8 | 9.0 | 7.0 | 6.1 | 5.8 |
| $b$ | 25.3 | 19.2 | 14.4 | 17.1 | 13.9 |
| $s$ | 59.6 | 68.5 | 68.0 | 65.6 | 68.1 |
| $f'$ | 38 | 46 | 44 | 42 | 44 |
| $m'$ | 40 | 38 | 42 | 32 | 34 |
| $c'$ | 22 | 16 | 14 | 26 | 22 |
| $n'$ | 75 | 74 | 76 | 69 | 74 |

1. Basalt, average of 2 analyses; 2. Andesite-basalt, average of 8 analyses; 3. Andesite, average of 6 analyses (excluding Shirinki and Fuss Peak); 4. Basic andesite, Fuss Peak, average of 3 analyses; 5. Acid andesite, Shirinki, average of 2 analyses.

At present the apportionment of analyses according to islands and individual volcanoes is still not sufficient for any quantitative calculations for the arc as a whole. Only Alaid, which is a small island, and Kunashir, among the large ones, can be characterized quantitatively.

The average compositions of the analyzed lavas for the individual portions of the arc are presented in Tables 10-13, and the corresponding variation curves are plotted in Fig. 80; on this diagram the volcanoes of Shirinki and Fuss Peak in the western zone,

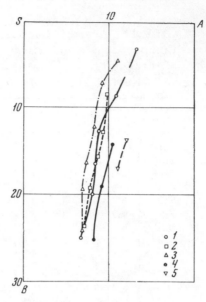

Fig. 80. Variation curves for the average values of the analyzed lavas: 1)
Points for the northern islands; 2) points
for the central islands; 3) points for the
southern islands; 4) points for the western zone; 5) Shirinki and Fuss Peak.

which have greater alkalinities than the western zone as a whole,
are plotted separately.

It is clear from the tables and from the figure in question that
the rocks from the North and Central Kurile Islands actually belong
to one and the same type, in spite of the marked difference in crustal
type beneath them — continental in the first case, suboceanic (actually, oceanic) in the second. The variation curves for the rocks of
these two portions of the arc fall between the Pelée and Lassen
Peak lines.

The variation curve for the lavas of the South Kurile Islands
runs somewhat to the left of these and coincides with the Pelée-
type curve.

The volcanoes of the western zone have more alkalic lavas,
and most of them are close to the Lassen Peak type; the lavas of
Shirinki and Fuss Peak have a still more alkalic character and are
similar to the Yellowstone type.

TABLE 12. Average Chemical
Compositions of the Lavas
of the Central Kurile Islands

| Components | 1 | 2 | 3 | 4 | 5 |
|---|---|---|---|---|---|
| $SiO_2$ | 51.6 | 54.2 | 56.9 | 61.0 | 66.4 |
| $TiO_2$ | 1.1 | 1.0 | 1.0 | 1.1 | 0.7 |
| $Al_2O_3$ | 17.9 | 18.2 | 18.0 | 16.6 | 15.6 |
| $Fe_2O_3$ | 3.6 | 3.5 | 3.1 | 3.8 | 2.8 |
| FeO | 5.8 | 5.5 | 5.3 | 3.4 | 3.2 |
| MnO | 0.2 | 0.2 | 0.2 | 0.2 | 0.2 |
| MgO | 6.1 | 4.3 | 3.5 | 2.8 | 1.7 |
| CaO | 10.3 | 9.3 | 7.8 | 6.5 | 4.6 |
| $Na_2O$ | 2.6 | 2.9 | 3.3 | 3.5 | 4.0 |
| $K_2O$ | 0.8 | 0.9 | 0.9 | 1.1 | 0.8 |
| $a$ | 7.1 | 8.0 | 8.8 | 9.4 | 10.0 |
| $c$ | 8.7 | 8.5 | 8.0 | 6.5 | 6.0 |
| $b$ | 23.7 | 19.4 | 15.9 | 13.0 | 8.6 |
| $s$ | 60.5 | 64.1 | 67.3 | 71.1 | 75.4 |
| $f'$ | 38 | 45 | 51 | 52 | 65 |
| $m'$ | 42 | 39 | 38 | 37 | 33 |
| $c'$ | 20 | 16 | 11 | 11 | 2 |
| $n'$ | 82 | 82 | 84 | 82 | 88 |

1. Basalt, average of 4 analyses; 2. Andesite-basalt, average of 11 analyses; 3. Basic andesite, average of 12 analyses; 4. Acid andesite, average of 13 analyses; 5. Dacite, average of 3 analyses.

These variation curves for the lavas of the Kurile volcanoes slope somewhat more steeply than the variation curves for Zavaritskii's lava types, so that we can speak only tentatively of the coincidence of a given Kurile curve and curves of the Pelée or other types.

Usually the appearance of calc-alkalic, "orogenic" lavas is considered to be the result of the contamination of alkalic olivine basalts by sialic material. If this is true, in the Kurile Island we should anticipate the development of a distinct longitudinal zonation corresponding to the differences in crustal structure along the arc. However, the longitudinal zonation that is shown by the chemical

TABLE 13.  Average Chemical
Compositions of the Lavas
of the South Kurile Islands

| Compo-nents | 1 | 2 | 3 | 4 | 5 | 6 |
|---|---|---|---|---|---|---|
| $SiO_2$ | 51.1 | 55.1 | 58.1 | 61.1 | 67.5 | 71.0 |
| $TiO_2$ | 1.1 | 0.9 | 0.9 | 0.8 | 0.7 | 0.6 |
| $Al_2O_3$ | 17.5 | 17.7 | 17.0 | 17.1 | 15.7 | 14.8 |
| $Fe_2O_3$ | 3.7 | 3.6 | 4.2 | 4.0 | 2.5 | 1.3 |
| $FeO$ | 7.6 | 5.4 | 4.5 | 3.7 | 2.5 | 2.2 |
| $MnO$ | 0.2 | 0.2 | 0.2 | 0.2 | 0.1 | 0.1 |
| $MgO$ | 5.2 | 4.7 | 3.8 | 2.7 | 1.4 | 0.8 |
| $CaO$ | 10.2 | 9.1 | 7.6 | 6.3 | 5.0 | 3.5 |
| $Na_2O$ | 2.6 | 2.6 | 2.6 | 3.0 | 3.5 | 4.1 |
| $K_2O$ | 0.8 | 0.7 | 1.1 | 1.1 | 1.1 | 1.6 |
| $a$ | 7.1 | 7.0 | 7.5 | 8.4 | 9.3 | 11.1 |
| $c$ | 8.5 | 8.7 | 7.9 | 7.6 | 5.9 | 4.2 |
| $b$ | 24.0 | 19.5 | 16.4 | 12.3 | 7.2 | 4.6 |
| $s$ | 60.4 | 64.8 | 68.2 | 71.7 | 77.6 | 80.1 |
| $f'$ | 45 | 45 | 50 | 59 | 64 | 70 |
| $m'$ | 37 | 42 | 40 | 39 | 33 | 30 |
| $c'$ | 18 | 13 | 10 | 2 | 3 | 0 |
| $n'$ | 82 | 86 | 78 | 80 | 82 | 80 |

1. Basalt, average of 8 analyses; 2. Andesite-basalt, average of 5 analyses; 3.
Basic andesite, average of 13 analyses;
4. Acid andesite, average of 12 analyses;
5. Dacite, average of 14 analyses; 6.
Liparite, average of 2 analyses.

analyses is very weak; the change is, at most, of only "half a type"
and, what is much more important, this zonation does not correlate
at all with the change in structure of the crust.  Moreover, the volcanoes on the most strongly contrasting types of crust, continental
in the North Kurile Islands and suboceanic (actually, oceanic) in the
Central Kurile Islands, actually belong to one and the same petrochemical type, whereas the volcanoes of the main and western zones
which are separated from one another by 10-20 km and lie on the
same type of crust (continental) erupt lavas that are rather distinct
in their petrochemical relations.  These facts, based on indisputable

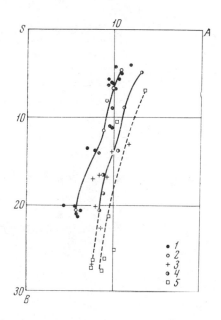

Fig. 81. 1) Analysis points for Akan-Shiretoko chain of volcanoes; 2) average values for lavas of the Akan-Shiretoko chain; 3) analysis points for the lavas of the Daisetsu-Tokachi chain of volcanoes; 4) average values for lavas of the Daisetsu-Tokachi chain; 5) analysis points for lavas of Rishiri Volcano.

geophysical and geochemical data, force us to question whether contamination and the assimilation of crustal rocks play any significant role in the volcanic processes and to look for a source of volcanism below the crust, in the upper part of the mantle.

However, in order to check on this contemplated conclusion, we must examine the characteristic petrochemical features of the volcanic rocks in other volcanic arcs as well as in the Pacific Ocean area and in the adjacent parts of the continents. The next chapter will be devoted to this examination.

But before turning to the other volcanic regions, let us examine the chemical characteristics of the lavas from the very southern end of the Kurile arc.

## TABLE 14. Numerical Characteristics of the Lavas of the Kurile–Zone Volcanoes on Hokkaido

| Rock composition and locality | Numerical characteristics | | | | | | | |
|---|---|---|---|---|---|---|---|---|
| | $a$ | $c$ | $b$ | $s$ | $f'$ | $m'$ | $c'$ | $n'$ |
| Akan–Shiretoko chain | | | | | | | | |
| Two-pyroxene andesite, Shiretoko Volcano | 10.2 | 6.4 | 12.2 | 71.2 | 59 | 34 | 7 | 77 |
| Two-pyroxene andesite, Rausu | 9.8 | 6.4 | 11.2 | 72.6 | 60 | 32 | 8 | 78 |
| Andesite-basalt, Kutcharo | 4.3 | 13.4 | 20.0 | 62.3 | 51 | 33 | 16 | 86 |
| " | 5.8 | 8.2 | 21.2 | 64.8 | 58 | 27 | 15 | 90 |
| Andesite, Kutcharo | 9.7 | 6.8 | 11.1 | 72.4 | 61 | 33 | 6 | 77 |
| Rhyolitic ignimbrite, Kutcharo | 11.6 | 3.7 | 4.0 | 80.7 | 61 | 37 | 2 | 76 |
| Rhyolitic pumice, Kutcharo | 10.2 | 3.4 | 4.2 | 82.2 | 69 | 29 | 2 | 77 |
| Andesite-basalt, Masshu | 5.6 | 10.6 | 20.1 | 63.7 | 48 | 34 | 18 | 92 |
| " | 6.2 | 7.7 | 20.6 | 65.5 | 56 | 27 | 17 | 90 |
| Andesite, Masshu | 6.8 | 8.5 | 13.7 | 71.0 | 60 | 29 | 11 | 90 |
| Dacitic pumice, Masshu | 9.2 | 5.5 | 8.1 | 77.2 | 67 | 35 | 8 | 86 |
| Rhyolite, Masshu | 9.5 | 3.5 | 5.8 | 81.2 | 63 | 29 | 8 | 83 |
| Andesite, Kamuishu | 8.3 | 8.3 | 14.0 | 69.4 | 54 | 34 | 12 | 81 |
| Andesite, Atsonupuri | 9.4 | 5.3 | 11.3 | 74.0 | 60 | 27 | 13 | 87 |
| Dacitic pumice, Nakajima | 10.1 | 5.0 | 6.8 | 78.1 | 62 | 35 | 3 | 80 |
| Dacite, Nakajima | 10.0 | 5.2 | 6.3 | 78.5 | 64 | 33 | 3 | 81 |
| " | 9.9 | 4.3 | 6.1 | 79.7 | 66 | 33 | 1 | 79 |
| Dacitic pumice, Atsonupuri | 10.6 | 3.6 | 5.6 | 80.2 | 64 | 27 | 9 | 79 |
| Rhyolite, Atsonupuri | 10.2 | 2.9 | 4.9 | 81.3 | 62 | 37 | 1 | 79 |
| Andesite-basalt, Akan | 5.8 | 9.6 | 20.8 | 63.8 | 50 | 33 | 17 | 87 |
| Dacitic ignimbrite, Akan | 9.9 | 4.8 | 9.0 | 76.3 | 62 | 27 | 11 | 78 |
| Andesite, Meakan | 7.9 | 8.9 | 13.8 | 69.4 | 54 | 41 | 5 | 78 |

## TABLE 14 (Continued)

| Rock composition and locality | Numerical characteristics | | | | | | | |
|---|---|---|---|---|---|---|---|---|
| | $a$ | $c$ | $b$ | $s$ | $t'$ | $m'$ | $c'$ | $n'$ |
| Daisetsu-Tokachi chain | | | | | | | | |
| Andesite-basalt, Daisetsu | 8.7 | 7.8 | 18.5 | 65.0 | 44 | 44 | 12 | 75 |
| Andesite, Daisetsu | 9.2 | 6.8 | 16.7 | 67.3 | 46 | 41 | 13 | 73 |
| " " | 8.4 | 7.5 | 16.5 | 67.6 | 45 | 38 | 17 | 74 |
| " " | 9.9 | 5.8 | 13.9 | 70.4 | 46 | 38 | 16 | 75 |
| Dacite, Daisetsu | 11.3 | 5.6 | 8.9 | 74.2 | 54 | 40 | 6 | 74 |
| Andesite-basalt, Chubetsu | 7.9 | 7.1 | 20.0 | 65.0 | 44 | 37 | 19 | 69 |
| Basalt, Tokachi | 7.6 | 8.8 | 26.8 | 56.8 | 45 | 38 | 17 | 73 |
| Andesite-basalt, Tokachi | 8.5 | 7.9 | 22.6 | 61.0 | 45 | 38 | 17 | 79 |
| Andesite, Tokachi | 7.7 | 10.3 | 17.0 | 65.0 | 53 | 42 | 5 | 72 |
| " " | 11.7 | 4.5 | 13.1 | 70.7 | 51 | 33 | 16 | 70 |
| Rhyolitic ignimbrite, Tokachi | 13.2 | 1.7 | 4.8 | 80.3 | 61 | 21 | 18 | 64 |
| Rishiri Island | | | | | | | | |
| Basalt | 7.7 | 8.2 | 26.7 | 57.4 | 34 | 50 | 16 | 89 |
| " | 10.1 | 6.5 | 25.2 | 58.2 | 38 | 36 | 26 | 86 |
| " | 9.1 | 6.5 | 26.2 | 58.2 | 32 | 47 | 21 | 84 |
| " | 8.9 | 5.8 | 27.6 | 57.7 | 36 | 43 | 21 | 90 |
| Andesite-basalt | 9.6 | 7.6 | 21.2 | 61.6 | 38 | 40 | 22 | 88 |
| Andesite | 10.4 | 6.2 | 10.7 | 72.7 | 47 | 40 | 13 | 85 |
| Dacite | 13.5 | 4.0 | 7.0 | 75.5 | 61 | 36 | 3 | 82 |

TABLE 15. Average Chemical
Compositions for Lavas
of the Akan–Shiretoko
Chain (Hokkaido)

| Compo-nents | 1 | 2 | 3 | 4 |
|---|---|---|---|---|
| $SiO_2$ | 53.1 | 61.0 | 69.4 | 73.4 |
| $TiO_2$ | 1.1 | 0.8 | 0.5 | 0.5 |
| $Al_2O_3$ | 18.0 | 17.0 | 14.9 | 13.3 |
| $Fe_2O_3$ | 3.4 | 2.8 | 2.3 | 1.5 |
| FeO | 7.6 | 4.8 | 2.3 | 1.8 |
| MnO | 0.2 | 0.1 | 0.1 | 0.1 |
| MgO | 3.6 | 2.4 | 1.3 | 0.8 |
| CaO | 10.4 | 6.8 | 4.5 | 3.0 |
| $Na_2O$ | 2.2 | 3.2 | 3.7 | 4.0 |
| $K_2O$ | 0.4 | 1.1 | 1.3 | 1 6 |
| $a$ | 5.6 | 8.9 | 10.0 | 10.9 |
| $c$ | 9.8 | 7.2 | 4.9 | 3 2 |
| $b$ | 20.5 | 12.5 | 6.8 | 4.6 |
| $s$ | 64.1 | 71.4 | 78.3 | 81.3 |
| $f'$ | 53 | 57 | 60 | 63 |
| $m'$ | 31 | 33 | 32 | 28 |
| $c'$ | 16 | 10 | 8 | 9 |
| $n'$ | 90 | 81 | 81 | 79 |

1. Andesite-basalt, average of 5 analyses;
2. Andesite, average of 7 analyses; 3.
Dacite, average of 6 analyses; 4. Rhyolite,
average of 4 analyses.

   The Kurile chain of volcanoes extends farther south than Kunashir, passing onto the island of Hokkaido. Here the volcanoes form
two volcanic zones: the Akan–Shiretoko zone, which immediately
adjoins Kunashir, and the Daisetsu–Tokachi zone. In the back part
of the arc, closer to the Sea of Japan, lies Rishiri Island, which
also belongs to the Kurile arc (Katsui, 1961).

   Katsui (1961) has recently published a summary of the analyses
of these volcanoes. We will not present the actual analyses, but will
give only the results of their recalculation (Table 14) and the average values separately for each zone (Tables 15–16). These data are
represented graphically in Fig. 81. The rocks of the Akan–Shiretoko
chain are similar in their petrochemical relations to the rocks of

TABLE 16.  Average Chemical
Compositions for Lavas
of the Daisetsu-Tokachi
Chain (Hokkaido)

| Compo-nents | 1 | 2 | 3 | 4 | 5 | 6 |
|---|---|---|---|---|---|---|
| $SiO_2$ | 47.6 | 54.3 | 57.9 | 61.6 | 65.1 | 72.7 |
| $TiO_2$ | 1.6 | 1.0 | 0.7 | 0.7 | 0.6 | 0.3 |
| $Al_2O_3$ | 18.2 | 17.3 | 17.1 | 15.8 | 16.3 | 13.4 |
| $Fe_2O_3$ | 5.0 | 2.7 | 3.5 | 3.8 | 2.6 | 2.6 |
| FeO | 7.6 | 6.8 | 4.5 | 3.4 | 2.6 | 0.8 |
| MnO | 0.2 | 0.2 | 0.1 | 0.1 | 0.1 | 0.1 |
| MgO | 5.8 | 4.7 | 3.8 | 2.9 | 2.1 | 0.6 |
| CaO | 10.7 | 8.8 | 7.9 | 6.1 | 4.8 | 2 2 |
| $Na_2O$ | 2.7 | 2.9 | 2.9 | 3.3 | 3.8 | 3.9 |
| $K_2O$ | 0.8 | 1.3 | 1.6 | 2.3 | 2.0 | 3 4 |
| $a$ | 7.6 | 8.4 | 8.8 | 10.6 | 11.3 | 13.2 |
| $c$ | 8.8 | 7.5 | 7.2 | 5.4 | 5.6 | 1.7 |
| $b$ | 26.8 | 20.6 | 16.5 | 13.6 | 8.9 | 4.8 |
| $s$ | 56.8 | 63.5 | 67.5 | 70.4 | 74.2 | 80.3 |
| $f'$ | 45 | 44 | 45 | 48 | 54 | 61 |
| $m'$ | 38 | 40 | 39 | 36 | 40 | 21 |
| $c'$ | 17 | 16 | 16 | 16 | 6 | 18 |
| $n'$ | 83 | 77 | 74 | 69 | 74 | 64 |

1. Basalt, 1 analysis; 2. Andesite-basalt, average of 3 analyses; 3. Basic andesite, average of 3 analyses; 4. Acid andesite, average of 2 analyses; 5. Dacite, 1 analysis; 6. Rhyolite, 1 analysis.

the South Kurile Islands, and their variation curve is close to that of the Pelée type.  The rocks of the Daisetsu-Tokachi chain have a much more alkalic character and are similar to the Lassen Peak type; they are more closely similar to the western zone of the Kurile Islands.  The rocks of the island of Rishiri have a still more alkalic character.  The Akan-Shiretoko and Daisetsu-Tokachi chains, which have such different petrochemical characteristics, rest on the same type of crust.  If we judge from the gravimetric data (Tsuboi, 1954) and from the character of the dispersion of the surface waves (Santo, 1963), the crust on Hokkaido belongs to the continental type.

In view of the Hokkaido volcanoes, the independence of the rock chemistry from the crustal type is even more clearly expressed; the volcanoes of the extreme northern and extreme southern parts of the arc are situated on crust of the same, continental type, but they erupt somewhat different lavas; on the other hand, the difference in character of the crust in the North and Central Kuriles, or in the South Kuriles and on Hokkaido exerts no influence on the chemical character of the lavas.

The differences in alkalinity of the lavas on Hokkaido are determined only by their positions; the alkalinity increases in moving from the outer to the inner part of the arc.

*Chapter VII*

# Characteristic Petrochemical Features
# of the Lavas of the Other Island Arcs
# and of Intracontinental
# and Oceanic Volcanoes

## ISLAND AND VOLCANIC ARCS

A system of island arcs envelops the whole continent of Asia on its eastern side and extends southeastward as far as New Zealand (Fig. 82).

The Kurile-Kamchatka arc is joined to the Japanese North Honshu arc in Hokkaido. In central Honshu the chain of arcs, which has been single up to this point, bifurcates; one system of arcs continues to follow along the southeast edge of the continent (Ryukyu and Philippine arcs), the other goes off due south into the ocean (Izu-Bonin, Marianas, Yap, and Palau arcs), separating the inner, deep-water Philippine Sea from the Pacific Ocean proper. In the area of western New Guinea the two arc systems rejoin.

The arcs of islands located to the north and northeast of Australia have an extremely interesting characteristic: the abyssal trenches of New Britain, Bougainville (Solomon Islands), and the New Hebrides are located on the continental rather than on the oceanic side of the corresponding arcs; as a result, the focal surfaces of the deep earthquakes dip toward the ocean rather than toward the continent, as is the usual case.

The northernmost island arc, the Aleutian, unites the volcanoes of the Asiatic continent with the North American continent. Along the Pacific coast of both Americas there are no island arcs in the usual sense of the word, but their analogs are chains of volcanoes, volcanic arcs, that border the whole eastern shore of the Pacific Ocean. The volcanoes of South America are joined to the Antarctic

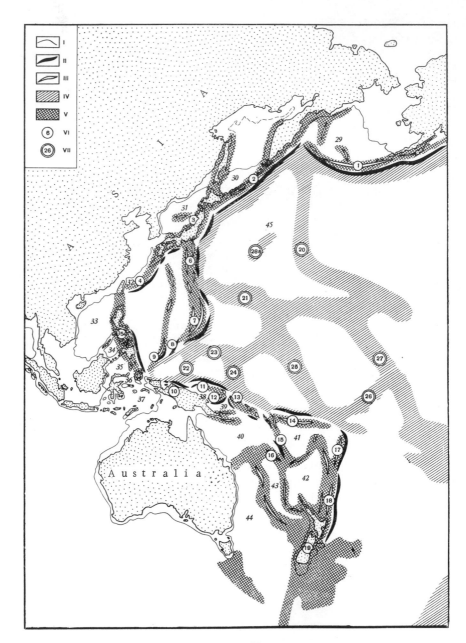

Fig. 82

Fig. 82. Sketch of the topography of the floor of the western Pacific Ocean: I) Edge of the continental shelf; II) abyssal oceanic trenches; III) axes of submarine mountain ridges of the transition zone; IV) oceanic rises on which are situated mountain ridges or groups of submarine volcanoes; V) island arcs; VI) island arcs and their associated trenches (1-19): 1) Aleutian; 2) Kurile-Kamchatka; 3) Japan; 4) Nansei; 5) Philippine; 6) Izu-Bonin; 7) Marianas; 8) Yap; 9) Palau; 10) New Guinea; 11) West Melanesian; 12) New Britain; 13) Solomon Islands; 14) East Melanesian; 15) New Hebrides; 16) New Caledonia; 17) Tonga; 18) Kermadec; 19) New Zealand; VII) ocean rises (20-28): 20) Hawaiian; 21) Marcus-Wake; 22) Eauripik; 23) Caroline; 24) Kapingamirangi; 25) Marshall Islands; 26) Tokelau and Cook Islands; 27) Line Islands; 28a) northwestern Pacific mountains; Numbers without circles, basins in the transition zone (29-45): 29) Bering Sea; 30) Sea of Okhotsk; 31) Sea of Japan; 32) East China Sea; 33) South China Sea; 34) Sulu Sea; 35) Celebes Sea; 36) Molucca Sea; 37) Banda Sea; 38) Bismarck Sea; 39) Solomon Sea; 40) Coral Sea; 41) north basin of Fiji Sea; 42) south basin of Fiji Sea; 43) New Caledonian basin; 44) Tasman Sea; 45) Philippine oceanic basin (Udintsev, 1960).

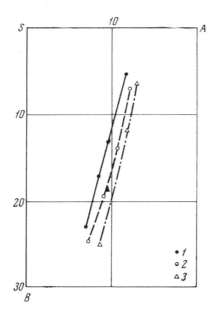

Fig. 83. Diagram of the average compositions of the Kamchatka lavas: 1) East Kamchatka; 2) Central Kamchatka Depression; 3) Sredinnyi Ridge.

islands and to Graham Land by the arc of the South Sandwich Islands. Terror and Erebus Volcanoes in the Ross Sea can be thought
of as a continuation of this unbroken volcanic belt which, through the
submarine New Zealand Sill, is locked into a complete ring at the New
Zealand volcanoes. The whole Pacific is embraced within a belt of
present-day volcanoes, which form the celebrated "Pacific Ring of
Fire."

A second zone of island and volcanic arcs extends in an almost east-west direction from Indonesia through Burma, Iran, and
the Caucasus to the Mediteranean Sea.

The vast majority of present-day and Quaternary subaerial
volcanoes are confined to the island and volcanic arcs.

## 1. Kamchatka

The volcanic zone of Kamchatka is the direct northward extension of the Kurile volcanic zone. Judging from the data of deep
seismic sounding in the adjacent water areas of the Pacific Ocean
and the Sea of Okhotsk and according to the gravimetric data on the
peninsula itself, the crust in Kamchatka belongs to the continental
type. The junction of continental and oceanic crust occurs along
the northern part of the abyssal Kurile-Kamchatka Trench, which
in the region of Cape Kamchatskii intersects the Aleutian Trench.
The northenmost modern volcano (Shiveluch) occurs approximately
at the point of intersection.

Seismologic work by the author (Gorshkov, 1956) established
the existence of a deep-seated magmatic hearth beneath the Klyuchevskaya Volcano group. Its depth was found to be on the order of
60-80 km, which corresponds to the upper mantle. Later a similar
result was obtained by other investigators in the Avachinskii Volcano group (Fedotov and Farberov, 1962). The work of Fedotov
(1964) established that the subcrustal part of the mantle in the Kamchatka region, as in the Kurile Islands, is characterized by a lower
than usual seismic velocity. Estimates of the conductive heat flow
made by Polyak indicate approximately double the value of the average for the earth as a whole.

The Kamchatka volcanoes can be divided into three zones in
passing from the east coast into the interior. These zones are
superposed on older structures.

## TABLE 17.  Average Compositions of the Quaternary and Recent Lavas of Kamchatka

| Compo-nents | 1 | 2 | 3 | 4 | 5 | 6 | 7 | 8 | 9 | 10 | 11 | 12 |
|---|---|---|---|---|---|---|---|---|---|---|---|---|
| $SiO_2$ | 51.5 | 56.0 | 59.6 | 65.9 | 52.4 | 54.7 | 59.8 | 64 9 | 50.9 | 55.5 | 61 5 | 66.3 |
| $TiO_2$ | 1.0 | 0.8 | 0.7 | 0.6 | 1 0 | 0.9 | 0 6 | 0.4 | 1.2 | 0.9 | 0.9 | 0.5 |
| $Al_2O_3$ | 18.1 | 18,3 | 17.8 | 16.8 | 16 6 | 17.8 | 17.3 | 17.6 | 17.2 | 17.8 | 16 6 | 16.7 |
| $Fe_2O_3$ | 4.8 | 4.4 | 3.6 | 2.4 | 4.2 | 3.8 | 4.1 | 2.3 | 5.0 | 2.6 | 3.4 | 3.2 |
| $FeO$ | 5.2 | 4.1 | 3.6 | 2.4 | 5.2 | 4.8 | 2.5 | 2.0 | 4.7 | 5.4 | 3.3 | 1.2 |
| $MnO$ | 0.3 | 0.1 | 0.1 | 0.2 | 0.2 | 0.2 | 0.1 | 0.1 | 0.1 | 0.2 | 0.1 | 0.1 |
| $MgO$ | 6.0 | 4.2 | 3.1 | 1.6 | 6.9 | 4.9 | 3.6 | 1.6 | 6.9 | 5.2 | 2.3 | 1.4 |
| $CaO$ | 9.7 | 8.2 | 6.8 | 4.3 | 9.8 | 8.4 | 6.7 | 5.1 | 9.7 | 7.6 | 5.8 | 3.9 |
| $Na_2O$ | 2.5 | 2.9 | 3.2 | 4.0 | 2.4 | 3.2 | 3.8 | 4.2 | 3.1 | 3.4 | 3.9 | 4.1 |
| $K_2O$ | 0.9 | 1.0 | 1.5 | 1.8 | 1.3 | 1.3 | 1.5 | 1.8 | 1.2 | 1.4 | 2.2 | 2.6 |
| $a$ | 7.0 | 8.2 | 9.5 | 11.6 | 7.2 | 9.1 | 11.6 | 12.0 | 8.6 | 9.6 | 11.8 | 12.9 |
| $c$ | 9.0 | 8.5 | 7.5 | 5.3 | 7.5 | 7.5 | 6.4 | 5.9 | 7.2 | 7.2 | 5.3 | 4.8 |
| $b$ | 23.0 | 17.1 | 13.2 | 7.8 | 24.7 | 19.4 | 14.0 | 7.0 | 25.0 | 18.6 | 11.8 | 6.4 |
| $s$ | 61.0 | 66.2 | 69.8 | 75.3 | 60.6 | 64.0 | 69.0 | 75.1 | 59.2 | 64.6 | 71.1 | 75.9 |
| $a'$ | — | — | — | 7 | — | — | — | — | — | — | — | — |
| $f'$ | 41 | 47 | 51 | 58 | 35 | 42 | 43 | 56 | 35 | 41 | 52 | 62 |
| $m'$ | 45 | 43 | 41 | 35 | 47 | 43 | 44 | 39 | 47 | 48 | 33 | 38 |
| $c'$ | 14 | 10 | 8 | — | 18 | 15 | 13 | 5 | 18 | 11 | 15 | 0 |
| $n'$ | 80 | 81 | 77 | 77 | 74 | 79 | 79 | 78 | 79 | 78 | 73 | 70 |

**1.** Basalt of East Kamchatka, average of 29 analyses; **2.** Andesite-basalt of East Kamchatka, average of 30 analyses; **3.** Andesite of East Kamchatka, average of 30 analyses; **4.** Dacite of East Kamchatka, average of 14 analyses; **5.** Basalt of the Central Kamchatka Depression, average of 28 analyses; **6.** Andesite-basalt of the Central Kamchatka Depression, average of 32 analyses; **7.** Andesite of the Central  Kamchatka Depression, average of 34 analyses; **8.** Dacite of the Central Kamchatka Depression, average of 4 analyses; **9.** Basalt of the Sredinnyi Ridge, average of 12 analyses; **10.** Andesite-basalt of the Sredinnyi Ridge, average of 5 analyses; **11.** Andesite of the Sredinnyi Ridge, average of 4 analyses; **12.** Dacite of the Sredinnyi Ridge, average of 16 analyses.

1.  Volcanic zone of the Kamchatka east coast, including the volcanoes from Kambal'nyi on the south to the Gamchenskii group on the north.

2.  Zone of the Central Kamchatka Depression, in which are included the Klyuchevskaya Volcano group and Shiveluch Volcano.

3.  The volcanoes of the Sredinnyi Ridge, which runs along the central axis of the peninsula.

Many investigators have examined the petrochemistry of the individual regions and of Kamchatka as a whole. The average values for all types of rocks from the individual zones, on the basis of 260 modern analyses, are given by Naboko (1960). She pointed out that the plotted points representing the individual analyses for each zone were badly scattered, but that in general the alkalinity of the rocks increases from the east coast to the Sredinnyi Ridge.

We have used the average values from the paper by Naboko; these, recalculated to 100% for the main petrogenic elements, are represented in Table 17, and the corresponding points and variation curves are plotted in Fig. 83. From the table and diagram in question, it is seen that all three lines are parallel to one another and that the alkalinity increases significantly as we go from the east coast to the inner part of the peninsula.

The variation curve for lavas from the eastern zone falls between the Pelée and Lassen Peak curves, the curve for the Central Kamchatka Depression coincides rather well with the Lassen Peak curve, and the curve for the lavas of the Sredinnyi Ridge for the most part falls between the Lassen Peak and Yellowstone curves.

The three lines are not parallel to the corresponding curves plotted by Zavaritskii but as in the Kurile Islands they have a somewhat steeper slope and so cut across the Zavaritskii variation curves.

We see that the general character of the distinctive petrochemical features of the lavas of the Kurile Islands and of Kamchatka are similar.

## 2. Japan

The volcanoes of Japan are placed in four volcanic arcs by Japanese authors, and in these arcs separate zones are recognized. (Ishikawa and Katsui, 1959).

1. The Kurile arc, which ends on Hokkaido, whose rocks have been considered in the preceding chapter, and to which we need not return.

2. The North Honshu zone.

3. The volcanoes of the Fossa Magna and of the Izu-Bonin Islands (Fuji zone).

TABLE 18.  Average Lava
Compositions for Volcanoes
of the Nasu Zone

| Components | 1 | 2 | 3 | 4 | 5 | 6 |
|---|---|---|---|---|---|---|
| $SiO_2$ | 51.7 | 55.8 | 58.9 | 61.6 | 66.7 | 72.5 |
| $TiO_2$ | 0.8 | 0.9 | 0.9 | 0.8 | 0.6 | 0.5 |
| $Al_2O_3$ | 17.8 | 17.5 | 16.7 | 16.9 | 16.2 | 14.5 |
| $Fe_2O_3$ | 3.3 | 2.9 | 3.0 | 3.0 | 2.4 | 0.9 |
| FeO | 7.3 | 6.1 | 5.2 | 4.3 | 2.6 | 1.9 |
| MnO | 0.1 | 0.1 | 0.1 | 0.1 | 0.1 | 0.1 |
| MgO | 6.2 | 4.6 | 4.0 | 2.5 | 1.6 | 0.7 |
| CaO | 10.5 | 9.0 | 7.6 | 6.5 | 4.8 | 3.2 |
| $Na_2O$ | 1.9 | 2.3 | 2.6 | 3.2 | 3.8 | 3.8 |
| $K_2O$ | 0.4 | 0.9 | 1.0 | 1.1 | 1.2 | 1.9 |
| $a$ | 5.0 | 6.5 | 7.3 | 8.7 | 10.4 | 10.9 |
| $c$ | 9.8 | 8.7 | 7.6 | 7.2 | 5.7 | 3.8 |
| $b$ | 24.2 | 19.1 | 16.2 | 12.3 | 7.3 | 4.2 |
| $s$ | 61.0 | 65.7 | 68.9 | 71.8 | 76.6 | 81.1 |
| $a'$ | — | — | — | — | — | 12 |
| $f'$ | 42 | 45 | 47 | 57 | 61 | 61 |
| $m'$ | 44 | 42 | 42 | 35 | 37 | 27 |
| $c'$ | 14 | 13 | 11 | 8 | 2 | — |
| $n'$ | 89 | 78 | 79 | 81 | 82 | 76 |

**1.** Basalt, average of 12 analyses; **2.** Andesite-basalt, average of 12 analyses; **3.** Basic andesite, average of 19 analyses; **4.** Acid andesite, average of 13 analyses; **5.** Dacite, average of 5 analyses; **6.** Rhyolite, average of 2 analyses.

4.  The volcanoes of southwestern Japan.

Each of these regions is coupled with a corresponding abyssal trench.  In passing from the ocean to the islands across the abyssal trench a characteristic picture of the gravity anomalies, similar to that in the Kurile Islands, is observed.

Judging from the existing seismic data (Usami et al., 1958; Matsuzawa et al., 1960; Mikumo et al., 1961) the crustal structure is close to continental on the main Japanese island, Honshu, but the "granitic" layer has only a small thickness (4 km or less).  The

TABLE 19.  Average Lava
Compositions for Volcanoes
of the Chokai Zone

| Compo-<br>nents | 1 | 2 | 3 | 4 |
|---|---|---|---|---|
| $SiO_2$ | 54.1 | 56.9 | 61.1 | 66.1 |
| $TiO_2$ | 0.9 | 0.9 | 0.7 | 0.7 |
| $Al_2O_3$ | 16.8 | 18.6 | 17.1 | 16.4 |
| $Fe_2O_3$ | 4.8 | 3.8 | 3.5 | 2.6 |
| FeO | 5.4 | 4.3 | 3.6 | 2.3 |
| MnO | 0.2 | 0.2 | 0.1 | 0.1 |
| MgO | 4.8 | 3.5 | 2.9 | 1.8 |
| CaO | 8.7 | 7.6 | 6.1 | 4.5 |
| $Na_2O$ | 3.2 | 3.1 | 3.3 | 3.8 |
| $K_2O$ | 1.1 | 1.1 | 1.6 | 1.7 |
| $a$ | 8.9 | 8.8 | 9.8 | 10.9 |
| $c$ | 6.2 | 8.3 | 6.7 | 5.5 |
| $b$ | 22.0 | 15.2 | 12.2 | 7.6 |
| $s$ | 62.9 | 67.7 | 71.3 | 76.0 |
| $a'$ | — | — | — | 2 |
| $f'$ | 43 | 51 | 52 | 59 |
| $m'$ | 37 | 41 | 41 | 39 |
| $c'$ | 20 | 8 | 7 | — |
| $n'$ | 82 | 81 | 76 | 77 |

1.  Andesite-basalt, average of 2 analyses;
2.  Andesite, average of 9 analyses; 3.  Acid
andesite, average of 8 analyses; 4.  Dacite,
average of 3 analyses.

total crustal thickness is 25–28 km.  The velocity in the subcrustal part of the mantle is less than normal, equaling 7.7 km/sec.

In the Japan region 58 measurements of heat flow have been made, 39 of these on land and 19 at sea (Ueda and Horai, 1964).  The eastern shores of the Japanese islands and the adjacent parts of the Pacific Ocean are characterized by somewhat lower heat-flow values, but western Japan and the region of the Fossa Magna transverse fracture, to which areas modern volcanism is confined, are distinguished by higher than usual (approximately doubled) conductive heat flow.

The problems of the petrochemistry of the volcanic rocks have attracted a great deal of attention from the Japanese volcanologists

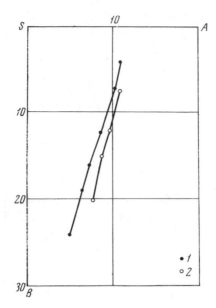

Fig. 84. Diagram of the average lava composition of North Honshu: 1) Nasu zone; 2)
Chokai zone.

and petrologists. The higher than usual alkalinity of the volcanic
rocks along both shores of the Sea of Japan was pointed out by
Tomita (1935) as much as 30 years ago when he distinguished the
"Circum-Japan Sea alkaline province."

In 1954 Katsui, on the basis of a rather limited number of
chemical analyses, pointed out that in northern Honshu the inner
(Chokai) zone was distinguished from the outer (Nasu) zone by its
higher alkalinity (Katsui, 1954). Later this conclusion was sub-
stantiated by a large amount of data (Kawano et al., 1961).

Kuno recognized three petrographic provinces in Japan: (1)
tholeiitic, on the eastern margin of Japan; (2) alkaline, along the
west coast of the islands; and (3) calc-alkaline, situated in the middle part of the country (Kuno, 1959).

In Kuno's opinion, the first two series developed from corresponding primary magmas formed by the partial melting of the
peridotite substratum at depths of less than 200 km in the first case
and of more than 200 km in the second. In neither case has the
magma undergone contamination. The calc-alkaline (hypersthenic)

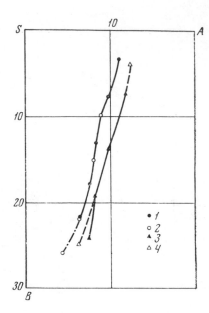

Fig. 85. Diagram of the average compositions in the Fuji arc: 1) Continental part
of the Izu-Hakone zone; 2) oceanic part
of the Izu-Hakone zone; 3) continental part
of the Fuji zone; 4) oceanic part of the
Fuji zone.

series, in Kuno's opinion, is formed from magma generated at a
depth of about 200 km and then contaminated by sialic crustal material. Subsequently Kuno (1960) added to the two "primary" magmas still another, "high-alumina" magma.

The main difference between the tholeiitic and the calc-alkaline series Kuno sees as the presence of microlites of pigeonite
in the former and hypersthene in the latter. As a result Kuno ex-
plains the existence, on one and the same volcano, of lavas of the
pigeonitic and hypersthenic series (and the common interlayering
of these rocks) by nourishment from different magmatic hearths
(Kuno, 1964).

Kuno's ideas differ from those of Sugimura, who came to deny
the volcanic arcs and zones distinguished earlier. In Sugimura's
opinion, only two island arcs exist in Japan: the first stretches
from Kamchatka through the Kurile Islands, northeastern Japan,

TABLE 20. Average Lava Compositions of the Izu–Hakone Zone

| Components | 1 | 2 | 3 | 4 | 5 | 6 | 7 | 8 | 9 |
|---|---|---|---|---|---|---|---|---|---|
| $SiO_2$ | 54.1 | 58.2 | 63.4 | 66.6 | 75.5 | 51.9 | 54.4 | 61.3 | 66.2 |
| $TiO_2$ | 0.8 | 0.8 | 0.8 | 0.8 | 0.4 | 1.0 | 1.1 | 1.0 | 0.7 |
| $Al_2O_3$ | 17.2 | 16.3 | 15.2 | 15.7 | 12.7 | 15.9 | 15.9 | 15.2 | 15.2 |
| $Fe_2O_3$ | 2.9 | 2.5 | 2.1 | 2.3 | 1.3 | 3.5 | 3.1 | 1.8 | 2.6 |
| $FeO$ | 7.0 | 6.3 | 5.4 | 3.2 | 1.3 | 9.7 | 9.8 | 7.8 | 3.6 |
| $MnO$ | 0.2 | 0.2 | 0.2 | 0.1 | 0.1 | 0.2 | 0.2 | 0.2 | 0.1 |
| $MgO$ | 5.0 | 3.7 | 2.6 | 1.1 | 0.5 | 5.0 | 3.7 | 2.3 | 1.8 |
| $CaO$ | 9.6 | 8.4 | 6.2 | 5.4 | 2.6 | 10.7 | 8.8 | 6.6 | 5.5 |
| $Na_2O$ | 2.4 | 2.8 | 3.1 | 3.8 | 4.2 | 1.7 | 2.5 | 3.3 | 3.6 |
| $K_2O$ | 0.6 | 0.8 | 1.0 | 1.0 | 1.4 | 0.4 | 0.5 | 0.5 | 0.7 |
| $a$ | 6.3 | 7.5 | 8.3 | 9.8 | 11.0 | 4.3 | 6.3 | 8.0 | 8.9 |
| $c$ | 8.7 | 7.3 | 6.0 | 5.6 | 2.8 | 8.7 | 7.7 | 6.3 | 5.8 |
| $b$ | 21.6 | 17.8 | 13.0 | 7.7 | 3.3 | 26.0 | 21.9 | 15.0 | 9.8 |
| $s$ | 63.4 | 67.4 | 72.7 | 76.9 | 82.9 | 61.0 | 64.1 | 7.7 | 75.5 |
| $f'$ | 44 | 48 | 54 | 64 | 68 | 49 | 56 | 61 | 57 |
| $m'$ | 41 | 35 | 34 | 24 | 24 | 33 | 29 | 26 | 32 |
| $c'$ | 15 | 17 | 12 | 12 | 8 | 18 | 15 | 13 | 11 |
| $n'$ | 87 | 84 | 82 | 85 | 82 | 87 | 89 | 91 | 88 |

Continental portion: 1. Andesite-basalt, average of 8 analyses; 2. Basic andesite, average of 14 analyses; 3. Acid andesite, average of 10 analyses; 4. Dacite, average of 3 analyses; 5. Rhyolite, average of 4 analyses. Oceanic portion: 6. Basalt, average of 28 analyses; 7. Andesite-basalt, average of 7 analyses; 8. Andesite, average of 2 analyses; 9. rhyolite, average of 7 analyses.

and the Shichito Islands to the Marianas; the second, from Kyushu through the Ryukyu Islands to Taiwan (Sugimura, 1960). The increase in alkalinity within each arc Sugimura relates to the increase in depth of earthquakes.

The author of the present work has examined the distinctive geomorphic and petrochemical features of the individual island arcs and chains, the existing geophysical data, and also the new data on the morphology of the abyssal trenches that are so closely related to the volcanic arcs. As a result, he came to the conclusion that the original ideas as to the existence of several island arcs are in general correct and require only insignificant changes in their details.

TABLE 21. Average Lava Compositions of the Fuji Zone

| Components | 1 | 2 | 3 | 4 | 5 | 6 |
|---|---|---|---|---|---|---|
| $SiO_2$ | 50.4 | 52.7 | 6^.8 | 69.1 | 51.2 | 76.9 |
| $TiO_2$ | 1.4 | 1.1 | 0.8 | 0.4 | 0.9 | 0.1 |
| $Al_2O_3$ | 17.6 | 19.0 | 16.8 | 14.4 | 17.6 | 13.8 |
| $Fe_2O_3$ | 3.6 | 2.5 | 3.6 | 2.1 | 5.6 | 0.6 |
| $FeO$ | 8.3 | 6.8 | 3.2 | 2.1 | 4.5 | 0.5 |
| $MnO$ | 0.2 | 0.2 | 0.1 | 0.1 | — | 0.1 |
| $MgO$ | 5.2 | 4.3 | 3.5 | 1.4 | 7.0 | 0.3 |
| $CaO$ | 9.8 | 9.6 | 6.4 | 4.1 | 10.2 | 1.1 |
| $Na_2O$ | 2.7 | 3.1 | 3.4 | 3.8 | 2.5 | 4.1 |
| $K_2O$ | 0.8 | 0.7 | 1.4 | 2.5 | 0.5 | 2.5 |
| | | | | | | |
| $a$ | 7.4 | 8.1 | 9.8 | 11.8 | 6.2 | 12.4 |
| $c$ | 8.4 | 9.2 | 6.6 | 3.5 | 8.8 | 2.8 |
| $b$ | 24.3 | 19.2 | 13.7 | 7.4 | 25.0 | 4.0 |
| $s$ | 59.9 | 63.5 | 69.9 | 77.3 | 60.0 | 80.8 |
| $a'$ | — | — | — | — | — | 70 |
| $f'$ | 47 | 43 | 46 | 51 | 37 | 19 |
| $m'$ | 37 | 38 | 44 | 31 | 48 | 11 |
| $c'$ | 16 | 15 | 10 | 18 | 15 | — |
| $n'$ | 83 | 88 | 79 | 69 | 89 | 71 |

Continental portion: **1.** Basalt, average of 12 analyses; **2.** Andesite-basalt, average of 7 analyses; **3.** Andesite, average of 2 analyses; **4.** Dacite, average of 3 analyses. Oceanic portion: **5.** Basalt, average of 2 analyses; **6.** Rhyolite, average of 3 analyses.

## North Honshu Arc

The North Honshu arc extends for 800 km from southwestern Hokkaido through the central and western parts of Honshu to the region of the Fossa Magna, where it intersects the Fuji zone. This arc is divided into two zones: the outer Nasu zone, where present-day activity is greatest, and the inner Chokai zone, which extends from Oshima-Oshima Island in the Sea of Japan southward along the shore of the Sea of Japan.

In Tables 18 and 19 the average values and the results of re-calculation of the analyses are given for each zone, and in Fig. 84 the petrochemical diagram is presented. In the recalculation 85 new analyses (Kawano et al., 1961) were used. In conformity with the earlier data of Japanese authors (Katsui, 1954; Ishikawa and

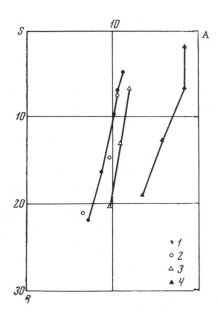

Fig. 86. Diagram of the average lava compositions for the Ryukyu arc: 1) Continental portion of the outer zone; 2) oceanic portion of the outer zone; 3) inner zone; 4) alkalic zone of Kyushu.

Katsui, 1959), it is seen in Fig. 84 that the lavas of the Chokai zone are actually more alkalic. The variation curve for the Chokai-zone lavas is close to that for the Pelée type; that for the Nasu-zone lavas is close to the Lassen Peak type.

Fuji Zone. The Fuji volcanic zone runs along the Fossa Magna, which cuts across the island of Honshu from its northern to its southern shore; it extends farther into the Pacific Ocean along the Izu-Shichito chain. This Nampo island arc is accompanied on the east by the abyssal Izu-Bonin Trench which at its northern end joins the Japan Trench, the two intersecting at a small angle, and on the south it is separated from the Marianas Trench by the Marcus-Nekker Ridge. No direct determinations of the crustal structure are known to us from this region, but judging from the gravimetric data this island arc rests on crust of the oceanic type. Oceanic crust is developed both in the Pacific Ocean and in the Philippine Sea.

The Fuji zone is divided into two subzones, northern and southern (Ishikawa and Katsui, 1959). Considering that the chemistry of the lavas in these two "subzones" is different and that the northern is an "inner" zone and the southern an "outer" zone, we regard it as expedient to recognize two independent zones here: (1) the outer or Izu-Hakone zone extends from Hakone Volcano through the Izu-Shichito Islands to the Bonin Islands (the latter are composed of Paleogene or older rocks and are not considered here); (2) the inner Fuji zone proper extends from Yatsugatake Volcano at the very center of Honshu through Fuji and Akagi Volcanoes on seaward to the islands of Niijima and Kozu.

Examination of the petrochemistry of these zones is of interest, because both zones begin on land and extend into the ocean, having very different types of crustal structure in the different parts.

As a result of this, it is interesting to examine the continental and oceanic portions of both zones separately. All the analyses used in the following recalculations were taken from the compilation "Chemical Composition of the Volcanic Rocks of Japan" (1962).

In Table 20 the average values corresponding to the continental and oceanic portions of the Izu-Hakone zone are presented, and in Table 21 are presented the same data for the Fuji zone. The results of the recalculations are depicted in Fig. 85, on which an extremely interesting feature shows up. As in the other arcs, a transverse zonation is apparent — the inner (Fuji) zone has a more alkalic character than the outer (Izu-Hakone) zone; the former is close to the Pelée type, the latter, to the Lassen Peak type. Longitudinal zonation is not displayed in either of these two zones; the continental part of each variation curve coincides completely with the oceanic part, and at the ends of the curves they extend one another. In this arc the independence of the petrochemistry of the volcanic rocks from variations in crustal type is very clearly displayed.

Southwestern Japan

In southwestern Japan several volcanic zones are recognized: the Norikura, Daisen, and Ryukyu zones.

The Norikura zone is situated in back of the Fuji zone in central Honshu. The volcanoes of this zone form the Hida Range,

which extends across the island to the Sea of Japan. This range is
not parallel to the Fuji zone, but runs at a small angle to it. Analytical data on the volcanoes of this zone are almost nonexistent
and do not permit us to make a confident petrochemical analysis.
Most probably the Norikura zone must be thought of as one of the
rear Fuji zones.

The D a i s e n   z o n e ,  according to Ishikawa and Katsui, extends from the island of Honshu along the Japan-Sea coast and then
turns southwestward, passing over onto the west coast of Kyushu.

It seems to us that the Daisen zone comes to an end on Honshu
and that the lavas of the west coast of Kyushu form an independent
zone which is a back zone relative to the Ryukyu zone.

Conceived of in this way, the Daisen zone includes several
dormant volcanoes that extend into southwestern Honshu, along the
coast of the Sea of Japan. This zone is accompanied by the small,
not very deep Kyushu Trench located between the Izu-Shichito and
Ryukyu Islands. The depth of this trench is not great: it does not
exceed 6000 m. The seismic activity usually related to abyssal
trenches and volcanic chains is not displayed here.

All this allows us to suggest that the Kyushu Trench has been
"tied off"; the deep-seated processes that give rise to earthquakes,
volcanism, and the development of a trench ceased here rather recently and the trench has begun to be filled with sediments.

In all probability an outer and an inner chain can be recognized here also; however, chemical-analytical information on the
volcanoes of the Daisen zone is almost non-existent, and it is not
yet possible to carry out a petrochemical analysis here.

The R y u k y u   a r c  begins on the north coast of Kyushu
(Futago Volcano) and stretches southward across the whole island,
including within it the Aso and Aira Calderas and continuing still
farther along the chain of Ryukyu (Nansei) Islands that separate the
abyssal Philippine Sea from the shallow East China Sea.

To the east the abyssal Ryukyu or Nansei Trench extends along
the Ryukyu Islands. We know of no direct data on the structure of
the crust in this region. Starting from general considerations, we
can suggest that the Nansei Trench separates the oceanic crust of
the Philippine Sea from continental or transitional crust in the eastern margin of Asia.

## TABLE 22.  Average Lava Compositions for the Ryukyu Arc

| Compo-nents | 1 | 2 | 3 | 4 | 5 | 6 | 7 | 8 | 9 | 10 | 11 | 12 | 13 | 14 | 15 |
|---|---|---|---|---|---|---|---|---|---|---|---|---|---|---|---|
| $SiO_2$ | 50.7 | 57.5 | 63.4 | 68.2 | 72.1 | 54.4 | 60,3 | 68.4 | 53.1 | 61.8 | 66.4 | 52.0 | 57.3 | 62.9 | 71.0 |
| $TiO_2$ | 1.4 | 0.9 | 0.7 | 0.6 | 0.4 | 0.8 | 0,7 | 0.4 | 1.0 | 0.9 | 0.7 | 2.5 | 1.2 | 0.7 | 0.2 |
| $Al_2O_3$ | 19.1 | 17.5 | 16.5 | 15.3 | 15.4 | 17.5 | 16,5 | 14.9 | 18.0 | 16.0 | 16.4 | 17.8 | 18.2 | 18.1 | 14.3 |
| $Fe_2O_3$ | 5.1 | 2.4 | 2.7 | 1.7 | 0.8 | 2.9 | 2,7 | 1.6 | 3.5 | 3.3 | 2.1 | 4.8 | 4.3 | 1.9 | 2.3 |
| FeO | 4.8 | 5.5 | 3.7 | 3.4 | 1.6 | 6.3 | 4,5 | 3.3 | 6.0 | 3.6 | 2.2 | 5.6 | 3.6 | 2.6 | 0.8 |
| MnO | 0.1 | 0.1 | 0.1 | 0.1 | 0.1 | 0.1 | 0,1 | 0.1 | 0.1 | 0.1 | 0.1 | 0.2 | 0.2 | 0.3 | 0.1 |
| MgO | 5.3 | 3.7 | 2.2 | 1.2 | 0.6 | 5.2 | 2,9 | 1.3 | 4.3 | 2.7 | 1.3 | 4.0 | 2.2 | 0.6 | 0.1 |
| CaO | 10.2 | 8.0 | 5.5 | 4.0 | 3.0 | 9.7 | 7,4 | 4.4 | 9.1 | 5.9 | 4.3 | 6.2 | 4.7 | 3.3 | 0.4 |
| $Na_2O$ | 2.4 | 2.9 | 3.3 | 3.6 | 3.4 | 2.3 | 3,2 | 3.6 | 3.3 | 3.7 | 4.0 | 4.4 | 4.8 | 5.5 | 6.5 |
| $K_2O$ | 0.9 | 1.5 | 1.9 | 1.9 | 2.7 | 0 8 | 1,6 | 2.0 | 1.6 | 2.0 | 2.5 | 2.4 | 3.5 | 4.1 | 4.3 |
| $a$ | 6.9 | 8.7 | 10.1 | 10.6 | 11.2 | 6,5 | 9.6 | 10.6 | 9.8 | 11.1 | 12.0 | 13.5 | 15.9 | 18.4 | 18.4 |
| $c$ | 9.1 | 7.6 | 6.2 | 4.8 | 3.7 | 8,7 | 6.3 | 4.5 | 7.4 | 5.1 | 4.7 | 5.4 | 4.3 | 2.9 | 1.7 |
| $b$ | 21.7 | 16.3 | 9.7 | 6.9 | 4.8 | 21,0 | 14.6 | 7.4 | 20.2 | 13.0 | 6.7 | 18.8 | 12.7 | 6.6 | 1.6 |
| $s$ | 62.3 | 67.4 | 74.0 | 77.7 | 80.3 | 63,8 | 69.4 | 77.7 | 62.6 | 70.8 | 76.6 | 62.3 | 67.1 | 72.1 | 78.3 |
| $a'$ | — | — | — | 2 | 36 | — | — | — | — | — | — | — | — | — | — |
| $f'$ | 42 | 46 | 60 | 69 | 44 | 41 | 47 | 60 | 44 | 49 | 58 | 51 | 58 | 66 | 80 |
| $m'$ | 43 | 39 | 39 | 29 | 20 | 43 | 34 | 30 | 37 | 35 | 33 | 37 | 30 | 16 | 6 |
| $c'$ | 15 | 15 | 1 | — | — | 16 | 19 | 10 | 19 | 16 | 9 | 12 | 12 | 18 | 14 |
| $n'$ | 80 | 74 | 73 | 74 | 65 | 80 | 75 | 72 | 76 | 74 | 71 | 73 | 73 | 67 | 66 |

Outer Zone. Continental portion: **1.** Basalt, average of 2 analyses; **2.** Andesite, average of 20 analyses; **3.** Acid andesite, average of 57 analyses; **4.** Dacite, average of 27 analyses; **5.** Rhyolite, average of 4 analyses. Oceanic portion: **6.** Andesite-basalt, average of 5 analyses; **7.** Andesite, average of 13 analyses; **8.** Dacite, average of 8 analyses. Inner Zone: **9.** Andesite-basalt, average of 7 analyses; **10.** Andesite, average of 14 analyses; **11.** Dacite, average of 3 analyses. Alkalic Zone of Kyushu: **12.** Aphyric basalt, 1 analysis; **13.** Trachyandesite, 1 analysis; **14.** Soda trachyte, average of 3 analyses; **15.** Riebeckite trachyte, average of 2 analyses.

The volcanoes of this arc are active ones; here we know of deep-focus earthquakes that form a focal zone dipping beneath the continent, and the volcanoes continue their intensive activity. The calderas on Kyushu are known to be young (Holocene) centers for widespread ignimbrite fields. The principal interest in this zone stems from the fact that it is the second, continent-ward volcanic zone relative to the outer Izu-Hakone and Marianas arcs.

The continental portion of the Ryukyu arc may be divided into two subzones (or independent zones):  an outer zone of the Ryukyus proper, which includes Kirishima Volcano and the Aira Caldera along with Sakurajima Volcano, and an inner zone, the chain of volcanoes from Futago to Aso Caldera (Aso zone).

To characterize petrochemically the rocks of this zone we have used more than 170 analyses. It must be pointed out that the analyses are very nonuniformly characteristic of the volcanoes: some volcanoes are represented by only two or three analyses; others (e.g., Sakurajima), by several dozen.

For material distributed in this way, a simple arithmetic average of all the analyses from the zone may characterize only the one volcano with the predominant number of analyzed rocks. In order to avoid this, we calculated average values for each rock type present for each volcano and then derived from these the average values for the separate zones as a whole, giving equal weight to the averages for each volcano.

The average lava compositions for the Ryukyu arc, calculated in this way, are shown in Table 22 and in Fig. 86. It can be seen from the table and figure that the continental portion of the outer Ryukyu zone and its oceanic extension have uniform petrochemical characteristics and that their variation curves actually coincide. The inner (Aso) zone, as in the other cases, has a somewhat more alkalic character.

The variation curve for the outer zone falls between the Pelée and Lassen Peak lines; that of the inner zone, between the Lassen Peak and Yellowstone lines. However, both curves, outer and inner, have steeper slopes and intersect the Pelée and Lassen Peak curves in their upper parts.

As we can see, the special position of the Ryukyu arc is not reflected in its characteristic petrochemical features. The chemistry of the rocks of this "interior" arc is the same as that of the "exterior" Izu-Hakone arc. In other words, those processes that led to the formation of the Izu-Hakone arc did not exert any influence on the Ryukyu arc.

In the back part of the Ryukyu arc there is one more active volcano, Unzen. The few analyses of its lavas do not indicate a still higher alkalinity which, as it were, violates the established principle of gradually increasing alkalinity toward the inner part of a volcanic arc. However, it is possible that there is a systematic error in the alkali determinations, because the petrographic descriptions of these rocks emphasize that the principal mafic phenocrysts are hornblende and biotite. This situation generally is characteristic of rocks with a higher than usual alkalinity.

Still farther west, in the Tsushima Strait and on the very coast of Kyushu, lie small eroded island volcanoes of Quaternary age. A notable feature of these volcanoes is the strongly alkalic character of their lavas, ranging from trachybasalts to trachytes. Analytical data on the islands of Kakara, Madara, and Matsu are also presented in Table 22, and their variation curve is shown along with the curve for the Ryukyu arc in Fig. 86. We can see that the curves for the lavas of western Kyushu, correlative with their high alkalinity, fall far to the right of the curves of all the island arcs and have a somewhat gentler slope. This curve coincides with the Etna-type curve, according to Zavaritskii, which separate the calc-alkalic type from the purely alkalic.

We will encounter such a type of curve again when we examine the lavas of the intracontinental volcanoes.

In concluding the petrochemical analysis of the volcanic rocks of Japan, let us note that in spite of the complexity of the geologic structure of this country and even of the differences in type of crustal structure (from oceanic to continental) along some of the volcanic arcs, all the arcs show very similar petrochemical characteristics:

1. The rocks of all the volcanic arcs and zones belong to the calc-alkalic series.

2. Longitudinal petrochemical variation is generally absent. Even in passing from the continental portions of zones to the oceanic portions, the type of variation curve remains the same.

3. Transverse petrochemical variation is very clearly expressed. A clear increase in alkalinity is displayed even in the first ten or so kilometers in passing from the outer zone to the inner one in any arc. The outer arc usually belongs to the Pelée type or has a transitional character between the Pelée and Lassen Peak types. The inner arc belongs, correspondingly, to the Lassen Peak type or to a type intermediate between the Lassen Peak and Yellowstone types.

TABLE 23. Average Lava Compositions
for the Lavas of Some Island Arcs
in the Southwestern Pacific Ocean

| Com-po-nents | Marianas Islands | | New Guinea | | Tonga Islands | |
|---|---|---|---|---|---|---|
| | 1 | 2 | 3 | 4 | 5 | 6 |
| $SiO_2$ | 51.0 | 53.6 | 56.2 | 59.8 | 54.8 | 66.6 |
| $TiO_2$ | 1.1 | 0.9 | 1.2 | 1.3 | 1.0 | 0.9 |
| $Al_2O_3$ | 17.9 | 20.5 | 16.3 | 16.7 | 14.3 | 12.2 |
| $Fe_2O_3$ | 3.6 | 2.4 | — | 2.9 | 3.0 | 1.8 |
| $FeO$ | 7.4 | 6.5 | 4.2 | 2.3 | 8.0 | 6.4 |
| $MnO$ | 0.2 | 0.2 | 0.1 | 0.1 | 1.8 | 1.2 |
| $MgO$ | 4.7 | 2.4 | 8.0 | 4.0 | 4.8 | 1.3 |
| $CaO$ | 10.9 | 10.2 | 8.0 | 6.4 | 10.1 | 5.6 |
| $Na_2O$ | 2.5 | 2.7 | 2.3 | 4.1 | 1.7 | 2.9 |
| $K_2O$ | 0.7 | 0.6 | 3.8 | 2.4 | 0 5 | 1.1 |
| $a$ | 6.6 | 7.5 | 10.3 | 12.4 | 4.4 | 7.9 |
| $c$ | 9.2 | 11.1 | 5.5 | 4.9 | 7.4 | 4.1 |
| $b$ | 23.5 | 15.6 | 21.0 | 14.4 | 25.0 | 13.3 |
| $s$ | 60.7 | 65.8 | 63.2 | 68.3 | 63.2 | 74.7 |
| $f'$ | 46 | 58 | 19 | 33 | 47 | 65 |
| $m'$ | 35 | 28 | 62 | 47 | 33 | 16 |
| $c'$ | 19 | 14 | 19 | 20 | 20 | 19 |
| $n'$ | 86 | 87 | 48 | 72 | 84 | 81 |

1. Basalt, average of 2 analyses; 2. Andesite,
average of 2 analyses; 3. Trachyandesite, 1
analysis; 4. Andesite, average of 12 analyses;
5. Basalt, average of 2 analyses; 6. Andesite,
average of 2 analyses.

In the back part of the Ryukyu arc even purely alkalic rocks
are present.

4. In the break-up of volcanic arcs, as A. N. Zavaritskii (1946)
pointed out, volcanic activity dies out in the cut-off portions. The
Daisen zone is "tied off" in this way, squeezed between the Ryukyu
and Fuji arcs.

## 3. Island Arcs of the Southwest Pacific

The island arcs of the southwestern Pacific make up one of
the most interesting regions. Here, to the south of Japan, the arcs

TABLE 24. Average Compositions
of Lavas from Volcanoes of the
"Anomalous Arcs" of the
Southwestern Pacific

| Compo-nents | New Britain | | New Hebrides | |
|---|---|---|---|---|
|  | 1 | 2 | 3 | 4 |
| $SiO_2$ | 48.7 | 61.0 | 50.1 | 55.4 |
| $TiO_2$ | 0.8 | 1.0 | 1.0 | 1.3 |
| $Al_2O_3$ | 18.4 | 14.8 | 17.4 | 18.4 |
| $Fe_2O_3$ | 4.2 | 2.8 | 4.0 | 1.8 |
| $FeO$ | 6.6 | 4.1 | 7.9 | 6.3 |
| $MnO$ | 0.2 | 0.1 | 0.4 | 0.1 |
| $MgO$ | 5.5 | 2.0 | 5.0 | 2.6 |
| $CaO$ | 12.5 | 8.6 | 10.2 | 8.0 |
| $Na_2O$ | 2.6 | 3.6 | 2.3 | 3.8 |
| $K_2O$ | 0.5 | 2.0 | 1.7 | 2.3 |
| $a$ | 6.5 | 10.6 | 7.6 | 11.9 |
| $c$ | 9.4 | 4.5 | 8.1 | 6.7 |
| $b$ | 26.4 | 15.4 | 25.0 | 15.5 |
| $s$ | 57.7 | 69.5 | 59.3 | 65.9 |
| $f'$ | 40 | 41 | 46 | 50 |
| $m'$ | 37 | 21 | 35 | 29 |
| $c'$ | 23 | 38 | 19 | 21 |
| $n'$ | 89 | 73 | 67 | 72 |

1. Basalt, average of 2 analyses; 2. Andesite, average of 4 analyses; 3. Basalt,
Ambrym Volcano, 1 analysis; 4. Andesite, Yasour Volcano, 1 analysis.

bifurcate, to be united again near New Guinea. North of New Guinea
and between it and New Zealand the system of island arcs again bifurcates, and right here are located the "anomalous arcs," whose
abyssal trenches lie not on the side toward the ocean, but toward
the continent of Australia, while the focal planes of deep-focus
earthquakes dip toward the ocean, not toward the continent.

Study of the volcanism, in particular of the petrochemical characteristics of the lavas, is of exceptional interest in this area. Unfortunately, chemical analyses of volcanic rocks from this area are
numbered literally in single numbers and a full petrochemical analysis is impossible. Only qualitative characterization can be made,
and this will change significantly as new information becomes available

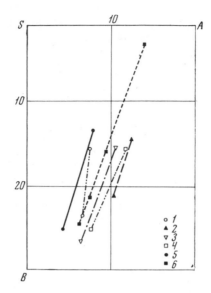

Fig. 87. Diagram of the lava composi-
tions of the island arcs of the southwest
Pacific and New Zealand: 1) Marianas
Islands; 2) New Guinea; 3) New Britain;
4) New Hebrides; 5) Tonga Islands; 6)
New Zealand.

The existing rock analyses from the Marianas arc and New
Guinea are presented in Table 23, and in Table 24 are given the analyses from the "anomalous arcs," New Britain and the New Hebrides
Islands. The recalculated data are presented in Fig. 87, from which
it can be seen that the rocks of all these arcs belong to calc-alkalic
varities and vary somewhat in alkalinity.

The variation curve for lavas of the Marianas arc is very steep,
almost vertical (this may be due to the small number of analyses)
and is situated between the lines for the Pelée and Lassen Peak
types.

The curve for the New Guinea lavas coincides nicely with the
curve of the Yellowstone type; its slope is gentler than that of the
examples considered earlier.

The variation curve for the anomalous arc of the Solomon Islands coincides with the Lassen Peak curve and is parallel to the
New Guinea curve. The line for the New Hebrides lavas passes

TABLE 25. Average Chemical
Compositions of New
Zealand Lavas

| Compo-nents | 1 | 2 | 3 | 4 | 5 |
|---|---|---|---|---|---|
| SiO$_2$ | 51.3 | 57.3 | 60.0 | 75.2 | 45.4 |
| TiO$_2$ | 0.9 | 0.7 | 0.7 | 0.2 | 2.0 |
| Al$_2$O$_3$ | 17.4 | 15.7 | 16.4 | 13.4 | 14.0 |
| Fe$_2$O$_3$ | 2.8 | 1.7 | 1.4 | 0.8 | 2.9 |
| FeO | 7.1 | 6.2 | 5.1 | 1.1 | 9.7 |
| MnO | — | 0.1 | 0.1 | 0.1 | 0.2 |
| MgO | 6.1 | 6.2 | 4.4 | 0.2 | 11.2 |
| CaO | 11.5 | 8.4 | 7.0 | 1.4 | 10.3 |
| Na$_2$O | 2.4 | 2.6 | 3.2 | 4.4 | 3.2 |
| K$_2$O | 0.5 | 1.1 | 1.7 | 3.2 | 1.1 |
| a | 6.3 | 7.4 | 9.4 | 13.8 | 8.3 |
| c | 8.9 | 6.6 | 6.2 | 1.6 | 4.8 |
| b | 24.4 | 21.2 | 15.9 | 2.2 | 36.3 |
| s | 60.4 | 64.8 | 68.5 | 82.4 | 50.6 |
| a' | — | — | — | 6 | — |
| j' | 38 | 35 | 39 | 79 | 31 |
| m' | 44 | 49 | 47 | 15 | 49 |
| c' | 18 | 16 | 14 | — | 20 |
| n' | 89 | 78 | 75 | 68 | 81 |

1. Basalt, average of 3 analyses; 2. Andesite-basalt, average of 6 analyses; 3. Andesite, average of 4 analyses; 4. Rhyolite, average of 21 analyses; 5. Auckland basalt, average of 13 analyses.

between the lines of the Lassen Peak and Yellowstone types, but it has a somewhat gentler slope than the curves of the adjacent arcs (because of the small number of analyses?). In each case, in spite of the small number of analyses, the membership of all the analyzed lavas in the different types of the calc-alkalic family is quite definitely shown.

## 4. Tonga Islands and New Zealand

The Tonga and Kermadec Islands, along with New Zealand, form the southernmost volcanic arc in the western Pacific. On the east this region is bounded by the almost rectilinear Tonga-Kerma-

dec Trench, which is up to 10,882 m deep. This trench is the natural boundary between the complex region of Melanesia to the west
and the deep-water, but simpler basin of the Pacific Ocean to the
east.

On the north the Tonga Trench bends sharply to the west, separating the oceanic islands of Samoa from the Tonga arc, and then
it rapidly pinches out. To the south this trench passes into the
Kermadec Trench; near the North Island of New Zealand this in
turn passes into the relatively shallow Hikurangi Trench, which
pinches out near the south end of the North Island. The chain of
modern volcanoes comes to an end at this same point.

The focal surface of the deep-focus earthquakes dips rather
steeply to the west, and shocks as deep as 700 km are known here.

To the east of the Tonga Trench lies crust of the normal
oceanic type with "normal" seismic velocities in the subcrustal
parts of the mantle of 8.1-8.2 km/sec. Immediately beneath the
islands (in the Tofua Trough), at a depth of 12 km, a velocity of 7.6
km/sec has been observed, which is anomalously low for the mantle
(Raitt et al., 1957). Thus, the volcanic islands of Tonga are located directly on oceanic crust, while the velocity in the subcrustal
mantle (as in other arcs) has a lower value than usual.

The crust in New Zealand is close to being continental, but
it has a somewhat smaller than usual thickness of 20-25 km. The
velocity in the subcrustal parts of the mantle is also low, down to
7.5 km/sec (Eiby, 1958).

The available analyses for the Tonga Islands are given in
Table 23, and the more numerous analyses of the New Zealand lavas
are recalculated to averages in Table 25. The analytical data for
the Tonga Islands and New Zealand are presented in Fig. 87. The
rocks in question belong to the calc-alkalic series; the Tonga Island lavas belong to the Pelée type, the New Zealand lavas are similar to the Lassen Peak type. Both variation curves parallel the corresponding curves of Zavaritskii.

The somewhat greater alkalinity of the New Zealand lavas apparently is explained by their greater distance from the abyssal
trench. Still farther west, in the Auckland region, there is a large
group of cones of basaltic composition. The more acid representatives are absent here, but the basalts are strongly alkalic, corre-

TABLE 26.  Average Chemical Analyses of Lavas
of the Aleutian Islands

| Components | 1 | 2 | 3 | 4 | 5 | 6 | 7 | 8 | 9 | 10 | 11 | 12 | 13 | 14 | 15 | 16 | 17 |
|---|---|---|---|---|---|---|---|---|---|---|---|---|---|---|---|---|---|
| $SiO_2$ | 52.3 | 53.8 | 58.4 | 63.4 | 69.8 | 48.5 | 53.4 | 59.5 | 63.9 | 77.0 | 51.6 | 53.9 | 59.1 | 65.4 | 73.9 | 46.6 | 61.0 |
| $TiO_2$ | 0.9 | 0.9 | 0.8 | 0.7 | 0.4 | 0.9 | 0.8 | 0.6 | 0.7 | 0.1 | 1.2 | 1.4 | 1.0 | 0.8 | 0.2 | 1.4 | 0.5 |
| $Al_2O_3$ | 18.3 | 18.6 | 18.0 | 16.8 | 15.3 | 18.0 | 18.2 | 17.4 | 16.8 | 13.7 | 17.4 | 17.4 | 16.6 | 15.7 | 13.5 | 18.8 | 18.6 |
| $Fe_2O_3$ | 2.9 | 3.0 | 2.9 | 1.8 | 1.4 | 4.5 | 4.4 | 4.1 | 3.0 | 0.5 | 3.6 | 2.8 | 2.5 | 1.9 | 0.9 | 6.2 | 3.9 |
| $FeO$ | 6.5 | 5.8 | 4.1 | 3.9 | 2.2 | 6.3 | 4.8 | 3.1 | 2.4 | 0.5 | 6.7 | 7.2 | 5.6 | 4.1 | 1.6 | 5.1 | 1.2 |
| $MnO$ | 0.2 | 0.2 | 0.2 | 0.2 | 0.1 | 0.2 | 0.2 | 0.2 | 0.2 | 0.1 | 0.2 | 0.2 | 0.2 | 0.2 | 0.1 | 0.2 | 0.1 |
| $MgO$ | 5.5 | 4.4 | 3.1 | 1.8 | 1.0 | 6.1 | 4.2 | 2.6 | 2.5 | 0.3 | 5.6 | 3.8 | 2.9 | 0.9 | 0.1 | 5.5 | 1.1 |
| $CaO$ | 10.2 | 8.9 | 7.5 | 4.7 | 3.4 | 11.8 | 9.4 | 7.3 | 4.9 | 1.1 | 10.1 | 8.8 | 6.6 | 3.6 | 1.3 | 12.0 | 6.5 |
| $Na_2O$ | 3.2 | 3.3 | 3.6 | 4.5 | 4.4 | 2.7 | 3.4 | 3.6 | 3.9 | 4.2 | 2.9 | 3.4 | 3.9 | 4.5 | 4.3 | 2.6 | 4.2 |
| $K_2O$ | 0.9 | 1.1 | 1.4 | 2.2 | 2.0 | 1.0 | 1.2 | 1.6 | 1.7 | 2.5 | 0.9 | 1.1 | 1.6 | 2.9 | 4.1 | 1.6 | 2.9 |
| $a$ | 8.5 | 9.2 | 10.2 | 13.0 | 12.5 | 7.7 | 9.6 | 10.5 | 11.1 | 12.2 | 7.9 | 9.2 | 11.0 | 13.7 | 14.6 | 8.5 | 13.9 |
| $c$ | 8.2 | 8.2 | 7.2 | 4.7 | 3.8 | 8.4 | 7.7 | 6.6 | 5.8 | 1.2 | 7.9 | 7.3 | 5.7 | 3.5 | 1.4 | 8.7 | 5.8 |
| $b$ | 22.8 | 19.4 | 14.2 | 9.5 | 5.4 | 27.1 | 20.0 | 13.7 | 9.4 | 4.2 | 24.0 | 19.8 | 15.0 | 8.1 | 2.6 | 26.7 | 8.8 |
| $s$ | 60.5 | 63.4 | 68.4 | 72.8 | 78.3 | 56.8 | 62.7 | 69.2 | 73.7 | 82.4 | 60.2 | 63.7 | 68.3 | 74.7 | 81.4 | 56.2 | 71.5 |
| $a'$ | — | — | — | — | — | — | — | — | — | 63 | — | — | — | — | — | — | — |
| $f'$ | 39 | 45 | 47 | 57 | 63 | 38 | 44 | 49 | 52 | 25 | 41 | 48 | 51 | 70 | 87 | 40 | 53 |
| $m'$ | 42 | 40 | 38 | 33 | 31 | 39 | 36 | 33 | 46 | 12 | 40 | 33 | 33 | 19 | 8 | 36 | 21 |
| $c'$ | 19 | 15 | 15 | 10 | 6 | 23 | 20 | 18 | 2 | — | 19 | 22 | 16 | 11 | 5 | 24 | 26 |
| $n'$ | 84 | 81 | 79 | 76 | 77 | 80 | 81 | 77 | 78 | 73 | 83 | 83 | 79 | 71 | 62 | 71 | 69 |

Rat Islands:  1. Basalt, average of 4 analyses; 2.  Andesite-basalt, average of 7 analyses; 3.  Andesite, average of 17 analyses; 4.  Dacite, average of 7 analyses; 5. Rhyolite, average of 2 analyses.  Andreanof Islands:  6. Basalt, 1 analysis; 7. Andesite-basalt, average of 9 analyses; 8.  Andesite, average of 8 analyses; 9.  Dacite, 1 analysis; 10. Rhyolite, 1 analysis.  Fox Islands:  11. Basalt, average of 8 analyses; 12.  Andesite-basalt, average of 6 analyses; 13.  Andesite, average of 6 analyses; 14.  Dacite, average of 3 analyses; 15.  Rhyolite, average of 3 analyses; Bogosloff Island:  16. Basalt, 1 analysis; 17.  Dacite, 1 analysis.

sponding approximately to the Highwood type $(a = 8.3,\ b = 36.3)$, i.e., transverse chemical zonation is displayed very clearly in New Zealand.

## 5. Aleutian Islands

Let us turn now to the Aleutian Islands, the link between Asia and America.  This arc has a rather considerable length of more than 1700 km.  On the south the islands are bordered by the abyssal Aleutian Trench, which pinches out eastward as it approaches the

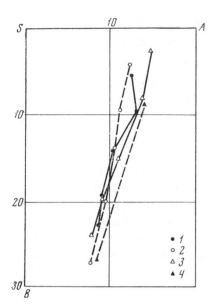

Fig. 88. Diagram of the average compositions of the Aleutian lavas: 1) Rat Islands;
2) Andreanof Islands; 3) Fox Islands; 4)
Bogosloff Island.

continent and which has been filled to a considerable degree with
sediments (Gates and Gibson, 1956).

Seismic studies have established that the Aleutian arc is situated on crust of oceanic type (Shor, 1962). The Aleutian Islands are
one of the most seismic regions of the world; the earthquake focal
zone dips northward, beneath the Bering Sea, whose crust is also of
oceanic type.

The island chain is divided geographically into four parts:
the Near Islands, Rat Islands, Andreanof Islands, and Fox Islands.
The volcanoes of the Aleutian Islands are located mainly to the north
of the arc axis, i.e., close to its inner edge. The volcanoes began
to be studied only after World War II, and at present only a small
part of the islands and their volcanoes have been described in the
bulletins of the U.S. Geological Survey. About 80 chemical analyses
of the volcanic rocks have been published. Unfortunately, the accuracy of the analyses does not always inspire confidence; the results of duplicate alkali determinations (Coats, 1952) often show dif-

ferences of up to 1%, and this may produce substantial differences in petrochemical interpretations.

In conformity with the geographic division, we have divided the existing analyses of the Aleutian lavas into three groups. The average values for the groups are presented in Table 26, while the results of the recalculations are displayed in Fig. 88. From the data presented it is easily seen that the volcanoes of the Aleutian arc reveal a clear longitudinal zonation in petrochemistry.

The variation curve for the lavas of the Rat Islands (which are closest to Kamchatka) coincides approximately with the curve of the Lassen Peak type in its lower part (from basalt to andesite), but the dacites from these islands are clearly enriched in alkalies, causing the curve to bend distinctly to the right, while the most acid rocks, the rhyolites, contain less alkalies and the curve bends to the left, forming a broad arc. What causes such a nonuniform distribution of the alkalies is not clear.

In any case assimilation is not called for, since this part of the arc lies wholly on crust of the oceanic type.

The variation curve for the lavas of the middle part of the arc (Andreanof Islands) has a more normal character; through most of its length it coincides with the line of the Lassen Peak type, deviating from this only near the top. Such a trend is characteristic of the other arcs of southeast Asia.

Finally, the curve for the lavas of the Fox Islands, which are closest to the American continent, is located between the curves for the Lassen Peak and Yellowstone types and runs parallel to them.

The points representing the lavas of Bogosloff, which lies on the inner side of the arc, show, as in other situations, the more alkalic character of the lavas of this volcano.

Thus, in the Aleutian arc a very distinct longitudinal petrochemical variation is shown. The variation curves for the lavas of the western part of the arc are similar in their trend to the curves for the island arcs of southeast Asia, while the variation curves of the eastern part have a somewhat different slope, one which is also displayed by the volcanoes of the American continent. We have already encountered a similar trend in the curves from several arcs in the southwestern Pacific.

## 6. Eastern Part of the "Pacific Ring"

To the east the Aleutian Islands volcanic zone passes into the Alaska Peninsula. All the tectonic structures of Alaska farther to the east turn to the southeast. In contrast to the continent of Asia, island arcs are absent along the American coast, but the volcanic arcs that extend along the eastern shore of the Pacific Ocean from Alaska on the north to southern Chile on the south are completely analogous to them.

The volcanologic, geologic, and geophysical investigation of this vast area is extremely nonuniform. Naturally North America has been studied best of all; the main contribution to the study of the volcanoes of the USA has been made by the scientists of the University of California (Coombs et al., 1960, and others). Study of the volcanoes of Central Ameria has even been begun by them (Mooser et al., 1958).

The South American volcanoes have been studied very poorly, and in spite of the fact that the most widespread volcanic rock of the island arcs, andesite, is named after the Andes of South America, chemical analyses of the lavas of this region are very scarce and most were made during the last century.* Thus a petrochemical analysis of the South American lavas is impossible at present, and we will limit our analysis to some areas of North and Central America.

Volcanoes of the USA. Volcanic cones within the territory of the United States extend along the crest of the Cascade Range, which is analogous to the inner, volcanic ridge of an island arc. Eruptions of volcanoes in the United States are very rare, and in general volcanic activity seems to be dying out.

Going to the north, it is possible to trace the transition of the volcanic zone of the North American continent into the volcanic chain of the Aleutian island arc.

The nonvolcanic Coast Ranges, which extend continuously along the Pacific shore of North America, are analogous to the outer, non-volcanic chain of islands in the double island arcs. It is possible to trace the passage of the Coast Ranges into the outer islands of the Aleutian arc (e.g., Kodiak Island).

---

*Thus, for example, in the catalog of Chilean volcanoes (Casterano, 1963), many of the analyses are dated 1850.

The interior valleys that come between the Coast Ranges and the Cascade Range are analogs of the depressions (straits) between the inner and outer island chains of the double island arcs.

A broad zone of interior ridges and valleys extends to the east of the Cascade Range; here belong the massive basalt outpouring of the Columbia River, Snake River, and so on.

In the basin of the Snake River is located the Craters of the Moon National Monument, with its several dozen recent scoria cones. Still farther east lies the chain of the Rocky Mountains, where the famous Yellowstone National Park is located, with its Eocene-Miocene volcanism. The Rocky Mountains drop away to the prairies of the Great Plains, which is a purely continental structure. At the very boundary between the Rocky Mountains and the Great Plains is a region of mainly Tertiary volcanism, the Highwood Mountains. Apparently there are also manifestations of Quaternary volcanism here, though rather limited in scale (Larsen, 1941).

The thickness of the earth's crust varies across the trend of the main structures within rather broad limits (Eaton, 1963; Pakiser, 1963). Beneath the Coast Ranges and the Valley of California it is 20-25 km; beneath the Sierra Nevada it thickens to as much as 50 km, and the "root" is much wider than the ridge itself. In the Basin and Range Province the crustal thickness again decreases to 20-25 km. Beneath the Great Plains the crust thickens to 40-45 km.

Over broad areas from the Rocky Mountains to the Coast Ranges and locally along the Pacific coast itself, seismic velocities in the subcrustal part of the mantle have a lower than usual value of 7.6-7.9 km/sec. Right here, in the eastern part of the Basin and Range Province, anomalously low velocity values were first observed in the subcrustal parts of the mantle (Berg et al., 1960). In the Great Plains region and locally along the Pacific coast, the velocities in the subcrustal part of the mantle have "normal" values of 8.0-8.2 km/sec.

American workers relate the low seismic velocities in the western United States to the fact that the East Pacific Rise supposedly extends through there, and beneath it the velocities are anomalously low (Menard, 1960). However, as we have seen, low velocities are characteristic of all the island and volcanic arcs. In our opinion, the same general principle holds for North America.

## TABLE 27. Average Lava Compositions for the Volcanoes of the USA and Mexico

| Components | 1 | 2 | 3 | 4 | 5 | 6 | 7 | 8 | 9 | 10 | 11 | 12 | 13 | 14 |
|---|---|---|---|---|---|---|---|---|---|---|---|---|---|---|
| $SiO_2$ | 53.7 | 61.4 | 68.6 | 74.0 | 51.8 | 61.5 | 69.2 | 74.9 | 52,3 | 48.7 | 53.6 | 55.6 | 60.3 | 48.3 |
| $TiO_2$ | 0.9 | 0.8 | 0.3 | 0.2 | 0.6 | 0.5 | 0.2 | 0.2 | 2,5 | 0.7 | 0.8 | 1.0 | 0.8 | 2.0 |
| $Al_2O_3$ | 18.2 | 17.2 | 16.2 | 14.2 | 18.4 | 17.1 | 16.2 | 14.3 | 14,3 | 12.9 | 15.6 | 18.3 | 17.4 | 13.3 |
| $Fe_2O_3$ | 3.8 | 2.4 | 1.2 | 0.7 | 4.2 | 2.8 | 1.8 | 1.0 | 2,2 | 4.9 | 3.9 | 1.6 | 1.4 | 5.1 |
| $FeO$ | 4.6 | 3.3 | 2.1 | 1.3 | 4.9 | 2.8 | 1.2 | 0.6 | 13,2 | 4.5 | 4.5 | 5.8 | 4.4 | 6.6 |
| $MnO$ | 0.1 | 0.1 | 0.1 | — | 0.2 | 0.1 | 0.1 | 0.1 | 0,4 | 0.1 | 0.1 | 0.1 | 0.1 | 0.2 |
| $MgO$ | 5.6 | 3.1 | 1.3 | 0.4 | 6.4 | 3.2 | 1.3 | 0.3 | 2,3 | 9.7 | 4.9 | 5.5 | 3.7 | 9.4 |
| $CaO$ | 8.9 | 6.1 | 3.3 | 1.3 | 10.0 | 6.5 | 3.3 | 1.1 | 6,7 | 12.0 | 7.5 | 7.2 | 6.2 | 9.9 |
| $Na_2O$ | 3.3 | 4.0 | 4.0 | 3.7 | 2.8 | 3.8 | 4.1 | 3.4 | 3,7 | 2.7 | 3.4 | 3.9 | 4.0 | 3.4 |
| $K_2O$ | 0.9 | 1.6 | 2.9 | 4.2 | 0.7 | 1.7 | 2.6 | 4.3 | 2,4 | 3.8 | 5.7 | 1.0 | 1.7 | 1.8 |
| $a$ | 8.7 | 11.4 | 13.0 | 13.8 | 7.4 | 10.8 | 12.7 | 12.9 | 11,4 | 10.8 | 14.8 | 10.2 | 11.4 | 9.6 |
| $c$ | 8.1 | 5.9 | 4.0 | 1.6 | 8.1 | 6.1 | 3.9 | 1.2 | 3,8 | 2.8 | 6.1 | 7.2 | 6.0 | 3.6 |
| $b$ | 20.5 | 11.9 | 5.6 | 4.3 | 23.0 | 13.0 | 5.6 | 4.5 | 23,0 | 34.2 | 21.2 | 18.0 | 13.4 | 32.9 |
| $s$ | 62.7 | 70.8 | 77.4 | 80.3 | 67.5 | 70.1 | 77.8 | 81.4 | 61,8 | 52.2 | 57.9 | 64.6 | 69.2 | 53.9 |
| $a'$ | — | — | 7 | 36 | — | — | 14 | 59 | — | — | — | — | — | — |
| $f'$ | 38 | 44 | 52 | 49 | 37 | 41 | 47 | 31 | 65 | 23 | 34 | 39 | 41 | 31 |
| $m'$ | 48 | 43 | 41 | 15 | 47 | 44 | 39 | 10 | 16 | 45 | 37 | 52 | 47 | 45 |
| $c'$ | 14 | 13 | — | — | 16 | 15 | — | — | 19 | 32 | 29 | 9 | 12 | 24 |
| $n'$ | 84 | 79 | 68 | 57 | 85 | 78 | 70 | 54 | 71 | 51 | 48 | 85 | 78 | 68 |

Cascade Mountains: **1.** Andesite-basalt, average of 13 analyses; **2.** Andesite, average of 29 analyses; **3.** Dacite, average of 14 analyses; **4.** Rhyolite, average of 14 analyses. Lassen Peak: **5.** Basalt, average of 5 analyses; **6.** Andesite, average of 14 analyses; **7.** Dacite, average of 8 analyses; **8.** Rhyolite, average of 6 analyses. Craters of the Moon: **9.** Basalt, 1 analysis. Highwood Mountains: **10.** Phonolite, average of 2 analyses; **11.** Analcite phonolite, 1 analysis. Mexico: **12.** Andesite-basalt, Paricutin Volcano, 1943-44 lava, average of 5 analyses; **13.** Andesite, Paricutin Volcano, 1952 lava, 1 analysis; **14.** Basalt, San Martin Volcano, 1 analysis.

In any case, no changes in the manifestations of volcanism dependent on the presence of a new structure (the oceanic rise) are observed in the western United States.

No abyssal trench is recognized in the topography of the ocean floor along the North American coast; however, it has been established by deep seismic sounding that there is an abyssal trench which has been completely filled with unconsolidated sediments

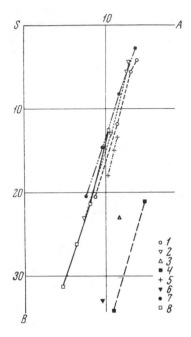

Fig. 89. Diagram of the average compositions of the lavas of the North and Central American volcanoes: 1) Cascade Mountains; 2) Lassen Peak; 3) Craters of the Moon; 4) Highwood Mountains; 5) Paricutin Volcano; 6) San Martin Volcano; 7) El Salvador; 8) Nicaragua.

(Hodgson, 1963). Clearly the active processes that produce a trench have already stopped here, and it has rapidly filled with sediment. The dying out of the deep-seated processes is also displayed in the dying out of volcanic activity and in the fact that here, in contrast to the rest of the marginal part of the Pacific Ocean, deep-focus earthquakes and the corresponding focal zone are absent. We note that south of California volcanic activity becomes more apparent; there an abyssal trench and zones of deep-focus earthquakes are again found.

In the western United States and Canada one can see the transition from an "active" boundary between the continent and an ocean "of the Pacific type" to a passive "Atlantic-type" boundary.

The petrochemistry of Cenozoic volcanism in the western United States has been examined in considerable detail by Burri (1926) and by Zavaritskii (1950). On a Zavaritskii diagram we can easily see that with increasing distance from the Pacific coast the alkalinity of the rocks increases considerably; it changes from the calc-alkalic Lassen Peak type through the Yellowstone type to the alkalic rocks of the Maros-Highwood type. The names of all these types are taken from geologic localities in the western United States [except "Maros," named after the Maros or Mures River in Hungary].

We will examine only the recent analyses of Quaternary lavas (published mainly since 1940). The average compositions of the Cascade lavas (Coombs et al., 1960) are given in Table 27; in that table are also given separate data for Lassen Peak. Those values are very similar, and the variation curves (Fig. 89) coincide well with Zavaritskii's Lassen Peak curve.

To the east of the Cascade Range Quaternary volcanic activity is rather weakly displayed.

Several dozen small cinder cones of recent origin lie within Craters of the Moon National Monument in the Basin and Range Province (Snake River basin). There is only one analysis of a basalt from the Craters of the Moon (Coombs and Howard, 1960). Its point (see Table 27 and Fig. 89) falls on the variation curve of the Etna type, which correlates well with the position of this volcano in the inner part of the Cascade volcanic arc.

The Highwood volcanic region, on the boundary between the Rocky Mountains and the Great Plains, is one of the easternmost regions that are farthest within the continent from the Cascade volcanic arc. In the Highwood Mountains volcanic activity occurred mainly in Eocene-Miocene times; however, the possibility is not excluded that small-scale eruptions occurred at the beginning of the Quaternary (Larsen, 1941). Analyses of phonolites of supposedly Quaternary age are given in Table 27 and in Fig. 89. As in the older Tertiary rocks, these also belong to the Maros-Highwood type.

As we can see, the results of recalculation of the recent analyses of the Quaternary rocks lead to the same results: with increasing distance into the continent from the Pacific Ocean the alkalinity of the rocks clearly increases from the Lassen Peak type to the Maros-Highwood type. A distinct decrease in the scale of activity from Tertiary to Quaternary times is not reflected in the petrochemical characteristics of the rocks.

Volcanoes of Mexico. A distinctive feature that requires special consideration is displayed in the distribution of the Mexican volcanoes. Unfortunately, analyses of the lavas from the Mexican volcanoes are very scarce and, except for Paricutin, almost all were made in the nineteenth century.

Near the coast of southern Mexico, from approximately 20° North latitude, extends the abyssal Central American Trench, with depths of up to 6000 m. However, the volcanoes of Mexico are not situated parallel to the trench, as in all the other volcanic chains, but in a transverse region running from the Pacific coast on the west (Ceboruco Volcano) to the Gulf of Mexico in the east (San Martin

Volcano). On the trend of this chain to the west, in the Pacific Ocean, lie the volcanic islands of Revilla Gigedo and a series of submarine volcanoes, while still farther to the west runs the Clarion fracture zone.

Recent oceanologic investigations in the eastern Pacific Ocean have established the existence of a system of almost east-west fracture zones. On the basis of magnetometric data American investigators state that horizontal movement has taken place along these fracture zones with displacements of hundreds of kilometers. However, no traces of such large-scale movements have been detected on the continental extensions of these fractures, and the scale of the movements, if they actually have occurred, seems greatly exaggerated to us.

Nevertheless, it is impossible not to see some influence of the Clarion fracture zone in the distribution of the Mexican volcanoes.* The rather numerous analyses of the Paricutin lavas (see Table 27), although they show some scatter, in general coincide well with the trend of the island and volcanic arc variation curves. Paricutin Volcano is located in the western (Pacific) half of the volcanic chain, and the variation curve for its lavas (Fig. 89) is located between the Lassen Peak and Yellowstone curves.

As mentioned earlier, analyses of the lavas of the other Mexican volcanoes are not numerous, and we have not used them. However, an isolated analysis of a basalt from the easternmost volcano, San Martin, made in 1927, can be used to characterize the change in petrochemistry along the chain. The point plotted for its analysis (Fig. 89) falls in the field of purely alkalic basalts, on a Highwood-type curve, i.e., in Mexico as in the other cases that were examined, an increase in alkalinity is observed in passing from the outer to the inner edge of the volcanic arc.

Thus, although the Clarion fracture zone exerts the greatest influence on the location of the Mexican volcanic chain, the petrochemistry of the volcanic rocks does not depend on the superposition of this structure, but is subject to the same general principles that were found in the other island and volcanic arcs.

Looking ahead, we can point out that the volcanoes of the Revilla Gigedo Islands, which lie on this same fracture zone, but on the

---

* There are no volcanoes on the continental extensions of the other fracture zones.

TABLE 28. Average Compositions of Lavas of Central America

| Compo-nents | 1 | 2 | 3 | 4 | 5 | 6 | 7 | 8 |
|---|---|---|---|---|---|---|---|---|
| $SiO_2$ | 52.1 | 59.5 | 67.7 | 72.8 | 49.3 | 49.8 | 50.3 | 59.3 |
| $TiO_2$ | 1.1 | 0.6 | 0.6 | 0.6 | 0.6 | 1.4 | 1.6 | 0.7 |
| $Al_2O_3$ | 18.8 | 16.6 | 16.6 | 14.3 | 15.8 | 17.3 | 19.1 | 17.9 |
| $Fe_2O_3$ | 3.8 | 3.4 | 2.2 | 1.3 | 3.7 | 4.5 | 4.7 | 2.1 |
| FeO | 6.0 | 5.7 | 2.1 | 0.8 | 7.3 | 5.6 | 5.0 | 5.5 |
| MnO | 0.3 | 0.2 | 0.1 | 0.1 | 0.2 | 0.1 | 0.1 | 0.2 |
| MgO | 4.3 | 2.7 | 1.2 | 0.4 | 9.4 | 6.3 | 4.3 | 2.3 |
| CaO | 10.0 | 6.5 | 3.3 | 1.9 | 11.5 | 11.8 | 11.0 | 6.9 |
| $Na_2O$ | 2.6 | 3.2 | 3.6 | 3.6 | 1.8 | 2.5 | 2.6 | 3.9 |
| $K_2O$ | 1.0 | 1.6 | 2.6 | 4.2 | 0.4 | 0.7 | 1.3 | 1.2 |
| $a$ | 7.5 | 9.6 | 11.6 | 13.6 | 4.6 | 6.3 | 8.0 | 10.3 |
| $c$ | 9.4 | 6.5 | 3.8 | 2.3 | 8.2 | 8.7 | 9.3 | 7.2 |
| $b$ | 20.5 | 14.6 | 8.3 | 2.8 | 31.3 | 26.2 | 21.4 | 12.7 |
| $s$ | 62.6 | 69.3 | 76.3 | 81.3 | 55.9 | 58.8 | 61.3 | 69.8 |
| $a'$ | — | — | 30 | 15 | — | — | — | — |
| $f'$ | 47 | 60 | 46 | 63 | 32 | 36 | 43 | 57 |
| $m'$ | 37 | 32 | 24 | 22 | 50 | 42 | 36 | 32 |
| $c'$ | 16 | 8 | — | — | 18 | 22 | 21 | 11 |
| $n'$ | 79 | 76 | 68 | 56 | 87 | 89 | 75 | 84 |

Salvador: 1. Andesite-basalt, average of 4 analyses; 2. Andesite, average of 3 analyses; 3. Dacite, average of 7 analyses; 4. Rhyolite, average of 2 analyses. Nicaragua: 5. Basalt, 1947 flow from Cerro Negro, 1 analysis; 6. Basalt, average of 4 analyses; 7. Andesite-basalt, average of 6 analyses; 8. Andesite, Coseguina Volcano, 1835 bomb, 1 analysis.

other side of the abyssal trench, belong to a very different, oceanic petrochemical type. The case in question clearly and conclusively indicates that, although deep tectonic fractures may determine the appearance of a chain of volcanoes, the chemistry of the lavas does not depend on the tectonics.

Volcanoes of Central America. The volcanoes of Central America are situated in chains that parallel the Central American Trench. There are recent analyses for the volcanoes of El Salvador and Nicaragua. The recalculations of the analyses (Table 28, Fig. 89) show that all the lava flows belong to a type similar to that of Lassen Peak.

New information has been obtained in the last few years on a number of South American volcanoes, but no new chemical analyses have yet been published. Judging from the microscopic descriptions and from the old analyses, there are no distinguishing features between the lavas of the South American Andes and the lavas of other volcanic arcs. Apparently the same regularities are displayed there.

In examining the distinctive petrochemical features of the volcanic rocks of the "Pacific ring of fire"; we see that in all the island and volcanic arcs the same characteristics are displayed. The most distinct characteristic is the gradual increase in the alkalinity of the rocks as one moves from the outer edge of the volcanic arc toward the interior of the continent.

Let us see if this characteristic is displayed in Indonesia, which belongs to a very different, east-west-trending (Tethyan) system of volcanic arcs.

## 7. Indonesian Volcanoes

In the region north of New Guinea the Indonesian island arc approaches the "Pacific ring of fire;" from there it goes through the island of Halmahera and the Lesser Sunda Islands to Java and Sumatra. To the north of Sumatra the arc passes through the Nicobar and Andaman Islands into Burma. Indonesia is one of the regions in which the concentration of volcanoes is the greatest in the world. Just the active volcanoes number 128 here. A large portion of the active volcanoes (65) are situated on Java and Sumatra; we will examine only these two islands.

The Indonesian island arc is double, and the volcanoes are confined to the inner arc. On the south the arc is enclosed by the Java Trench, whose depth reaches 7450 m. Strong gravity anomalies were first found in the Indonesian arc, where they are associated with the Java Trench. Negative anomalies in the trench region are interpreted as a peculiar "sucking down" of the earth's crust, and on this basis a geotectonic hypothesis was constructed (Umbgrove, 1952; Kuenen, 1952; Bemmelen, 1957) which now has only historical interest. With the abyssal trench is linked a zone of deep-focus earthquakes whose focal plane dips to the north, toward the con-

TABLE 29.  Average Lava Compositions of Indonesian Volcanoes

| Compo-nents | 1 | 2 | 3 | 4 | 5 | 6 | 7 |
|---|---|---|---|---|---|---|---|
| $SiO_2$ | 51.0 | 52.9 | 55.8 | 61.6 | 50.3 | 57.5 | 51.6 |
| $TiO_2$ | — | — | — | — | 1.0 | 0.7 | — |
| $Al_2O_3$ | 18.2 | 19.2 | 19.1 | 17.4 | 18.2 | 18.3 | 18.5 |
| $Fe_2O_3$ | 4.8 | 3.7 | 4.7 | 4.2 | 4.6 | 4.0 | 3.2 |
| FeO | 5.9 | 5.4 | 3.4 | 2.2 | 4.3 | 2.6 | 7.7 |
| MnO | 0.2 | 0.2 | 0.2 | 0.1 | 0.2 | 0.2 | 0.2 |
| MgO | 5.5 | 4.1 | 2.9 | 2.5 | 3.7 | 1.8 | 5.7 |
| CaO | 9.8 | 9.3 | 8.1 | 5.7 | 9.3 | 5.7 | 9.3 |
| $Na_2O$ | 3.0 | 3.5 | 3.6 | 3.7 | 3.6 | 4.0 | 2.7 |
| $K_2O$ | 1.6 | 1.7 | 2.2 | 2.6 | 4.8 | 5.1 | 1.1 |
| $a$ | 9.1 | 10.5 | 11.5 | 12.1 | 15.3 | 16.6 | 7.8 |
| $c$ | 7.9 | 8.1 | 7.5 | 5.8 | 4.8 | 4.3 | 8.8 |
| $b$ | 23.8 | 19.6 | 15.4 | 11.3 | 21.4 | 12.0 | 23.2 |
| $s$ | 59.2 | 61.8 | 65.6 | 70.8 | 58.5 | 67.1 | 60.2 |
| $a'$ | — | — | — | — | — | — | — |
| $f'$ | 42 | 44 | 50 | 52 | 39 | 51 | 45 |
| $m'$ | 40 | 37 | 33 | 37 | 30 | 25 | 43 |
| $c'$ | 18 | 19 | 17 | 11 | 31 | 24 | 12 |
| $n'$ | 75 | 76 | 72 | 70 | 53 | 54 | 80 |

| Compo-nents | 8 | 9 | 10 | 11 | 12 | 13 | 14 |
|---|---|---|---|---|---|---|---|
| $SiO_2$ | 54.1 | 62.1 | 68.3 | 52.9 | 55.3 | 61.5 | 73.8 |
| $TiO_2$ | — | — | — | — | — | — | — |
| $Al_2O_3$ | 18.6 | 18.6 | 16.1 | 16.4 | 17.7 | 17.5 | 14.3 |
| $Fe_2O_3$ | 3.7 | 2.8 | 1.9 | 2.8 | 3.1 | 3.4 | 1.4 |
| FeO | 5.9 | 3.0 | 2.2 | 6.5 | 5.6 | 3.2 | 0.7 |
| MnO | 0.1 | 0.1 | 0.1 | 0.1 | 0.1 | 0.1 | — |
| MgO | 4.4 | 1.7 | 1.0 | 7.9 | 5.0 | 2.8 | 0.5 |
| CaO | 8.7 | 5.8 | 3.4 | 8.8 | 8.5 | 6.1 | 1.9 |
| $Na_2O$ | 3.5 | 3.9 | 4.4 | 3.3 | 3.2 | 3.4 | 3.7 |
| $K_2O$ | 1.0 | 2.0 | 2.6 | 1.3 | 1.5 | 2.0 | 3.7 |
| $a$ | 9.4 | 11.8 | 13.4 | 9.0 | 9.5 | 10.6 | 13.1 |
| $c$ | 8.0 | 6.8 | 3.9 | 6.2 | 7.3 | 6.5 | 2.2 |
| $b$ | 19.6 | 8.8 | 5.6 | 25.8 | 19.8 | 11.8 | 3.5 |
| $s$ | 63.0 | 72.6 | 77.1 | 59.0 | 63.4 | 70.1 | 81.2 |
| $a'$ | — | — | — | — | — | — | 23 |
| $f'$ | 47 | 61 | 67 | 33 | 41 | 52 | 53 |
| $m'$ | 39. | 34 | 31 | 51 | 43 | 40 | 24 |
| $c'$ | 14 | 5 | 2 | 16 | 16 | 8 | — |
| $n'$ | 83 | 75 | 71 | 79 | 76 | 72 | 60 |

Central Java:  1. Basalt, average of 5 analyses; 2. Andesite-basalt, average of 16 analyses; 3. Andesite, average of 23 analyses; 4. Andesite-dacite, average of 8 analyses.  Muriah Volcano:  5. Phonolite, average of 2 analyses; 6. Leucite phonolite, 1 analysis. Krakatau Volcano:  7. Basalt, average of 2 analyses; 8. Andesite-basalt, average of 4 analyses; 9. Dacite, average of 2 analyses; 10. Rhyolite, average of 10 analyses. Sumatra:  11. Basalt, average of 9 analyses; 12. Andesite-basalt, average of 4 analyses; 13. Andesite, average of 23 analyses; 14. Rhyolite, average of 20 analyses.

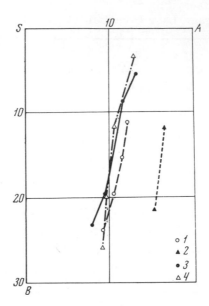

Fig. 90. Diagram of the average lava
compositions of the Indonesian volcanoes:
1) Java; 2) Muriah Volcano; 3) Krakatau
Volcano; 5) Sumatra.

tinent of Asia. Information on the structure of the crust is lacking.*
Beneath the large islands the crust is similar to the continental
type.

A petrochemical analysis of the lavas of the Indonesian volcanoes was made by Rittmann (1953); he showed that the alkalinity
of the rocks increases from the outer toward the inner part of the
arc. Let us see how this looks in terms of the Zavaritskii system.
The volcanoes of central Java (Table 29, Fig. 30) are similar to the
Yellowstone type. On the back side of the arc, on the north coast
of Java, is Muriah Volcano with its leucite-bearing lavas. As was
established by our investigations (Tazieff, Marinelli, and Gorshkov,
1966), this volcano has erupted within historic times, and the possibility of future eruptions is not eliminated. The variation curve
for its lavas falls in the fields of strongly alkalic lavas.

---

* At the Second International Oceanographic Congress (Moscow, 1966) R. W. Raitt reported that in the Banda Sea the crust was of oceanic type.

The lavas of Krakatau Volcano have a more calcic character than the lavas of the volcanoes of central Java; they belong to the Lassen Peak type.

The lavas of the Sumatra volcanoes are similar to this type; the variation curve for the Sumatra lavas has a steeper, "Asiatic" slope in contrast to the more gentle, "American"slope of the curves for Java and Krakatau.

As can be seen, there are no substantial differences between the Indonesian arc and the arcs of the circum-Pacific zone.

All of the island and volcanic arcs have several common, characteristic features:

The island and volcanic arcs lie on crust of diverse types, ranging from typically continental (Kamchatka and North America, for example) to typically oceanic or suboceanic (Marianas and Tonga Islands, for example). A common trait of all the arcs is the peculiar structure of the upper mantle that was noted earlier in the Kurile Islands: lower than normal seismic velocities in the subcrustal parts of the mantle and absence of the wave-guide or Gutenberg layer in the mantle. A characteristic distribution of gravity anomalies is found in all of them, and (except in North America) the island and volcanic arcs are accompanied by abyssal trenches and zones of deep-focus earthquakes. Heat flow in these regions has twice the average value for the earth as a whole.

The distinctive petrochemical features of the lavas of the island and volcanic arcs are rather similar. All the rocks belong to the calc-alkalic group, and the alkalinity of the rocks increases very significantly inward from the outer edge of the arc. In addition, even in the passage of an arc from the oceanic portion of the crust to the continental (e.g., in the Fuji arc), the petrochemical rock type remains the same. The "anomalous arcs" of the southwestern Pacific are similar in their petrochemical relationships to all the other arcs.

Island arcs sometimes are developed directly on oceanic crust, not along the margin of a continent, but at a distance of several hundred kilometers from it (e.g., the Bonin-Marianas arc). The trend of the variation curves for the lavas of the east Asiatic arcs is somewhat different from the trend of the curves for the American arcs;

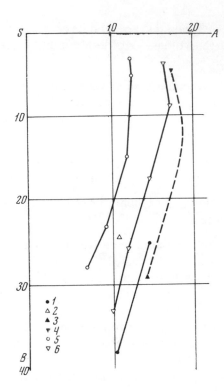

Fig. 91. Diagram of the average compositions
of lavas from some intracontinental volcanoes:
1) Uyun-Kholdonga Volcano; 2) Anyui Volcano; 3) Balgan-Tas Volcano; 4) Mayak Dome;
5) calc-alkalic lavas of eastern Asia; 6) alkalic lavas of eastern Asia.

on this basis one can recognize two groups of calc-alkalic rocks:
East Asiatic and American. The difference in the trends of the
variation curves does not depend on the type of structure or thickness of the crust. Thus, East-Asiatic slopes of variation curves
are found in Kamchatka, with its continental crust, and in the Marianas Islands, with their oceanic crust, while American-type curves
are found in the continental arc of Central America and in the oceanic
arc of the Tonga Islands.

## INTRACONTINENTAL VOLCANOES

We have already pointed out, in examining the petrochemistry
of the North American volcanic arc, that the volcanoes situated in

TABLE 30.  Average Lava Compositions
of Recent Volcanoes in Eastern Asia

| Compo-nents | 1 | 2 | 3 | 4 | 5 | Zavarit-skii values | 1 | 2 | 3 | 4 | 5 |
|---|---|---|---|---|---|---|---|---|---|---|---|
| SiO$_2$ | 44.5 | 51.8 | 52.6 | 46.3 | 68.5 | $a$ | 10.2 | 14.4 | 10.8 | 14.1 | 17.4 |
| TiO$_2$ | 2.6 | 2.4 | 1.8 | 3.2 | 0.5 | $c$ | 2.8 | 2.0 | 4.8 | 3.0 | 1.7 |
| Al$_2$O$_3$ | 12.6 | 14.3 | 15.5 | 15.6 | 16.0 | $b$ | 38.1 | 25.2 | 24.4 | 29.2 | 4.5 |
| Fe$_2$O$_3$ | 6.1 | 3.0 | 2.2 | 4.9 | 1.9 | $s$ | 48.9 | 58.4 | 60.0 | 53.7 | 76.4 |
| FeO | 4.6 | 6.1 | 7.8 | 5.6 | 0.6 | $f'$ | 24 | 32 | 38 | 32 | 48 |
| MnO | 0.2 | 0.1 | 0.1 | 0.2 | 0.1 | $m'$ | 51 | 44 | 44 | 44 | 37 |
| MgO | 12.3 | 6.9 | 6.6 | 7.9 | 1.0 | $c'$ | 25 | 24 | 18 | 24 | 15 |
| CaO | 10.8 | 6.8 | 7.7 | 8.7 | 2.0 | $n'$ | 40 | 51 | 73 | 74 | 70 |
| Na$_2$O | 2.5 | 3.5 | 3.7 | 4.9 | 5.7 | | | | | | |
| K$_2$O | 3.8 | 5.1 | 2.0 | 2.7 | 3.7 | | | | | | |

1. Most basic lava of the Mergen volcanoes; 2. Shikhlunite of the Mergen volcanoes; average of 9 analyses; 3. Basalt, Anyui Volcano, average of 6 analyses; 4. Trachy-basalt, Balgan-Tas Volcano, average of 13 analyses; 5. Trachyliparite, Mayak Dome, 1 analysis.

the back part of this arc erupt purely alkalic lavas (of the Maros-Highwood type) rather than calc-alkalic ones. We also find similar rocks in the back parts of the Ryukyu arc and on Muriah Volcano (on Java).

It remains for us to examine the phenomena of intracontinental volcanism in eastern Asia. In contrast to the cases just mentioned, the volcanoes of eastern Asia are not confined to the trends of contemporary island arcs, but form an irregular system.

In general, the present-day volcanoes on the continent of Asia are rather rare. Reliably recorded eruptions have taken place only in the Mergen group of volcanoes* in Manchuria in the eighteenth century. Here potash-rich basic rocks, shikhlunites, were erupted (Zavaritskii, 1939).

Detailed studies have been made rather recently on two young volcanic cones in northeastern Asia: Anyui Volcano in the valley of the Monnya River (Ustiev, 1958) and Balgan-Tas Cone in the basin of the Moma River (Rudich, 1964). In the latter region a somewhat

---

*There is rather vague information on recent eruptions in Tibet, but no data on the composition of the lavas.

TABLE 31. Average Chemical
Compositions of Lavas of the
Calc-Alkalic Series of the
Baikal Region and of Some Fields
in Eastern and Central Asia

| Components | 1 | 2 | 3 | 4 | 5 |
|---|---|---|---|---|---|
| $SiO_2$ | 48.4 | 55.8 | 57.9 | 67.1 | 76.8 |
| $TiO_2$ | 1.5 | 1.8 | 1.3 | 0.6 | 0.3 |
| $Al_2O_3$ | 15.7 | 13.4 | 16.8 | 16.3 | 12.0 |
| $Fe_2O_3$ | 6.5 | 5.0 | 4.3 | 3.0 | 1.5 |
| $FeO$ | 5.6 | 5.1 | 3.3 | 9.0 | 0.4 |
| $MnO$ | 0.2 | 0.2 | 0.1 | 0.1 | — |
| $MgO$ | 8.4 | 5.5 | 3 8 | 0.8 | 0.2 |
| $CaO$ | 9.6 | 7.9 | 5.6 | 4.2 | 0.8 |
| $Na_2O$ | 1.8 | 3.1 | 3.1 | 3.5 | 1.7 |
| $K_2O$ | 1.9 | 1.8 | 3.3 | 3.3 | 6.3 |
| $P_2O_5$ | 0.4 | 0.5 | 0.5 | 0.2 | — |
| | | | | | |
| $a$ | 6.9 | 9.2 | 11.8 | 12.5 | 12.3 |
| $c$ | 7.3 | 4.0 | 5.4 | 4.8 | 0.9 |
| $b$ | 28 0 | 23.2 | 14.9 | 5.3 | 3.2 |
| $s$ | 57.8 | 63.6 | 67.9 | 77 4 | 83.6 |
| $a'$ | — | — | — | — | 41 |
| $j'$ | 37 | 39 | 47 | 86 | 49 |
| $m'$ | 48 | 39 | 44 | 11 | 10 |
| $c'$ | 15 | 22 | 9 | 3 | — |
| $n'$ | 61 | 82 | 59 | 62 | 29 |

1. Olivine basalt, average of 51 analyses;
2. Andesite-basalt, average of 32 analyses; 3. Andesite, average of 12 analyses; 4. Dacite, average of 3 analyses;
5. Liparite, average of 3 analyses.

older trachyliparite dome (Mayak) has also been found. Data on the chemistry of the rocks of all these volcanoes are presented in Table 30 and are displayed graphically in Fig. 91.

Because the lavas of the young volcanoes are all alkalic basaltoids, it is rather difficult to judge the character of their differentiation; more acid differentiates are known only among the older lavas (middle and early Quaternary). In a recently published summary Belov (1963) gives average compositions for the calc-alkalic and alkalic rocks of the Baikal region and for some of the lava fields of eastern and central Asia; we have reproduced these in Tables 31 and 32 and in Fig. 91. Belov used 220 analyses in all.

TABLE 32.  Average Chemical
Compositions of Lavas of the
Alkali-Calcic Series of the
Baikal Region and of Some Fields
in Eastern and Central Asia

| Components | 1 | 2 | 3 | 4 | 5 * |
|---|---|---|---|---|---|
| $SiO_2$ | 46.6 | 49.4 | 54.6 | 62.6 | 70.2 |
| $TiO_2$ | 2.5 | 2.4 | 2.0 | 1.0 | 0.5 |
| $Al_2O_3$ | 14.0 | 15.9 | 16.5 | 16.7 | 14.7 |
| $Fe_2O_3$ | 5.6 | 4.4 | 5.0 | 3.4 | 2.3 |
| FeO | 7.2 | 6.4 | 3.5 | 1.1 | 0.9 |
| MnO | 0.2 | 0.1 | 0.1 | 0.1 | 0.1 |
| MgO | 8.2 | 6.5 | 4.0 | 1.5 | 0.4 |
| CaO | 10.0 | 8.1 | 5.6 | 4.0 | 1.4 |
| $Na_2O$ | 3.4 | 3.7 | 3.8 | 4.8 | 4.0 |
| $K_2O$ | 1.8 | 2.5 | 4.2 | 4.6 | 5.4 |
| $P_2O_5$ | 0.5 | 0.6 | 0.7 | 0.2 | 0.1 |
| | | | | | |
| $a$ | 9.8 | 11.8 | 14.5 | 17.1 | 16.3 |
| $c$ | 3.5 | 4.7 | 3.0 | 2.5 | 1.5 |
| $b$ | 33.2 | 25.8 | 17.5 | 8.9 | 3.6 |
| $s$ | 53.5 | 57.7 | 65.0 | 71.5 | 78.6 |
| $f'$ | 35 | 38 | 44 | 45 | 76 |
| $m'$ | 41 | 42 | 39 | 29 | 18 |
| $c'$ | 24 | 20 | 17 | 26 | 6 |
| $n'$ | 74 | 69 | 58 | 61 | 53 |

1.  Limburgitic basanitoids, average of 26
analyses; 2.  Basaltic trachybasalt, average of 46 analyses; 3.  Trachyandesitic
basalt, average of 30 analyses; 4.  Trachyte,
average of 9 analyses; 5.  Comendite, average of 9 analyses.

* An error of calculation that occurred in
the original has been corrected.

It will be seen from Tables 30-32 and from Fig. 91 that the calc-
alkalic continental lavas belong to the Yellowstone-Lassen Peak
type, the alkalic ones to the Highwood-Etna type.  The lavas of the
Mergen volcanoes and of Balgan-Tas are still more alkalic.

The variation curves for the lavas of the continental volcanoes
parallel those of the island arcs of eastern Asia.  In both cases they
are somewhat steeper than the corresponding curves of Zavaritskii
and cut across them.

In general, the trend of the variation curves for the lavas of intracontinental volcanoes coincides with that for the calc-alkalic lavas of the island and volcanic arcs; a continuous transition can be traced between them. As one goes from the outer edge of the island arc into the continent, more and more alkalic rocks are found. The most complete transition between the types can be seen in North America (from the Lassen Peak to the Highwood types). The change from calc-alkalic to alkalic rocks is also observed during the geologic development in the transition from the geosynclinal to the platform regime.

The crustal structure in the continents is more or less uniform; it is a thick (35 km or more), two-layered crust with a normal distribution of seismic velocities. In the subcrustal part of the mantle the seismic velocity is 8.0-8.2 km/sec, and at a depth of 120-200 km a layer with lower seismic velocities (the wave-guide or Gutenberg zone) is observed.

## INTRAOCEANIC VOLCANOES

Until recently the opinion was widespread that the lavas of the inner parts of the oceans were mainly alkalic olivine basalts that gave rise to the alkalic differentiation series. Recent studies have established that broad areas of the ocean floors are covered by characteristic oceanic tholeiitic basalts that are distinguished from the

TABLE 33. Average Composition
of Oceanic Tholeiites

| Components | Weight percent | Numerical characteristics |
|---|---|---|
| $SiO_2$ | 50.25 | $a = 6.4$ |
| $TiO_2$ | 1.56 | $c = 7.5$ |
| $Al_2O_3$ | 16.09 | $b = 27.9$ |
| $Fe_2O_3$ | 2.72 | $s = 58.2$ |
| $FeO$ | 7.20 | $f' = 34$ |
| $MnO$ | 0.19 | $m' = 42$ |
| $MgO$ | 7.02 | $c' = 24$ |
| $CaO$ | 11.82 | $n' = 96$ |
| $Na_2O$ | 2.81 | |
| $K_2O$ | 0.20 | |
| $P_2O_5$ | 0.15 | |

TABLE 34.  Average Chemical
Compositions of the Lavas
of the Hawaiian Islands

| Components | 1 | 2 | 3 | 4 | 5 |
|---|---|---|---|---|---|
| $SiO_2$ | 49.36 | 46.46 | 48.60 | 51.90 | 61.73 |
| $TiO_2$ | 2.50 | 3.01 | 3.16 | 2.57 | 0.50 |
| $Al_2O_3$ | 13.94 | 14.61 | 16.49 | 16.65 | 18.03 |
| $Fe_2O_3$ | 3.03 | 3.27 | 4.19 | 4.25 | 3.33 |
| FeO | 8.53 | 9.11 | 7.40 | 6.17 | 1.49 |
| MnO | 0.16 | 0.14 | 0.18 | 0.21 | 0.24 |
| MgO | 8.44 | 8.19 | 4.70 | 3.56 | 0.41 |
| CaO | 10.30 | 10.33 | 7.79 | 6.30 | 1.17 |
| $Na_2O$ | 2.13 | 2.92 | 4.43 | 5.22 | 7.42 |
| $K_2O$ | 0.38 | 0.84 | 1.60 | 2.01 | 4.19 |
| $P_2O_5$ | 0.26 | 0.37 | 0.69 | 0.93 | 0.17 |
| $a$ | 5.1 | 7.7 | 12.3 | 14.7 | 22.6 |
| $c$ | 6.6 | 5.8 | 5.1 | 4.0 | 0.9 |
| $b$ | 30.5 | 31.9 | 23.7 | 19.5 | 5.7 |
| $s$ | 57.8 | 54.6 | 58.9 | 61.8 | 70.8 |
| $f'$ | 35 | 36 | 47 | 50 | 79 |
| $m'$ | 46 | 43 | 34 | 31 | 12 |
| $c'$ | 19 | 21 | 19 | 19 | 9 |
| $n'$ | 90 | 84 | 80 | 80 | 73 |

1. Tholeiitic basalt, average of 181 analyses;
2. Alkalic olivine basalt, average of 28 analyses; 3. Hawaiite, average of 33 analyses; 4.
Mugearite, average of 13 analyses; 5. Soda
trachyte, average of 5 analyses.

continental tholeiites by their markedly lower content of potash,
which does not exceed 0.25% (Engel and Engel, 1964). In Table 33
the average composition of oceanic tholeiite is given, based on six
samples taken from different parts of the floor of the Pacific Ocean.

The oceanic tholeiites form the pedestals of all the intraoceanic islands and the alkalic olivine basalt-tracyte series
forms only a capping that does not exceed 3-5% of the total
volume of the island in question (Macdonald and Katsura, 1962;
Engel and Engel, 1964).

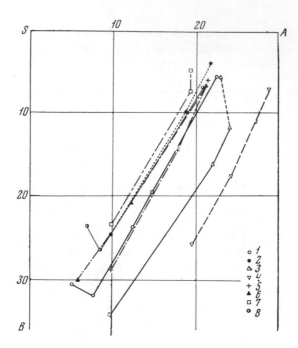

Fig. 92. Diagram of the average lava compositions of some intraoceanic volcanoes: 1) Hawaiian Islands; 2) Marquesas Islands; 3) Tahiti; 4) Rarotonga (Cook Islands); 5) Ponape (Eastern Caroline Islands); 6) Guadalupe Island; 7) San Benedicto Island (Revilla Gigedo Islands); 8) Galapagos Islands.

Acid differentiation products of the oceanic tholeiites are not known, but on the Hawaiian Islands an interlayering of the tholeiitic and olivine basalts is observed; transitional varieties are also known. Apparently the oceanic tholeiites give rise to the alkalic olivine basalts and later these in turn change, through a series of differentiates, to trachyte.

The intraoceanic islands form chains that extend for long distances, trending from southeast to northwest or west-northwest. The islands are more numerous in the southern half of the ocean (Polynesia).

The oceanic crust is thin (5 km, if the thickness of the water layer is neglected) and one-layered. The seismic velocities in the subcrustal parts of the mantle have normal values of 8.0-8.2 km/sec.

TABLE 35.  Average Chemical
Compositions of Lavas
of the Marquesas Islands

| Components | 1 | 2 | 3 |
|---|---|---|---|
| $SiO_2$ | 47.5 | 57.7 | 62.7 |
| $TiO_2$ | 3.7 | 1.2 | 0.7 |
| $Al_2O_3$ | 17.5 | 19.1 | 19.1 |
| $Fe_2O_3$ | 5.3 | 4.6 | 2.8 |
| FeO | 6.0 | 1.3 | 0.8 |
| MnO | 0.2 | 0.2 | 0.1 |
| MgO | 5.2 | 1.7 | 0.2 |
| CaO | 9.6 | 4.4 | 1.8 |
| $Na_2O$ | 3.0 | 5.8 | 6.2 |
| $K_2O$ | 2.0 | 4.0 | 5.6 |
| *a* | 9.7 | 18.8 | 21.8 |
| *c* | 7.1 | 3.6 | 1.9 |
| *b* | 24.6 | 10.1 | 4.0 |
| *s* | 58.6 | 67.5 | 72.3 |
| *f'* | 44 | 54 | 82 |
| *m'* | 36 | 28 | 11 |
| *c'* | 20 | 18 | 7 |
| *n'* | 70 | 69 | 63 |

**1.** Basalt, average of 11 analyses; **2.**
Andesine andesite, 1 analysis; **3.** Trachyte,
average of 3 analyses.

The low-velocity layer in the mantle beneath the oceans is very
much thicker than beneath the continents; it extends in depth from
60 to 200 km.

*Hawaiian Islands*

The Hawaiian Islands are one of the best-studied areas; many
petrographic theories have arisen as a result of their study.  This
makes it natural to begin our review of oceanic petrochemistry with
the Hawaiian Islands.

According to the results of geophysical investigations (Shor,
1960; Eaton, 1962), the crust in the region of the Hawaiian Islands
has a normal oceanic structure, but directly beneath the islands the
M discontinuity has been pushed down and the crust thickened be-

TABLE 36. Average Chemical
Compositions of the Lavas
of Tahiti

| Components | 1 | 2 | 3 | 4 |
|---|---|---|---|---|
| $SiO_2$ | 44.5 | 50.5 | 55.8 | 61 7 |
| $TiO_2$ | 3.5 | 2.5 | 1.3 | 0.9 |
| $Al_2O_3$ | 14.4 | 19.6 | 19.2 | 18.8 |
| $Fe_2O_3$ | 4.6 | 3.5 | 3.2 | 2.0 |
| FeO | 7.9 | 4.2 | 2.2 | 1.6 |
| MnO | 0.2 | 0.3 | 0.2 | 0.1 |
| MgO | 8.4 | 2.1 | 1.7 | 0.9 |
| CaO | 11.4 | 5.9 | 3.7 | 1.6 |
| $Na_2O$ | 3.5 | 7.3 | 7.8 | 7.0 |
| $K_2O$ | 1.6 | 4.1 | 4.9 | 5.4 |
| $a$ | 9.7 | 22.0 | 24.1 | 23.0 |
| $c$ | 4.4 | 2.1 | 0.5 | 0.9 |
| $b$ | 34.0 | 16.1 | 11.8 | 5.7 |
| $s$ | 51.9 | 59.8 | 63.6 | 70.4 |
| $f'$ | 33 | 46 | 43 | 55 |
| $m'$ | 41 | 22 | 24 | 27 |
| $c'$ | 26 | 32 | 33 | 18 |
| $n'$ | 77 | 74 | 71 | 67 |

1. Basanitoid, average of 9 analyses; 2. Ta-
hitite (trachyandesite), average of 7 analyses; 3. Phonolite, average of 3 analyses;
4. Phonolitic trachyte, average of 3 analyses.

cause of the piling up of the volcanic rocks. More detailed seismic,
gravimetric, and magnetometric studies (Furumoto and Woollard,
1965, and others) have shown that directly beneath the volcanoes there
are columnar bodies that reach through the crust and into the upper
mantle. The seismic velocity in these bodies is 7.7 km/sec. Apparently they represent the feeder zones of the volcanoes. The fact
that eruptions of the Hawaiian volcanoes are often preceded by earthquakes with hypocenter depths of 40-60 km agrees well with this;
the depth of the hypocenters gradually decreases, and when the earthquakes approach the surface an eruption begins (Eaton and Murata,
1960).

Chemically and petrochemically the lavas of the Hawaiian Islands have been studied repeatedly. Recently Macdonald and Katsura

TABLE 37. Chemical Compositions
of the Lavas of Rarotonga
(Cook Islands)

| Components | 1 | 2 | 3 |
|---|---|---|---|
| $SiO_2$ | 47.3 | 50.2 | 55.2 |
| $TiO_2$ | 3.3 | 2.0 | 0.8 |
| $Al_2O_3$ | 14.5 | 16.0 | 17.6 |
| $Fe_2O_3$ | 4.7 | 6.3 | 5.8 |
| $FeO$ | 6.0 | 3.3 | 1.0 |
| $MnO$ | 0.1 | — | — |
| $MgO$ | 3.2 | 1.7 | 1.4 |
| $CaO$ | 10.5 | 7.8 | 3.1 |
| $Na_2O$ | 6.5 | 8.6 | 9.4 |
| $K_2O$ | 3.7 | 4.1 | 5.7 |
| $a$ | 18.5 | 21.2 | 23.9 |
| $\bar{c}$ | 0.3 | 3.1 | 5.1 |
| $b$ | 26.8 | 17.6 | 7.2 |
| $s$ | 54.4 | 58.1 | 63.8 |
| $a + \bar{c}$ | 18.8 | 24.3 | 29.0 |
| $f'$ | 34 | 31 | 12 |
| $m'$ | 20 | 16 | 35 |
| $c'$ | 46 | 53 | 53 |
| $n'$ | 73 | 76 | 72 |

1. Alkali basalt, average of 2 analyses;
2. Murite, 1 analysis; 3. Phonolite, 1
analysis.

made 143 new chemical analyses with an estimate of the abundance
of the rocks and developed new average values on the basis of the
260 best analyses available (Macdonald and Katsura, 1964). These
values are given in Table 34 and the recalculated results are represented in Fig. 92. The variation curve for the lavas of the Hawaiian
Islands is considerably more gentle than the curves for lavas of the
island arcs and continental volcanoes, but the trend of the Hawaiian
curve, based on the new data, is not as gentle as on Zavaritskii's
diagram (1950, Fig. 87). Zavaritskii combined the dominant tholeiitic basalts and the alkalic differentiates of the subordinate olivine
basalts. This made the curve appear to be more gentle. The trend
of the Hawaiian variation curve has been revised for the present
work on the basis of the results of the more recent investigations
(Fig. 92).

*Polynesia*

There are rather numerous analyses of the lavas of Polynesia in the old reports of Lacroix (1927), and Lacroix (1928) and Chubb (1929) have published additional data on the Marquesas Islands.

In Lacroix's report analyses are not very numerous for many of the islands, and for others the analyses are very old; for this reason we have used only the analyses for the Marquesas Islands and for Tahiti (Society Islands), for which analyses are more numerous. The average values for the Marquesas Islands (19 analyses) and for Tahiti (22 analyses) are presented in Tables 35 and 36, respectively, and the results of the recalculations are presented in Fig. 92. From these data it can be seen that the variation curves for the Marquesas Islands and Tahiti run approximately parallel to the Hawaiian curve, though they differ in general alkalinity.

Let us look now at the petrochemistry of the lavas of Rarotonga (Cook Islands). There are only four analyses from here, but the rocks are saturated in alkalies and their examination is of systematic interest. The analytical and recalculated data are given in Table 37. In rocks that are not saturated in alkalies, recalculation according to the Zavaritskii method assigns all the alkalies as "aluminosilicate alkalinity" (the $a$ characteristic) and the remainder to the mafic minerals ($\bar{c}$ characteristic). Usually we work with rocks that are undersaturated in alkalies, and in plotting the aluminosilicate alkalinity on a diagram we take care of the total alkalinity at the same time. In working with alkali-oversaturated rocks we divide the total alkalinity into two parts ($a$ and $\bar{c}$), each of which is plotted separately on the right-hand side of the diagram. In this case the $a$ characteristic does not represent the total alkalinity, but only part of it. In comparing different rock series this creates considerable difficulty, because in individual cases the variation curves of the alkali-oversaturated rocks are plotted to the left of the curves for undersaturated rocks and the form of the variation curves is strongly modified.

We are interested primarily in the character of the changes produced during differentiation in the total alkalinity; thus, for the oversaturated rocks we will plot the value of $a + \bar{c}$ on the right side of the diagram, and not just the value of $a$. We think that doing it in this way is logically more valid.

The variation curve for the Rarotonga lavas, plotted in this way, is parallel to the curves of the other islands of Polynesia, but

TABLE 38. Average Chemical
Compositions of Lavas
of Ponape
(East Caroline Islands)

| Components | 1 | 2 | 3 |
|---|---|---|---|
| $SiO_2$ | 46.1 | 53.2 | 64.1 |
| $TiO_2$ | 3.3 | 1.8 | 0.5 |
| $Al_2O_3$ | 16.6 | 19.9 | 19.5 |
| $Fe_2O_3$ | 4.7 | 3.7 | 2.0 |
| FeO | 8.7 | 3.3 | 0.7 |
| MnO | 0.2 | 0.2 | 0.2 |
| MgO | 6.5 | 2.1 | 0.6 |
| CaO | 9.1 | 7.1 | 1.1 |
| $Na_2O$ | 3.5 | 6.2 | 7.4 |
| $K_2O$ | 1.3 | 2.5 | 3.9 |
| $a$ | 9.8 | 17.7 | 21.5 |
| $c$ | 6.3 | 4.5 | 1.3 |
| $b$ | 28.8 | 14.4 | 6.1 |
| $s$ | 55.1 | 63.1 | 71.1 |
| $a'$ | — | — | 46 |
| $f'$ | 43 | 45 | 39 |
| $m'$ | 40 | 25 | 15 |
| $c'$ | 17 | 30 | — |
| $n'$ | 79 | 78 | 75 |

1. Basalt, average of 4 analyses; 2.
Mugearite, 1 analysis; 3. Trachyte,
average of 2 analyses.

it is shifted significantly to the right of them, reflecting the higher than usual alkalinity of the rocks.

If we examine the total alkalinity in going from the Marquesas Islands across the island chains (from northeast to southwest) through Tahiti to Rarotonga, we will find that the alkalinity of the rocks increases. Whether this is the expression of some principle or whether it is a chance occurrence is still uncertain.

### East Caroline Islands

The Caroline Islands group in the western Pacific is divided genetically into two parts. The West Caroline Islands belong to the

Table 39.  Average Compositions
of Lavas of Guadalupe Island

| Components | 1 | 2 | 3 |
|---|---|---|---|
| $SiO_2$ | 49.1 | 50.6 | 62.4 |
| $TiO_2$ | 1.2 | 2.8 | 0.6 |
| $Al_2O_3$ | 15.0 | 17.9 | 18.1 |
| $Fe_2O_3$ | 3.3 | 4.8 | 3.2 |
| FeO | 5.7 | 5.2 | 1.4 |
| MnO | 0.2 | 0.2 | 0.2 |
| MgO | 7.7 | 4.4 | 0.9 |
| CaO | 12.7 | 8.1 | 2.0 |
| $Na_2O$ | 2.4 | 4.3 | 6.7 |
| $K_2O$ | 0.2 | 1.7 | 4.5 |
| $a$ | 5.6 | 12.2 | 21.1 |
| $c$ | 7.4 | 6.1 | 1.5 |
| $b$ | 30.0 | 20.8 | 6.7 |
| $s$ | 57.0 | 60.9 | 70.7 |
| $f'$ | 28 | 45 | 62 |
| $m'$ | 44 | 36 | 23 |
| $c'$ | 28 | 19 | 15 |
| $n'$ | 95 | 79 | 69 |

1. Tholeiitic basalt, experimental Mohole
near Guadalupe Island; 2. Alkalic olivine basalt, average of 4 analyses; 3. Trachyte, average of 2 analyses.

island-arc system; the East Caroline Islands are intraoceanic (and
in part coralline) islands.  There is a small number of modern analyses for the island of Ponape (Yagi, 1960).  The data on these is
presented in Table 38 and in Fig. 92.  As one can see, the Ponape
lavas are similar petrochemically to the Hawaiian lavas.  Their
variation curve is somewhat steeper and intersects the curve for
the Hawaiian lavas.

### Islands of the Eastern Pacific Ocean

Let us examine the petrochemistry of the volcanic rocks of
some of the islands of the eastern Pacific Ocean.

Guadalupe Island lies near the coast of Lower California,
from which it is separated by the abyssal Cedros Trench.  The is-

TABLE 40. Average Chemical
Compositions of the Lavas
of the Revilla Gigedo Islands

| Components | 1 | 2 | 3 | 4 |
|---|---|---|---|---|
| $SiO_2$ | 47.1 | 55.3 | 63.1 | 69.0 |
| $TiO_2$ | 2.5 | 1.6 | 0.6 | 0.2 |
| $Al_2O_3$ | 19.8 | 17.2 | 16.4 | 14.8 |
| $Fe_2O_3$ | 4 7 | 1.6 | 1.5 | 1 4 |
| FeO | 5.6 | 7.9 | 4 6 | 2.8 |
| MnO | 0.1 | 0.2 | 0.1 | 0.1 |
| MgO | 5.2 | 2.5 | 0.6 | 0 2 |
| CaO | 10 4 | 5.7 | 2.8 | 0.9 |
| $Na_2O$ | 3.5 | 5.7 | 6.4 | 6.1 |
| $K_2O$ | 1.1 | 2.3 | 3.9 | 4.5 |
| $a$ | 9.8 | 15.6 | 19.3 | 19.1 |
| $c$ | 8.9 | 3.7 | 2.4 | 0.2 |
| $b$ | 23.3 | 16.5 | 7.5 | 4.9 |
| $s$ | 58.0 | 64.2 | 70.8 | 75.8 |
| $f'$ | 42 | 55 | 74 | 73 |
| $m'$ | 39 | 25 | 13 | 5 |
| $c'$ | 19 | 20 | 13 | 22 |
| $n'$ | 83 | 79 | 72 | 68 |

1. Trachybasalt, average of 4 analyses;
2. Trachyandesite, 1 analysis; 3. Soda
trachyte, average of 4 analyses; 4. Soda
rhyolite, average of 3 analyses.

land is situated on crust of oceanic type at a distance of about 100 km from the continental slope. There are six modern analyses of the lavas of Guadalupe Island (Engel and Engel, 1964), the average values of which are given in Table 39; the recalculated results are plotted in Fig. 92.

The data on the composition of the tholeiitic basalt from the experimental Mohole drilled near Guadalupe on the ocean floor to a depth of 3566 m are also shown in Table 39 and Fig. 92 (Engel and Engel, 1961).

The rocks called olivine basalt (column 2 in Table 39) correspond most closely in their composition and according to the recalculated results to hawaiites or mugearites. The relatively small

TABLE 41. Analyses of Lavas
from the Galapagos Islands*

| Components | 1 | 2 | 3 |
|---|---|---|---|
| $SiO_2$ | 46.5 | 45.6 | 61.9 |
| $TiO_2$ | 1.7 | 1.8 | 0.2 |
| $Al_2O_3$ | 20.9 | 18.2 | 16.8 |
| $Fe_2O_3$ | 1.6 | 7.3 | 2.3 |
| FeO | 6.2 | 5.0 | 4.8 |
| MnO | 0.2 | 0.3 | 0.3 |
| MgO | 5.9 | 6.0 | 0.6 |
| CaO | 12.8 | 10.2 | 2.3 |
| $Na_2O$ | 2.6 | 3.2 | 7.2 |
| $K_2O$ | 0.4 | 0.8 | 3.2 |
| $a$ | 6.8 | 8.4 | 20.8 |
| $c$ | 11.5 | 8.1 | 1.0 |
| $b$ | 23.6 | 26.4 | 6.8 |
| $s$ | 58.1 | 57.1 | 71.4 |
| $f'$ | 34 | 44 | 71 |
| $m'$ | 44 | 39 | 10 |
| $c'$ | 22 | 17 | 19 |
| $n'$ | 91 | 85 | 77 |

1. Tholeiitic basalt (Engel and Engel, 1964); 2. Olivine basalt (Richardson, 1933); 3. Trachyte (Richardson, 1933).

*The figures are rounded off to the nearest 0.1%, but are not recalculated to 100%.

number of analyses does not allow us to judge with certainty the way in which the process of differentiation from tholeiitic basalts to alkalic basaltoids goes on — directly, as indicated by the points on the diagram, or through intermediate basic rocks that remain undiscovered. The Guadalupe variation curve almost coincides with the Marquesas Islands curve.

The island of San Benedicto (in the Revilla Gigedo group) lies not far from the coast of Mexico, from which it is separated by the abyssal Central American Trench. As mentioned in our discussion of the Mexican volcanoes, the Revilla Gigedo Islands lie on the trace of the Clarion fracture zone. If, as is assumed, the East Pacific Rise extends into the Gulf of California, then it may be as-

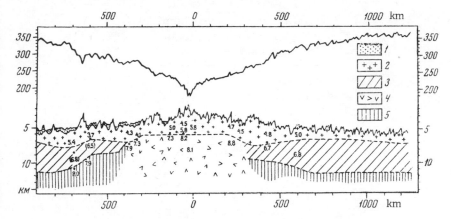

Fig. 93. Structure of the earth's crust and gravity anomalies in the area of the East Pacific Rise: 1) Unconsolidated sediment; 2) sedimentary layer; 3) "oceanic" ("basaltic") layer; 4) low-velocity region in the mantle; 5) upper mantle.

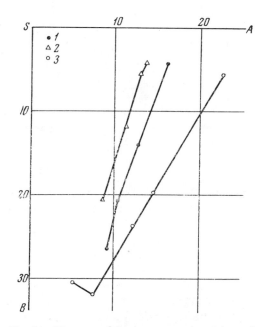

Fig. 94. Diagram of the average compositions of the lavas of Easter Island: 1) Easter Island; 2) Cascade Mountains of North America; 3) Hawaiian Islands.

## TABLE 42. Average Chemical Compositions of Lavas of Easter Island

| Components | 1 | 2 | 3 | 4 |
|---|---|---|---|---|
| $SiO_2$ | 48.4 | 54.0 | 61.4 | 72.8 |
| $TiO_2$ | 4.0 | 2.5 | 1.3 | 0.2 |
| $Al_2O_3$ | 15.4 | 15 1 | 14.7 | 13.3 |
| $Fe_2O_3$ | 4.7 | 5.3 | 3.3 | 2.0 |
| FeO | 8.7 | 8.1 | 6.4 | 1.8 |
| MnO | 0.2 | 0.2 | 0.2 | 0.1 |
| MgO | 4.7 | 2.9 | 1.3 | 0.1 |
| CaO | 9.4 | 6.7 | 4.8 | 0.8 |
| $Na_2O$ | 3.5 | 3 8 | 4.8 | 5.3 |
| $K_2O$ | 1.0 | 1.4 | 1.8 | 3.6 |
| $a$ | 9.2 | 10.5 | 12.9 | 16.1 |
| $c$ | 5.7 | 4 9 | 3.2 | 0.9 |
| $b$ | 26.3 | 20.6 | 14.0 | 4.3 |
| $s$ | 58.8 | 64.0 | 69.9 | 78.7 |
| $a'$ | — | — | — | 18 |
| $j'$ | 47 | 60 | 65 | 77 |
| $m'$ | 31 | 24 | 16 | 5 |
| $c'$ | 22 | 16 | 19 | — |
| $n'$ | 85 | 80 | 80 | 69 |

1. Basalt, average of 8 analyses; 2. Andesite-basalt, average of 2 analyses; 3. Andesite, average of 2 analyses; 4. Obsidian, average of 4 analyses.

sumed that the Revilla Gigedo Islands are located on the west flank of that rise, at a distance of about 200 km from its axis.

Data on the chemical composition of the San Benedicto lavas were kindly supplied to us by A. F. Richardson.* The average lava compositions are presented in Table 40, and the recalculated values are plotted in Fig. 92. As can be seen, the variation curve of the San Benedicto lavas is parallel to the curves of the other oceanic islands, being distinguished from them by a shift in the direction of somewhat smaller alkalinity.

---

*These were published in the Proceedings of the California Academy of Sciences, 1966, Vol. 33, No. 12.

The Galapagos Islands are located between the East
Pacific Rise and the coast of South America. The islands are composed of tholeiitic basalts with a "capping" of alkalic rocks. The
existing rather sparse analyses (Richardson, 1933; Engel and Engel,
1964) are given in Table 41 and are plotted in Fig. 92. It can be
seen from these data that the general course of differentiation here
and in the Hawaiian Islands is similar: from oceanic tholeiitic basalt "at right angles" to the general trend, to alkalic olivine basalt
and then to trachyte. The trend of the variation curve is parallel
to that of the other oceanic islands.

Easter (Pascua) Island, in contrast to the other intraoceanic islands, is located on the crest of the East Pacific Rise.
One of the major achievements of the IGY was the discovery of the
global system of oceanic ridges. The Pacific Ocean part of this system is traced through the southeastern part of the ocean toward the
Gulf of California; it is called the East Pacific Rise. The structure
of the earth's crust and upper mantle here is somewhat reminiscent
of that of an island arc: the seismic velocity in the subcrustal part
of the mantle has a low value (7.3-7.5 km/sec) and the Gutenberg
zone apparently is absent. The crust itself is thin, only 4 km (not
counting the thickness of the water layer), in the region of the rise.
In contrast to the nonseismic regions of the ocean floor, the oceanic
rise area is a seismic one, but the seismicity here is weaker than
in the island arcs, and the earthquakes do not exceed 60 km in depth.
On the ridge a weakly expressed, symmetrical minimum in the positive gravity anomaly (in its Bouguer reduction) is observed, from
+400-450 mgal on the ocean floor to +150-200 mgal at the rise axis.
In a narrow zone directly adjacent to the axis the heat-flow values
are 6-7 times greater than the average for the earth. In Fig. 93 a
sketch of the crustal and upper-mantle structure is presented, along
with the change in gravity anomalies in the area of the East Pacific
Rise (Talwani, 1965).

The chemical relationships of Easter Island were studied by
Bandy (1937), who also used the older data of Lacroix. The average
values for the important groups of lavas are given in Table 42, and
the recalculated results are plotted on a diagram in Fig. 94; on this
same diagram are plotted the variation curves for the lavas of the
Rocky Mountains [sic] of North America and for those of the Hawaiian Islands. As is clear from the diagram, the curve of the
Easter Island lavas is parallel to the curve of the continental rocks

and is different from the trend of the oceanic lavas. It coincides with the Yellowstone type.

The final chapter will be devoted to interpretations of the factual material presented and to the general conclusions.

*Chapter VIII*

# Evolution of Volcanism
# as a Reflection of the
# Evolution of the Upper Mantle

## TWO CLASSES OF VOLCANIC ROCKS

In examining the large volume of factual material on the chemistry of the lavas of the island and volcanic arcs, along with the lavas of the intraoceanic and intracontinental volcanoes, it has become very clear that all the variation curves can be assigned to one or the other of two groups that are distinguished by their slopes relative to the coordinate axes $SA$ and $SB$.

The variation curves of the calc-alkalic rocks of the island and volcanic arcs are approximately parallel to the variation curves of the alkalic rocks of the intracontinental volcanoes. The normal basalts of the island arcs ($Na_2O + K_2O = 2-3\%$) are generally replaced on the continental platforms by alkalic basalts and trachybasalts ($Na_2O + K_2O = 6-8\%$), while the acid dacites and rhyolites ($Na_2O + K_2O = 6-8\%$) are replaced by trachytes and phonolites ($Na_2O + K_2O = 12-14\%$). The initial stage of such a process of increasing alkalinity can be seen in any island arc where there are two parallel series of volcanoes, be it the Kurile Islands or the Aleutian Islands, Kamchatka or Japan. In the Ryukyu arc, on Java, and especially in the western United States a complete transition is traced from calc-alkalic to purely alkalic rocks.

The similar trend of the variation curves indicates that the differentiation processes in the lavas of the island-arc volcanoes and the intracontinental volcanoes are similar in character.

The change in composition of the lavas from calc-alkalic to alkalic also occurs during the process of geologic development leading from a geosynclinal to a platform regime.

Thus, for example, the geosynclinal Mesozoic lavas of north-eastern Asia are typical representatives of the calc-alkalic type, while the modern lavas in the same region are purely alkalic (Ustiev, 1958). Similar examples are known in other areas where folding has been completed.

Thus, from the existing petrochemical and geologic data it follows that the calc-alkalic lavas of the island arcs and the alkalic lavas of the intracontinental volcanoes form a single class of comagmatic rocks. The parent magma of this class is calc-alkalic. In going toward the interior of an island arc, and also with the passage of geologic time (with progressive development and completion of the process of folding) the alkalinity increases. This class of rocks we can call continental.

The variation curves for the lavas of the Asiatic island arcs and the volcanoes in the interior of Asia have a somewhat steeper slope than curves of the continental class in other parts of the Pacific rim. On this basis we can recognize two subclasses of continental rocks, "Asiatic" and "American." The boundary between these two subclasses occurs on the east between the Andreanof and Fox Islands in the Aleutian chain, and on the south between Sumatra and Java and farther on, somewhere to the north of New Guinea. The "Asiatic" subclass of rocks surrounds the continent of Asia like a belt up to 1200-1500 km wide. The other parts of the "Pacific ring of fire" belong to the "American" subclass.

The variation curves for most of the intraoceanic islands are also almost parallel to one another, but their trend differs significantly from the trend of the variation curves for the continental class of rocks. The slope of the oceanic variation curves indicates a more rapid rate of increase in alkalinity during differentiation in comparison with the rocks of the continental class.

The parent magma of the oceanic class of rocks is an oceanic tholeiite, for which a very low content of $K_2O$ (0.16-0.25) is characteristic. Oceanic tholeiite makes up broad areas of the ocean floor and the bases of all the intraoceanic islands. Acid differentiation products of the oceanic tholeiites are not known with certainty, but their transition to olivine basalts has been established (Macdonald and Katsura, 1964), and the latter give rise to a whole range of alkalic rocks on oceanic islands.

The petrochemical distinction of the Easter Island lavas from the lavas of the other intraoceanic islands is of great interest and importance, from our point of view. It was indicated earlier that the variation curve for the Easter Island rocks is parallel to the continental rather than the oceanic class of rocks.* We relate this distinction to the position of Easter Island on the axis of the East Pacific Rise.

The other islands that lie on the axes of midocean ridges possess the same characteristics in their lavas. These are St. Paul Island in the Indian Ocean and Iceland in the Atlantic Ocean. We will not introduce here either analytical data or diagrams, because this would involve the incorporation of material from the other islands of the Indian and Atlantic Oceans and from the territories surrounding them, which would increase considerably the scope of our work. These questions will be analyzed in detail in a paper that is being prepared on the petrochemistry of modern volcanism outside of the Pacific Ocean.

In Fig. 95 the trends of the variation curves and the areas that outline the continental and oceanic classes on an *ASB* diagram are indicated. Here we must note that the trend of the variation curves for the oceanic-island lavas appear to be less gently sloping than on the composite diagram of Zavaritskii (1950, Fig. 96). This is explained by the fact that Zavaritskii connected the alkalic differentiates of the oceanic lavas directly to the dominant tholeiitic basalts. The actual genetic relationship of the oceanic tholeiites, olivine basalts, and alkalic differentiates has become known only in the very last few years.

The slope of the variation curves relative to the *SB* axis is 15–18° for the continental rocks and about 30° for the oceanic ones.

In the area of basic and in part of intermediate compositions the oceanic and continental classes overlap one another to a considerable degree; the rocks are distinguished only by the trend of the variation curves. Thus when examining an insufficient amount of analytical data, and especially if the possibility of analytical errors is not taken into consideration, one cannot always subdivide the rocks in question into petrochemical affiliations. Apparently that is why this author's

---

* The "continental" character of the Easter Island rocks was noted earlier.

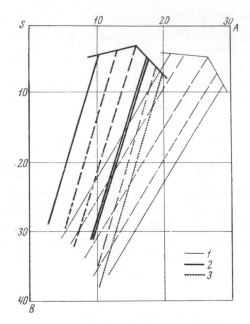

Fig. 95. Sketch showing disposition of the
variation curves for the oceanic and continental classes of rocks: 1) oceanic class
of rocks; 2) calc-alkalic rocks of the continental class; 3) alkalic rocks of the continental class.

idea of two classes of rocks was critized by Saltykovskii (1963), who
discussed figures consisting of points representing analyses on the
*ASB* plane and the variation curves drawn in these figures. In such
a construction even three or four analytical errors per hundred correct analyses will distort the actual picture considerably and will
lead to false conclusions. The source of the original data used by
Saltykovskii also seems a little strange: citing the works of Lacroix in 1927 and 1928, he for some reason does not use the groups
of islands for which there are rather numerous analyses (the Marquesas Islands or Tahiti, for example), but cites instead Mas-a-
Fuera with two(!) analyses and Raiatea with five analyses made at
the beginning of this century and published back in 1916. The data
presented by us, which is based on almost 2000 modern analyses,
very conclusively testifies to the existence of two classes of rocks,
oceanic and continental.

During analysis of the material we also at times got results

## TABLE 43. Silica and Alkali Contents of Lavas of Some Island Arcs Expressed as Percentages and as Molecular Ratios (M. R.)

| Compo-nents | 1 | | 2 | | 3 | | 4 | | 5 | | 6 | | 7 |
|---|---|---|---|---|---|---|---|---|---|---|---|---|---|
| | % | M.R. | % | M.R. | % | M.R. | % | M.R. | % | M.R. | % | M.R. | % |
| $SiO_2$ | 51.7 | 861 | 55.8 | 929 | 58.9 | 981 | 61.6 | 1026 | 66.7 | 1111 | 51.5 | 858 | 56.0 |
| $Na_2O$ | 1.9 | 31 | 2.3 | ·37 | 2.6 | 42 | 3.2 | 52 | 3.8 | 61 | 2.5 | 40 | 2.9 |
| $K_2O$ | 0.4 | 4 | 0.9 | 10 | 1.0 | 11 | 1.1 | 12 | 1.2 | 13 | 0.9 | 10 | 1.0 |

### TABLE 43 (Continued)

| Compo-nents | 7 | 8 | | 9 | | 10 | | 11 | | 12 | | 13 | |
|---|---|---|---|---|---|---|---|---|---|---|---|---|---|
| | M.R. | % | M.R. | % | M.R. | % | M.R. | % | M.R. | % | M.R. | % | M.R. |
| $SiO_2$ | 932 | 59.6 | 992 | 65.9 | 1097 | 50.9 | 848 | 55.5 | 924 | 61.5 | 1024 | 66.3 | 1104 |
| $Na_2O$ | 47 | 3.2 | 52 | 4.0 | 65 | 3.1 | 50 | 3.4 | 55 | 3.9 | 63 | 4.1 | 66 |
| $K_2O$ | 11 | 1.5 | 16 | 1.8 | 19 | 1.2 | 13 | 1.4 | 15 | 2.2 | 23 | 2.6 | 28 |

**1-5.** Average compositions of recent lavas of the Nasu zone (Honshu, Japan); **6-9.** Average composition of recent lavas of the volcanoes of East Kamchatka; **10-13.** Average compositions of recent lavas of the Sredinnyi Range on Kamchatka.

## TABLE 44. Silica and Alkali Contents of Lavas of Some Continental Volcanoes Expressed as Percentages and as Molecular Ratios (M.R.)

| Compo-nents | 1 | | 2 | | 3 | | 4 | | 5 | | 6 | | 7 | |
|---|---|---|---|---|---|---|---|---|---|---|---|---|---|---|
| | % | M.R. | % | M.R. | % | M.R. | % | M.R. | % | M.R. | % | M.R. | % | M.R. |
| $SiO_2$ | 46.6 | 776 | 49.4 | 823 | 54.6 | 909 | 62.6 | 1042 | 70.2 | 1169 | 44.5 | 741 | 51.8 | 863 |
| $Na_2O$ | 3.4 | 55 | 3.7 | 60 | 3.8 | 61 | 4.8 | 77 | 4.0 | 65 | 2.5 | 40 | 3.5 | 56 |
| $K_2O$ | 1.8 | 19 | 2.5 | 27 | 4.2 | 45 | 4.6 | 49 | 5.4 | 57 | 3.8 | 40 | 5.1 | 54 |

**1-5.** Average compositions of Cenozoic alkalic lavas of eastern Asia; **6-7.** Lavas of Uyun-Kholdonga (northeastern China).

## TABLE 45. Silica and Alkali Contents of Lavas of Some Oceanic Islands Expressed as Percentages and as Molecular Ratios (M.R.)

| Components | 1 | | 2 | | 3 | | 4 | | 5 | | 6 | | 7 | |
|---|---|---|---|---|---|---|---|---|---|---|---|---|---|---|
| | % | M.R. | % | M.R | % | M.R | % | M.R | % | M.R | % | M.R | % | M.R. |
| $SiO_2$ | 46.5 | 774 | 48.6 | 809 | 51.9 | 864 | 61.7 | 1028 | 47.5 | 791 | 57.7 | 961 | 62.7 | 1044 |
| $Na_2O$ | 2.9 | 47 | 4.4 | 72 | 5.2 | 84 | 7.4 | 119 | 3.0 | 48 | 5.8 | 94 | 6.2 | 100 |
| $K_2O$ | 0.8 | 9 | 1.6 | 17 | 2.0 | 21 | 4.2 | 45 | 2.0 | 21 | 4.0 | 42 | 5.6 | 59 |

### TABLE 45 (Continued)

| Components | 8 | | 9 | | 10 | | 11 | | 12 | | 13 | | 14 | |
|---|---|---|---|---|---|---|---|---|---|---|---|---|---|---|
| | % | M.R. | % | M.R. | % | M.R. | % | M.R. | % | M.R | % | M.R. | % | M.R. |
| $SiO_2$ | 47.3 | 788 | 50.2 | 836 | 55.2 | 919 | 48.4 | 806 | 54.0 | 899 | 61.4 | 1022 | 72.8 | 1212 |
| $Na_2O$ | 6.5 | 105 | 8.6 | 139 | 9.4 | 152 | 3.5 | 56 | 3.8 | 61 | 4.8 | 77 | 5.3 | 85 |
| $K_2O$ | 3.7 | 39 | 4.1 | 44 | 5.7 | 61 | 1.0 | 11 | 1.4 | 15 | 1.8 | 19 | 3.6 | 38 |

**1-4.** Average compositions of the lavas of the Hawaiian Islands; **5-7.** Average compositions of the lavas of the Marquesas Islands; **8-10.** Lavas of Rarotonga (Cook Islands); **11-14.** Average compositions of lavas of Easter Island.

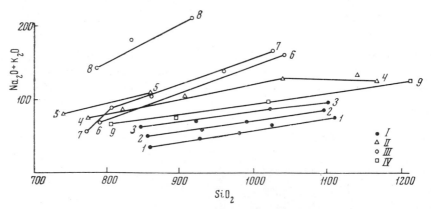

Fig. 96. Diagram of the relation between alkalies and silica in some lavas: 1) Island arcs; II) continental volcanoes; III) oceanic volcanoes; IV) East Pacific Rise; 1) Nasu zone; Japan; 2) East Kamchatka; 3) Sredinnyi Range, Kamchatka; 4) alkalic lavas of eastern Asia; 5) Uyun-Kholdonga Volcano; 6) Marquesas Islands; 7) Hawaiian Islands; 8) Cook Islands; 9) Easter Island.

that were contradictory or that deviated from "normal," but in all instances checking disclosed arithmetic errors or slips of the pen in calculating the average values or in their recalculation to Zavaritskii characteristics. This, finally, does not mean that no deviant types exist. One such "deviation" is the data from Easter Island, but this "deviation" only emphasized the regularities revealed by us, and about which we will speak further. Another "deviation" is apparently the lavas of Iwo Jima in the Bonin Island group. The scanty analyses of the recent lavas of the volcanoes of the Iwo Jima group indicate an oceanic character for the variation curve, although the islands are involved in the make-up of an island arc. Detailed examination of the submarine topography indicates that just at this point the island arc and the accompanying abyssal trench are broken by an oceanic structure, the submarine Marcus-Necker Ridge. It is very probable that Iwo Jima is genetically related to this oceanic structure rather than to the island arc. A marked change in the trend of the variation curves is caused by the process of assimilation of carbonate rocks (as at Vesuvius).

It is not yet certain, but it is theoretically possible that rocks of the oceanic class occur on relics of oceanic crust (and consequently also on relics of oceanic mantle) within the continental crust. In any case, any occurrence of "oceanic" volcanism on the continents or in the island arc or conversely, of "continental" volcanism in the expanses of the oceans, demands especially careful examination (and perhaps first of all a check on the chemical-analytical data).

As noted earlier, the two classes of rocks, continental and oceanic, are distinguished by the rate of increase in alkalinity during differentiation. It is possible, without recourse to Zavaritskii diagrams, to compare directly the change in total alkalinity with the change in acidity of the rocks. Because the atomic weights of potassium and sodium are very different, it is more convenient to consider their molecular quantities rather than the weight percentages. In most cases such a simplified variation diagram leads to the same results as examination of Zavaritskii diagrams.

In Tables 43-45 are presented the values of the silica and alkali content, along with the corresponding molecular quantities for some of the volcanic regions of the Pacific Ocean and its environs. In Fig. 96 these data are shown graphically.

It is not difficult to see that the variation lines relating the total alkalies to silica are straight segments for most of their lengths; for the island arcs and intracontinental volcanoes these are more or less parallel and are distinguished by their slopes from the intraoceanic family of lines which, in turn, are also almost parallel to one another. In this case the same two classes, oceanic and continental, are distinguished by the slopes of their lines. In the process the Easter Island lavas again are assigned to the continental type. In addition, just as in Fig. 95, the lines of the oceanic and continental types in part overlap one another.

It is possible to represent the data on Fig. 96 in algebraic form. A straight segment of any line can be represented by the formula

$$y = a + bx,$$

where $y$ is total alkalies (as a molecular ratio), $x$ the molecular ratio of silica, $a$ a value characteristic of the initial alkalinity, and $b$ the coefficient characteristic of the rate of increase in alkalinity (equal to the slope of the curve).

The continental and oceanic classes will be distinguished by the coefficient $b$: for the continental class of rocks it will fall in the range 0.06-0.20; for the oceanic class, in the range 0.30-0.40. The coefficient $a$ indicates the alkalinity of the rocks.

For convenience in graphic representation and in the recalculations the initial point on the abscissa is set at $SiO_2 = 700$, which corresponds approximately to 42% $SiO_2$. Under these conditions the calc-alkalic rocks have an $a$ coefficient in the range 10-60; when $a > 60$, continental rocks will be assigned to the alkalic series.

From Figs. 95 and 96 it can be seen that the fields of rocks belonging to the continental and oceanic classes in part overlap. Some conclusions follow from this that are very significant petrographically. Parent magmas of the continental and oceanic classes have very similar chemical compositions but give rise to different differentiation series. Thus, the alkalic rocks of intracontinental and intraoceanic volcanoes, which often have similar mineralogic and chemical compositions, are products of different differentiation series and ought not to be thought of as identical rocks. In the general petrography texts these rocks are not distinguished from one another, but in genetic petrography the continental and oceanic rocks ought to be distinguished.

There is no doubt that petrographic criteria for the distinction between the different genetic classes can be discovered by special study of them.

## VOLCANISM AND THE EARTH'S UPPER MANTLE

At present the basic difference in the structure and thickness of the earth's crust on the continents and beneath the oceans is well known. The oceanic solid crust has a thickness of 5-8 km and consists of one layer; the thickness of the continental crust averages 35 km and the crust itself is two-layered.

In both cases seismic velocities change abruptly at the crust-mantle boundary from 6.7-7.0 to 8.0-8.2 km/sec. Below this the velocities increase continuously, but at some depth in the upper mantle there is a rather thick layer in which the seismic velocities again decrease to 7.3-7.5 km/sec (Gutenberg, 1954). This low-velocity layer, often called the Gutenberg layer, begins at a depth of 60 km below the oceans, whereas beneath the continents its depth is twice as great, 120 km. The lower boundary in both cases is located at a depth of about 200 km.

The existence of the Gutenberg layer most probably is explained by the fact that at corresponding depths the melting temperature of the mantle material and its actual temperature almost coincide, i.e., the material is in a state close to melting (Daly's asthenosphere).

The difference in thickness and location of the asthenosphere layer beneath the continents and the oceans indicates that these two structures are different not only in the structure of the crust, but also in the structure of the upper mantle. The difference in structure extends to a depth of at least 400 km (Dorman et al., 1960). This difference in the deep structure of the continents and oceans is confirmed by observations on the motion of artificial satellites.

Beneath the island and volcanic arcs, as noted already, the structure of the upper mantle is again substantially different. In all cases where detailed investigations have been made, it has been established that just beneath the Mohorovicic discontinuity the mantle is characterized by a velocity of 7.5-7.8 km/sec and its density is somewhat lower than normal. This condition in the mantle extends to a depth of at least 100 km, and at the same time the wave guide or Gutenberg layer in the upper mantle is absent. Here the asthenosphere layer has, as it were, floated up to the lower boundary of the crust.

The low-velocity zone in the subcrustal mantle very clearly coincides with the zone of contemporary volcanism and does not depend on the structure of the crust. In some island arcs the crustal structure changes longitudinally from continental to oceanic or suboceanic, but the specificity of the structure of the subcrustal mantle remains the same over the entire extent.

Up to a certain point a similar picture is seen in the mid-ocean ridge areas, where it has been shown that below the Mohorovicic discontinuity there is a layer with relatively low seismic velocities. This was first observed in the North Atlantic about ten years aso, and later was confirmed in other parts of the Mid-Atlantic Ridge, in the Indian Ocean, and in the area of the East Pacific Rise. Thus, there is no doubt of the fact that the structure of the mantle has a special character beneath all the systems of midocean ridges.

The conductive heat flow in the island arcs and midocean ridges has a higher than normal value. This is expressed with special clarity in the midocean ridge areas. Numerous measurements indicate that, in a narrow zone immediately adjacent to the ridge axis, the heat-flow value is 6-7 times the average value (which is rather uniform over the whole surface of the planet). In the island-arc regions such measurements and estimates are not numerous, but they also indicate an apparently doubled heat-flow value.

As we can see, the island arcs and midocean ridges have a number of special traits which to some degree tie these two main volcanic structures together.

Finally, however, these two structures are by no means identical. To begin with, their volcanism is similar but not identical: in island-arc regions lavas mainly of the calc-alkalic type are erupted, while on the ocean ridges more alkalic varieties very likely are dominant.

The heat flow in island-arc areas apparently never reaches the high values that are characteristic of the oceanic ridges.

The velocities in the subcrustal parts of the mantle have lower than usual values in both structures, but judging from the gravimetric data (Talwani et al., 1965) the thickness of this layer is very much less beneath the oceanic ridges.

The character of the seismicity of the island arcs and the oceanic ridges is clearly different: in island-arc areas the depths of the earthquakes increase in passing from the ocean beneath the continent, where the strongest and deepest (up to 700 km) earthquakes occur. The oceanic ridge areas are also seismic (in contrast to the non-seismic parts of the ocean floor), but the seismicity is weak here, and the depth of the earthquakes does not exceed 60 km.

The distribution of gravity (Bouguer) anomalies is also substantially different: on the midocean ridges we find a weakly expressed symmetrical minimum in the positive anomalies, with a change from +400-450 mgal in the open ocean to +150-200 mgal over the ridge.

In an island arc area the picture is more complex. In going from the ocean to the island arc the Bouguer anomalies distinctly and rapidly decrease and may take on negative values over the abyssal trench. In the Puerto Rico area, for example, a change of over 500 mgal is observed in the gravity anomalies over a distance of 100 km. Between the abyssal trench and the island arc minor positive and negative anomalies are found, while on the continent the anomalies are almost zero (±50 mgal).

In examining the characteristics of volcanism, crustal structure, and upper-mantle structure, as well as from the characteristics of the geophysical fields, it can be seen that there are more similarities and fewer differences between the oceanic ridges and the island arcs than between these structures on the one hand and the oceanic and continental platforms on the other.

Apparently, similar but not completely identical processes go on in the upper mantle beneath the island arcs and the oceanic ridges.

In noting lower than usual seismic velocities in the subcrustal part of the mantle in some areas, Cook (1962) explained this phenomenon as a peculiar "blending" of the material of the mantle and crust. However, Cook's expression, "a mixture of crustal and mantle material," seems to us to be unfortunate in form and substance. It seems to us that the lower velocities in the subcrustal mantle in the island and volcanic arcs and in the midocean ridges reflects the special, "strained" physicochemical state of the material that is different from the "inert" mantle of the oceanic and continental platforms.

Usually all the varied types of lavas are explained in volcanology and petrology mainly by processes of contamination and assimilation. In particular the genesis of the calc-alkalic series of lavas is explained by the process of assimilation of the acid sialic material of the continental crust.

Earlier, during the petrochemical analysis of the lavas of the volcanoes of the Kurile Islands, it was noted that the calc-alkalic lavas were produced on crust of different types, both continental and suboceanic, where the processes of assimilation could not play any substantial role. This conclusion has been confirmed by abundant material from all the island and volcanic areas of the Pacific Ocean. Even in the western Aleutian Islands and on the Tonga and Marianas Islands, which are situated on oceanic crust, the volcanoes erupt only calc-alkalic lavas.

The calc-alkalic character of their volcanism and the already noted characteristics of the upper mantle structure are common to all the island and volcanic arcs. Thus, both the sources of nourishment for the volcanoes and the causes of their variety lie not in the crust but in the upper mantle.

More than ten years ago, on the basis of peculiarities in the transmission of transverse seismic waves in the region of the Klyuchevskaya group of volcanoes, the author established that the magmatic hearth of this group lies at a depth of 60-80 km (Gorshkov, 1956). Later similar depths for the magmatic sources of volcanoes were established by Fedotov and Farberov (1966) for the Avachinskii group of volcanoes on Kamchatka. The volcanic hearths in the Hawaiian Islands, judging from recent data (Macdonald, 1961; Eaton, 1962) lie at depths of 40-60 km, i.e., right in the upper mantle.

Both direct geophysical data and petrochemical analysis indicate that, with very rare exceptions, volcanism emerges everywhere as a "transcrustal" process. Everywhere the volcanic hearths lie beyond the crust, in the upper mantle. The composition of volcanic lavas actually does not depend on the composition of the rocks of the crust; in other words, the role of assimilation and contamination in the evolution of a magma is very slight. The sources of variation of the lavas are the magmas themselves and the processes that occur within them. Thus volcanic rocks can be thought of as being derived from the material of the upper mantle. In other words volcanism is, in its own way, an indicator of the composition and condition of the subcrustal mantle.

The existence of the two subclasses of continental rocks, "American" and "Asiatic," reflects some large-scale inhomogeneity of continental proportions in the upper mantle, apparently.

# SKETCH OF THE EVOLUTION OF VOLCANISM AND DEVELOPMENT OF THE CRUST

At present, perhaps, it is still too early to draw any concrete conclusions on the composition of and the processes within the mantle from the phenomena of volcanism. The relations that are apparent have too general a character, and further investigations may alter considerably some of the rather essential details.

In a very general way the following can be said: At present we can consider oceanic volcanism, which has its source in the asthenosphere at a depth of about 60 km, as primary. The relatively shallow occurrence of the asthenosphere layer brings about the rather widespread development of submarine volcanism on the oceanic platforms.

In some cases (the causes of which are not yet clear) a zone-melting type of process of deep-seated differentiation of mantle material sets in. This process has two branches: in some cases, when the process affects the very upper mantle, substantial reconstruction of the crust does not occur and formations of the oceanic-ridge type arise.

The asthenosphere layer in the upper mantle apparently moves upward, right to the Mohorovicic discontinuity; the layer of material with lower than usual seismic velocities "floats to the surface," so to speak. At the same time the heat flow strongly increases.

In other cases, if the process of differentiation takes place at greater depth, formations of the island-arc type arise, with consequent reconstitution of the crustal structure from the oceanic to the continental type. The inner asthenospheric or Gutenberg layer moves upward to the Mohorovicic discontinuity, that is, the layer of material with lower than usual seismic velocities again "floats

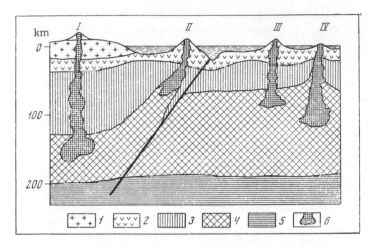

Fig. 97. Hypothetical section through the different types of primary magmatic regions: I) Intracontinental volcanoes; II) island-arc volcanoes; III) intraoceanic volcanoes; IV) volcanoes of the midocean ridges; 1) "granitic layer;" 2) "basaltic layer;" 3) subcrustal part of the mantle; 4) asthenospheric (Gutenberg) layer; 5) subasthenospheric layer of the upper mantle; 6) regions of magma generation and feeding conduits of the volcanoes.

to the surface." However, in this case the material "floats" up from greater depth.

A marked change takes place in the character of volcanism from oceanic to continental calc-alkalic. This change considerably anticipates in time the process of reconstitution of the crust, and the new island arc may be established directly on oceanic crust. The calc-alkalic lavas continue to be generated all during the course of deep-seated differentiation. As a rule a complete reconstitution of the crustal type is taking place at the same time. The process of differentiation, slowly dying out, apparently may continue to go on after the crustal reconstitution has come to an end; in any case, the calc-alkalic character of the lavas is rather commonly maintained in the younger mountain systems that are formed. With the end of the phase of calc-alkalic volcanism the "activated" layer of subcrustal mantle disappears and the upper boundary of the asthenosphere "sinks" under continental-platform conditions to a depth of about 120 km. The source of continental volcanism lies at a much greater depth than in the ocean, and thus the manifestations of continental volcanism are weaker.

On the basis of the gradual replacement in space of calc-alkalic lavas by alkalic continental ones, one can conjecture that the subcrustal asthenosphere layer of the island arcs may gradually pass into the deeper continental asthenosphere. Toward the ocean the boundary between the different types of asthenosphere is a sharp one that follows the focal plane of the earthquake foci (Fig. 97).

In the scientific literature the question is rather commonly discussed as to which comes first, volcanism or tectonism. In light of the concepts stated, we can say that volcanism, tectonics, and the type of crustal structure are paragenetically related and are caused by one common factor: the differentiation of the material of the upper mantle. In this relation, for example, the widespread phenomenon of acid volcanism and in particular, the widespread development of ignimbrites is caused not by anatexis or assimilation of a thick granitic crust, inasmuch as the granitic layer and the acid volcanic rocks are produced paragenetically at a specific stage in the development of the upper mantle.

It can scarcely be doubted that the composition of the subcrustal part of the mantle will be different in the areas of oceans, continents, island arcs, and oceanic ridges.

The thesis that volcanism is related directly to the upper mantle and is "transcrustal" in character seems very promising to us.

Actually, at present an ultradeep drilling project, "project Mohole," is being developed. The realization of this project will undoubtedly be a great contribution to geology and geophysics. However, we have hundreds of volcanoes — natural boreholes that reach to depths of dozens of kilometers. We still have not used the substantial information which the volcanoes furnish us, yet each sample of lava contains within it information on the composition and condition of the material at these depths which will scarcely be reached by drilling, at least during the lifetime of our scientific generation.

We must learn to extract this abundant information, but for this, special geologic, geophysical, and geochemical investigations will have to be set up to define the scientific concepts. According to the descriptive phrase of the past director of the Hawaiian Volcano Observatory, G. A. Macdonald, volcanoes are the windows through which we can look into the unattainable depths of the earth.

Actually, the study of volcanism can serve as one of the most powerful means of obtaining a knowledge of the interior of our planet. To realize this, purposeful geologic, geochemical, and geophysical investigations and widespread scientific coordination are necessary.

# References

Adams, W. M., and Furumoto, A. S. (1965), "A seismic refraction study of the Koolau volcanic plug," Pacific Science, Vol. 19, No. 3.

Afanas'ev, G. D. (1953), "To the problem of granite," Izv. Akad. Nauk SSSR, Ser. geol., No. 1.

Afanas'ev, G. D. (1960), "On the petrographic interpretation of geophysical data on the structure of the earth's crust," Izv. Akad. Nauk SSSR, Ser. geol., No. 7.

Afanas'ev, G. D. (1964), "Composition and structure of the earth's crust and the geologic-petrographic trends in this area," in: Geochemistry, Mineralogy, and Petrography, Izd. Nauka, Moscow.

Anderson, C. A. (1941), "Volcanoes of the Medicine Lake Highland, California," Univ. Calif. Publ., Bull. Dept. Geol. Sci., Vol. 25, No. 7.

Antipin (1775), "Report on the academy expedition of the nobleman Antipin in the Kurile Islands," Sobr. novostei (St. Petersburg), September, 49-50.

Atlas of Geographic Discoveries in Siberia and Northwestern America in the Seventeenth and Eighteenth Centuries (1964), Izd. Nauka, Moscow.

Aver'yanov, A. G., Veitsman, P. S., Gal'perin, E. I., Zverev, S. M., Zaionchkovskii, M. A., Kosminskaya, I. P., Krakshina, R. M., Mikhota, G. G., and Tulina, Yu. V. (1961), "Deep seismic sounding in the transition zone from the Asiatic continent to the Pacific Ocean in the period of the IGY," Izv. Akad. Nauk SSSR, Ser. geofiz., No. 2.

Aver'yanov, A. G., Veitsman, P. S., Gal'perin, E. I., Zverev, S. M., Kosminskaya, I. P., Krakshina, R. M., Mikhota, G. G., and Tulina, Yu. V. (1963), "Deep structure of the earth's crust in the transition zone from the Asiatic continent to the Pacific Ocean (on the basis of seismic data)" in: Geology and Metallogenesis of the Soviet Sector of the Pacific Ore Zone, Izd. Akad. Nauk SSSR, Moscow.

Balakina, L. M. (1962), "General regularities in the directions of the principal stresses active at the foci of earthquakes in the Pacific seismic zone," Izv. Akad. Nauk SSSR, Ser. geofiz., No. 11.

Bandy, M. C. (1937), "Geology and petrology of Easter Island," Bull. Geol. Soc. Am.,, Vol. 48, No. 11.

Baskin, S. I. (1952), "Travels of Evreinov and Luzhin in the Kurile archipelago," Izv. Vses. geog. obshchestvo, Vol. 84, No. 4.

Belousov, V. V. (1955), "On the geologic structure and development of the ocean basins," Izv. Akad. Nauk SSSR, ser. geol., No. 3.

Belousov, V. V., and Rudich, E. M. (1960), "On the place of island arcs in the history of the structural development of the earth," Sov. Geologiya, No. 10.

Belov, I. V. (1963), The Trachybasalt Formation of Pribaikal'e, Izd. Akad. Nauk SSSR.

Bemmelen, R. W. van (1947), "The Muriah Volcano (Central Java) and the origin of its leucite-bearing rocks," Nederlandse Akademie van Wetenschappen, Proc., Vol. 50, No. 6.

Bemmelen, R. W. van (1949), Geology of Indonesia, Netherlands Govt. Printing Office, The Hague.

Bemmelen, R. W. van (1954), Mountain Building, Nijhoff, The Hague.

Benson, W. N. (1941), "Cainozoic petrographic provinces in New Zealand and their residual magmas," Amer. J. Sci., Vol. 239, No. 8.

Bent, O. I. (1962), "New data on the material and chemical composition of the pumiceous rocks of the islands of Iturup and Kunashir (Kurile Islands)," Dokl. Akad. Nauk SSSR, Vol. 147, No. 6.

Berg, J. W., Jr., Cook, K. L. Narans, H. D., and Dolan, W. M. (1960), "Seismic investigation of crustal structure in the eastern part of the Basin and Range province," Bull. Seismol. Soc. Amer., Vol. 50, No. 4.

Bezrukov, P. L., and Udintsev, G. B. (1953), "New data on the geologic structure of the Far Eastern seas," Dokl. Akad. Nauk SSSR, Vol. 91, No. 2.

Bezrukov, P. L., Udintsev, G. B., Kanaev, V.F., and Zenkevich, N. L. (1958), "Submarine volcanoes and mountains in the Kurile Islands region," Trudy Lab. Vulkanologii Akad. Nauk SSSR, Vol. 13.

Bogoyavlenskaya, G. E., and Gorshkov, G. S. (1966), "Active volcanoes of the Central Kurile Islands" in: Recent Volcanism (Trudy Vses. Vulkanol. Soveshchaniya, Vol. 1), Izd. Nauka, Moscow.

Bowen, N. L. (1928), Evolution of the Igneous Rocks, Princeton Univ. Press, Princeton.

Brooks, J. A. (1962), "Seismic wave velocities in the New Guinea-Solomon Islands region," in: The Crust of the Pacific Basin (Am. Geophys. Union Geophys. Monograph, Vol. 6).

Broughton, W. R. (1804), A Voyage of Discovery to the North Pacific Ocean, London.

Burri, C. R. (1926), "Chemistry and provincial relationships of the young eruptive rocks of the Pacific Ocean and its margins," Schweiz. Min. Pet. Mitt., Vol. 6, No. 1.

Burri, C. R., and Sonder, R. A. (1936), "Contributions to the geology and petrography of the late Tertiary and recent volcanism in Nicaragua," Zeit. Vulkanol., Vol. 17, No. 1/2.

Byers, F. M., Jr. (1959), "Geology of Umnak and Bogoslof Islands, Aleutian Islands, Alaska," U.S. Geol. Surv. Bull., No. 1028-L.

Casertano, L. (1963), "Chilean continent," Catalogue of the Active Volcanoes of the World, Vol. 15.

Chayes, F. (1964), "A petrographic distinction between Cenozoic volcanism in and around the open oceans," J. Geophys. Research, Vol. 69, No. 8.

Chemekov, Yu. F. (1959), "Quaternary system in the Khabarovsk Krai and the Amur Oblast," in: Materials for the Quaternary Geology and Geomorphology of the USSR (Materialy VSEGEI, nov. ser., Vol. 2), Leningrad.

Chemekov, Yu. F. (1961), "Anthropogene stratigraphy and paleogeology of the Far East of the USSR"," in: Materials for the All-Soviet Conference for the Study of the Quaternary Period, Vol. 3.

Chernyi, I. (1871), "Journal or notes made by Cossack Captain Ivan Chernyi who visited the Kurile Islands as far as the nineteenth island, the route of the expedition and notes on the spacing of those islands, their inhabitants, and other subjects," in: Polonskii (1871).

Chernysheva, V. I. (1963), "Olivine basalt in the region of the north end of the submarine Hawaiian Ridge," Dokl. Akad. Nauk SSSR, Vol. 151, No. 6.

Chubb, L. J. (1930), "Petrography of the Marquesas Islands," Proc. Fourth Pacific Science Congress, Vol. 2B, 781-785.

Cisternas, A. (1961), "Crustal structure of the Andes from Rayleigh wave dispersion," Bull. Seismol. Soc. Am., Vol. 51, No. 3.

Clark, R. H. (1960), "Petrology of the volcanic rocks of Tongariro Subdivision," New Zealand Geol. Surv. Bull., No. 40.

Coats, R. R. (1952), "Magmatic differentiation in Tertiary and Quaternary volcanic rocks from Adak and Kanaga Islands, Aleutian Islands, Alaska," Bull. Geol. Soc. Am., Vol. 63, No. 5.

Coats, R. R. (1953), "Geology of Buldir Island, Aleutian Islands, Alaska," U.S. Geol. Surv. Bull., No. 989-A.

Coats, R. R. (1959), "Reconnaissance geology of Semisopochnoi Island, Aleutian Islands, Alaska," U.S. Geol. Surv. Bull., No. 1028-O.

Coats, R. R. (1962), "Magma type and crustal structure in the Aleutian arc," American Geophysical Union Monograph, No. 6.

Coats, R. R., Nelson, W. N., Lewis, R. Q., and Powers, H. A. (1961), "Geologic reconnaissance of Kiska Island, Aleutian Islands, Alaska," U.S. Geol. Surv. Bull., No. 1028-R.

Cook, J. (1782), A Voyage to the Pacific Ocean Performed under the Direction of Captains Cook, Clark, and Gore in the Years 1776-1780, Vols. 1-3, London.

Cook, K. L. (1962), "The problem of the mantle-crust mix: lateral inhomogeneity in the uppermost part of the earth's mantle," in: Advances in Geophysics, Vol. 9.

Coombs, H. A. (1939), "Mt. Baker, a Cascade volcano," Bull. Geol. Soc. Am., Vol. 50, No. 10.

Coombs, H. A., and Howard, A. D. (1960), "United States of America," Catalogue of the Active Volcanoes of the World, Part 9, Napoli.

Daly, R. A. (1933), Igneous Rocks and the Depths of the Earth, McGraw-Hill, New York.

Ditmar, K. von (1890), "Journeys and Life in Kamchatka in 1851-1855," Beiträge zur Kentniss des Russischen Reiches, Vol. 7.

Ditmar, K. von (1901), Journeys and Life in Kamchatka in 1851-1855 [in Russian], Part 1, St. Petersburg.

Dorman, J., Ewing, M., and Oliver, J. (1960), "Study of shear-velocity distribution in the upper mantle by mantle Rayleigh waves," Bull. Seismol. Soc. Am., Vol. 50, No. 1.

Dorshin, P. O. (1870), "On some volcanoes, their eruptions, and earthquakes in the former American possessions of Russia," Zap. Imp. SPB Mineral. Obshchest., ser. 2, No. 5.

Drewes, H., Fraser, G. D., Snyder, G. L., and Barnett, H. F., Jr. (1961), "Geology of Unalaska Island and adjacent insular shelf, Aleutian Islands, Alaska," U.S. Geol. Surv. Bull., No. 1028-S.

Eaton, J. P. (1962), "Crustal structure and volcanism in Hawaii," in: The Crust of the Pacific Basin (Amer. Geophys. Union Monograph, No. 6).

Eaton, J. P. (1963), "Crustal structure from San Francisco to Eureka, Nevada, from seismic-refraction measurements," J. Geophys. Research, Vol. 68, No. 20.

Eaton, J. P., and Murata, K. J. (1960), "How volcanoes grow," Science, Vol. 132, No. 3432.

Edwards, A. B. (1942), "Differentiation of the dolerites of Tasmania," J. Geol., Vol. 50, 451-480, 579-610.

Eiby, G. A. (1958), "The structure of New Zealand from seismic evidence," Geol. Rundschau, Vol. 47, No. 2.

Emery, K. O. (1958), "Shallow submerged marine terraces of Southern California," Bull. Geol. Soc. Am., Vol. 69, No. 1.

Engel, C. G., and Engel, A. E. J. (1961), "Composition of basalt cored in Mohole project," Bull. Am. Assn. Petrol. Geologists, Vol. 45.

Engel, C. G., and Engel, A. E. J. (1963), "Basalts dredged from the northeastern Pacific Ocean," Science, Vol. 140, No. 3573.

Engel, A. E. J., Engel, C. G., and Havens, R. H. (1965), "Chemical characteristics of oceanic basalts and the upper mantle," Bull. Geol. Soc. Am., Vol. 76, No. 7.

Érlikh, É. N. (1960), "On the evolution of Quaternary volcanism in the Sredinnyi Ridge zone of Kamchatka," Izv. Akad. Nauk SSSR, Ser. geol., No. 2.

Escher, O. (1858), "Otto Escher's expedition to the Amur," Petermann's Mitt., Vol. 4.

Evmeev, O. A. (1950), First Russian Geologists to Reach the Pacific Ocean, Moscow.

Fedorchenko, V. I. (1962), "Main stages in the postcaldera period of development of Golovnin Volcano (Kunashir)," Trudy SakhKNII, No. 12.

Fedotov, S. A. (1963), "On the absorption of transverse seismic waves in the upper mantle and the energy classification of nearby earthquakes from intermediate foci," Izv. Akad. Nauk SSSR, ser. geofiz., No. 6.

Fedotov, S. A. (1966), "Deep structure, properties of the upper mantle, and volcanic activity of the Kurile-Kamchatka island arc from seismologic data in 1964," in: Volcanism and Deep Structure of the Earth (Trudy Vtorogo Vses. Vulkanol. Soveshchaniya, Vol. 3), Izd. Nauka, Moscow.

Fedotov, S. A., and Farberov, A. I. (1966), "On the absorption of transverse seismic waves in the earth's crust and upper mantle in the region of the Avachinskii group of volcanoes," in: Volcanism and Deep Structure of the Earth (Trudy Vtorogo Vses. Vulkanol. Soveshchaniya, Vol. 3), Izd. Nauka, Moscow.

Fedotov, S. A., and Kuzin, I. P. (1963), "Velocity section of the upper mantle in the region of the South Kurile Islands," Izv. Akad. Nauk SSSR, ser. geofiz., No. 5.

Fedotov, S. A., Aver'yanova, V. N. Bagdasarova, A. M., Kuzin, I. P., and Tarakanov, R. Z. (1961), "Some results of the detailed study of the seismicity of the South Kurile Islands," Izv. Akad. Nauk SSSR, ser. geofiz., No. 5.

Fedotov, S. A., Bagdasarova, A. M., Kuzin, I. P., and Tarakanov, R. Z. (1963), "On the seismicity and deep structure of the southern Kurile island arc," Dokl. Akad. Nauk SSSR, Vol. 153, No. 3.

Fedotov, S. A., Kuzin, I. P., and Bobkov, M. F. (1964a), "Detailed seismologic investigation on Kamchatka in 1961-1962," Izv. Akad. Nauk SSSR, ser. geofiz., No. 9.

Fedotov, S. A., Matveeva, N. N., Tarakanov, R. Z., and Yanovskaya, T. B. (1964b), "On the velocities of longitudinal waves in the upper mantle in the region of Japan and the Kurile Islands," Izv. Akad. Nauk SSSR, ser. geofiz., No. 8.

Firsov, L. V. (1964), "Absolute age of the intrusive rocks of the islands of Kunashir and Urup (Kurile Islands)," Dokl. Akad. Nauk SSSR, Vol. 156, No. 4.

Fisher, N. H. (1957), "Melanesia," Catalogue of Active Volcanoes of the World, Part 5, Napoli.

Fisher, R. L. (1961), "Middle America Trench: topography and structure," Bull. Geol. Soc. Am., Vol. 72, No. 5.

Fisher, R. L., and Norris, R. M. (1960), "Bathymetry and geology of Sala y Gomez, southern Pacific," Bull. Geol. Soc. Am., Vol. 71, No. 4.

Fraser, G. D., and Barnett, H. F. (1959), "Geology of the Delarof and westernmost Andreanof Islands, Aleutian Islands, Alaska," U.S. Geol. Surv. Bull., No. 1028-I.

Furumoto, A. S., Thompson, N. J., and Woollard, G. P. (1965), "The structure of the Koolau Volcano from seismic refraction studies," Pacific Sci., Vol. 19, No. 3.

Furumoto, A. S., Woollard, G. P., et al., (1965), "Seismic refraction studies of the crustal structure of the Hawaiian archipelago," Pacific Sci., Vol. 19, No. 3.

Gainanov, A. G. (1955), "Pendulum determinations of the force of gravity in the Sea of Okhotsk and in the northwestern Pacific Ocean," Trudy Inst. Okeanol. Akad. Nauk SSSR, Vol. 12.

Gainanov, A. G. (1963), "On some results of gravimetric investigations in the Sea of Okhotsk, the Kurile-Kamchatka basin, and adjacent parts of the Pacific Ocean," in: Marine Gravimetric Investigations, Vol. 2.

Gainanov, A. G. (1964a), "On some characteristics of the structure of the earth's crust in the transition zones of the Pacific Ocean according to geophysical data," in: Geophysical Investigations, Vol. 1, Izd. MGU.

Gainanov, A. G. (1964b), "On the nature of the magnetic anomalies in the transition zones of the Pacific Ocean," Sov. Geologiya, No. 10.

Gainanov, A. G., and Smirnov, L. P. (1962), "Structure of the earth's crust in the transition region from the Asiatic continent to the Pacific Ocean," Sov. Geologiya, No. 3.

Gainanov, A. G., and Solov'ev, O. N. (1963), "On the nature of the magnetic anomalies in the transition region from the Asiatic continent to the Pacific Ocean," Dokl. Akad. Nauk SSSR, Vol. 151, No. 6.

Gainanov, A. G., and Ushakov, S. A. (1966), "Isostasy and the deep structure of the transition zone from the Asiatic continent to the Pacific Ocean in the region of the Kurile-Kamchatka basin," Dokl. Akad. Nauk SSSR, Vol. 158, No. 3.

Gainanov, A. G., Tulina, Yu. V., Kosminskaya, I. P., Zverev, S. M., Veitsman, P. S., and Solov'ev, O. N. (1965), "Complex interpretation of the data from geophysical investigations in the Sea of Okhotsk and the Kurile-Kamchatka zone," in: Results of Investigations during International Geophysical Projects (Seism. Issled., Vol. 6), Izd. Nauka, Moscow.

Gal'perin, E. I. (1958), "Study of the structure of the earth's crust in the transition area from the Asiatic continent to the Pacific Ocean," Inf. Byull. MGG, No. 5.

Gal'perin, E. I. Goryachev, A. V., and Zverev, S. M. (1958), "Investigations of the earth's crust in the transition area from the Asiatic continent to the Pacific Ocean," XII Razdel MGG, Seismologiya, No. 1.

Gates, O., and Gibson, W. (1956), "Interpretation of the configuration of the Aleutian Ridge," Bull. Geol. Soc. Am., Vol. 67, No. 2.

Golovnin, V. M. (1819), Condensed Memoirs of Fleet Captain-Lieutenant (now Captain of the First Rank) Golovnin, on his Voyage on the Sloop "Diana" to Describe the Kurile Islands, in 1811, St. Petersburg.

Gorkun, V. N., Rodionova, R. I., Fedorchenko, V. I., and Shilov, V. N. (1963), "On the distribution of some minor elements in the lavas of the northern part of Vernadskii Ridge on the island of Paramushir (Kurile Islands), in: Petrochemical Characteristics of Young Volcanism, Izd. Akad. Nauk SSSR.

Gorshkov, G. S. (1948a), "Names of volcanoes in the Kurile Islands," Izv. Vses. Geogr. Obshchestva, Vol. 80, No. 2.

Gorshkov, G. S. (1948b), "Sarychev Peak Volcano," Byull. Vulkanol. Stantsii, No. 15.

Gorshkov, G. S. (1954a), "Krenitsyn Peak," Byull. Vulkanol. Stantsii, No. 20.

Gorshkov, G. S. (1954b), "Volcanoes of Paramushir island and their condition in 1953," Byull. Vulkanol. Stantsii, No. 22.

Gorshkov, G. S. (1954c), "Chronology of eruptions of volcanoes in the Kurile chain," Trudy Lab. Vulkanol., No. 8.

Gorshkov, G. S. (1956), "On the deep magmatic hearth of Klyuchevskoi Volcano," Dokl. Akad. Nauk SSSR, Vol. 106, No. 4.

Gorshkov, G. S. (1957), "Catalog of the active volcanoes of the Kurile Islands," Byull. Vulkanol. Stantsii, No. 25.

Gorshkov, G. S. (1958a), "Active volcanoes of the Kurile island arc," in: Young Volcanism in the USSR (Trudy Lab. Vulkanol., No. 13).

Gorshkov, G. S. (1958b), "On some problems in the theory of volcanology," Izv. Akad. Nauk SSSR, ser. geol., No. 11.

Gorshkov, G. S. (1958c), "On some theoretical problems of volcanology," Bull. Volcanol., Ser. 2, Vol. 19.

Gorshkov, G. S. (1958d), "Kurile Islands," Catalogue of the Active Volcanoes of the World, Part 7, Napoli.

Gorshkov, G. S. (1960a), "Quaternary volcanism and petrochemistry of the modern lavas of the Kurile Islands," Dokl. Sov. Geologov na XXI Sessii Mezhdunar. Geol. Kong., Problema 13.

Gorshkov, G. S.(1960b), "Zavaritskii Caldera," Byull. Vulkanol. Stantsii, No. 30.

Gorshkov, G.S. (1961a), "On the petrochemistry of volcanic rocks in connection with the formation of island arcs," Publ. du Bureau Central Seismol. Internat., Ser. A, No. 22.

Gorshkov, G. S. (1961b), "Petrochemistry of volcanic rocks in relation to the formation of island arcs," Ann. di Geofis., Vol. 14, No. 2.

Gorshkov, G. S. (1961c), "Volcanic zone of the Kurile Islands," Proc. 9th Pacific Sci. Cong., Vol. 12.

Gorshkov, G. S. (1961d), "Welded tuff in Zavaritskii Caldera (Simushir island, Kurile Islands)," Trudy Lab. Vulkanol., No. 20.

Gorshkov, G. S. (1962a), "To the question of classification of some types of explosive eruptions," in: Questions of Volcanism, Izd. Akad. Nauk SSSR, Moscow.

Gorshkov, G. S. (1962b), "Petrochemical features of volcanism in relation to the types of the earth's crust," in: The Crust of the Pacific Ocean (Am. Geophys. Union Monograph, No. 6).

Gorshkov, G. S. (1963a), "Global characteristics of the petrochemistry of volcanic rocks and the basic structure of the earth," in: Petrochemical Characteristics of Young Volcanism [in Russian].

Gorshkov, G. S. (1963b), "Petrochemistry of volcanic rocks in relation to the formation of island arcs," in: General Questions of the Geology and Metallogenesis of the Pacific Ore Belt [in Russian], Moscow.

Gorshkov, G. S. (1964a), "Phenomena of volcanism and the upper mantle," in: Chemistry of the Earth's Crust [in Russian], Vol. 2, Moscow.

Gorshkov, G. S. (1964b), "Petrographic and chemical composition of the lavas of the Kurile volcanoes," in: Geology of the USSR [in Russian], Vol. 31, Part 1.

Gorshkov, G. S. (1965), "On the relation of volcanism and the upper mantle," Bull. Volcanol., Ser. 2, Vol. 28.

Gorshkov, G. S., and Bogoyavlenskaya, G. E. (1962), "On the petrography of the modern volcanic rocks of the Kurile island arc (North Kurile Islands)," Trudy Lab. Vulkanol., No. 21.

Gorshkov, G. S., and Naboko, S. I. (1962), "Modern volcanism of the Kamchatka-Kurile arc," in: Questions of Volcanism [in Russian].

Gorshkov, G. S., Markhinin, E. K., Fedorchenko, V. I., and Shilov, V. N. (1964), "Description of the volcanoes of the Kurile Islands," in: Geology of the USSR [in Russian], Vol. 31, Part 1.

Goryachev, A. V. (1960), "Some characteristic features of the recent tectonics of the Kurile island arc," Sov. Geologiya, No. 10.

Goryachev, A. V.(1962a), "On the relation of seismicity and modern volcanism in the Kurile-Kamchatka fold zone," Izv. Akad. Nauk SSSR, ser. geofiz., No. 11.

Goryachev, A. V. (1962b), "Principal regular features of the geotectonic regime of the Kurile-Kamchatka fold zone," Dokl. Akad. Nauk SSSR, Vol. 142, No. 1.

Goryachev, A. V. (1963), "Structural-tectonic regions of Kamchatka and the Kurile Islands," Dokl. Akad. Nauk SSSR, Vol. 153, No. 4.

Gubler, A. (1932), "The Kuriles," Mitt. d. Geogr.-Ethnogr. Gesells. Zürich, Vol. 32.

Gumennyi, Yu. K. (1962), "On the ore deposits of Kunashir island," Trudy SakhKNII, No. 10.

Gumennyi, Yu. K., and Neverov, Yu. L. (1961), "New data on the phenomena of active volcanism on Kunashir island," Trudy SakhKNII, No. 10.

Gutenberg, B. (1954), "Low-velocity layers in the earth's mantle," Bull. Geol. Soc. Am., Vol. 65.

Gutenberg, B. (1959), "The asthenosphere low-velocity layer," Ann. di Geophis., Rome.

Gutenberg, B., and Richter, K. (1941), Seismicity of the Earth (Geol. Soc. Am. Sp. Paper, 34).

Gzovskii, M. V. (1963), "Tectonophysics and the origin of magmas of different chemical composition," in: The Problem of Magma and the Genesis of Igneous Rocks [in Russian].

Hantke, G., and Parodi, A. (1966), "Colombia, Ecuador, and Peru," Catalogue of the Active Volcanoes of the World, Part 19, Rome.

Harada, D. (1934), "On the new cratered cone in Lake Midoriko on the island of Simushir (Central Kurile Islands)," Bull. Volcanol. Soc. Japan, Vol. 2, No. 1.

Herzen, R. P. von, and Maxwell, A. E. (1964), "Measurements of heat flow at the pre-
    liminary Mohole site off Mexico," J. Geophys. Research, Vol. 69, No. 4.

Herzen, R. P. von, and Uyeda, S. (1963), "Heat flow through the eastern Pacific Ocean
    floor," J. Geophys. Research, Vol. 68, No. 14.

Hess, H. H. (1948), "Major structural features of the Northwest Pacific," Bull. Geol.
    Soc. Am., Vol. 59, 417-445.

Hess, H. H., and Maxwell, J. C. (1953), "Major structural features of the south-west
    Pacific: a preliminary interpretation of H. O. 5484, Bathymetric chart, New
    Guinea to New Zealand," Proc. 7th Pacific Sci. Cong., Vol. 2.

Hibayashi, T. (1942), "Essays on the Kurile Islands," Geogr. J. [of Japan], Vol. 53, Nos.
    627, 630, 632; Vol. 54, No. 640.

Hobbs, W. H. (1953), "Origin of the lavas of the Pacific region," Proc. 7th Pacific Sci.
    Cong., Vol. 2.

Hodgson, J. H. (1963), "National report for Canada: Seismology and physics of the
    earth's interior, 1961-1962," Contr. Dominion Obs., Vol. 5, No. 19.

Homann, I. B. (1725), Great Atlas of the Entire World [in German], Nurenberg.

Ishikawa, T., and Katsui, Y. (1959), "Some considerations on the relation between the
    chemical character and the geographical position of the volcanic zones in
    Japan," J. Fac. Sci. Hokkaido Univ., Ser. IV, Vol. 10, No. 1.

Jimbo, K. (1892), General Geological Sketch of Hokkaido, with Special Reference to
    the Petrography, Sapporo.

Kamio, H. (1931), "Earthquake in Moroton Strait on Simushin in June 1920 and the
    eruption on Matua in January 1923," Geol. J. [of Japan], Vol. 38, No. 458.

Kanaev, V. F. (1960), "Geomorphologic observations on the Kurile Islands," Trudy
    Inst. Okeanologii Akad. Nauk SSSR, Vol. 32.

Kanaev, V. F. (1961), "New data on the geomorphology and vertical movements of
    the Kurile Islands chain," in: Materials for the All-Soviet Congress on the
    Study of the Quaternary Period [in Russian], Vol. 1, Moscow.

Kanaev, V. F., and Larina, N. I. (1959), "Submarine topography in the North Kuriles
    region," Trudy Inst. Okeanologii Akad. Nauk SSSR, Vol. 36.

Katsui, Y. (1954), "Chemical compositions of the lavas of the Chokai volcanic zone,
    Japan," Bull. Geol. Soc. Japan, Vol. 60, No. 704.

Katsui, Y. (1961), "Petrochemistry of the Quaternary volcanic rocks of Hokkaido and
    surrounding areas," J. Fac. Sci. Hokkaido Univ., Ser. IV, Vol. 11, No. 1.

Kawano, Y., Yagi, K., and Aoki, K. (1961), "Petrography and petrochemistry of the
    volcanic rocks of Quaternary volcanoes of northeastern Japan," Sci. Repts.
    Tohoku Univ., Ser. III, Vol. 7, No. 1.

Keller, F., Jr., Meuschke, J. L., and Alldredge, L. R. (1954), "Aeromagnetic surveys
    in the Aleutian, Marshall, and Bermuda Islands," Trans. Amer. Geophys. Union,
    Vol. 35, No. 4.

Kennedy, W. Q. (1933), "Trends of differentiation in basaltic magmas," Amer. J. Sci.,
    Vol. 25, No. 147.

Kennedy, W. Q., and Anderson, E. M. (1938), "Crustal layers and the origin of magmas,"
    Bull. Volcanol., Ser. 2, Vol. 3.

Khain, V. E. (1964), "Evolution of the earth's crust and the possible form of its rela-
    tion to processes in the upper mantle," Sov. Geologiya, No. 6.

Kondorskaya, N. V., and Postolenko, G. A. (1958), "Seismic activity in the Kurile-
    Kamchatka region in 1954-1955," Izv. Akad. Nauk SSSR, ser. geofiz., No. 9.
Kondorskaya, N. V., and Tarakanov, R. Z. (1961), "Kurile-Kamchatka earthquakes,"
    in: Earthquakes of the USSR [in Russian], Izd. Akad. Nauk SSSR.
Kondorskaya, N. V., and Tikhonov, V. I. (1960), "On the problem of the seismicity and
    tectonics of Kamchatka and the northern part of the Kurile chain," Dokl. Akad.
    Nauk SSSR, Vol. 130, No. 1.
Korsunskaya, G. V. (1948), "Volcanoes in the southern group of the Kurile Islands,"
    Izv. Vses. Geogr. Obshchestva, Vol. 80, No. 4.
Korsunskaya, G. V. (1958), Kurile Island Arc (A Physicogeographic Outline) [in Russian],
    Izd. Geografgiz, Moscow.
Kosminskaya, I. P. (1961), "Structure of the earth's crust in the deep basins of the Black,
    Japan, Caspian, Okhotsk, and Bering Seas," Byull. MOIP, otd. geol., No. 6.
Kosminskaya, I. P. (1963), "Study of the structure of the earth's crust in the USSR dur-
    ing the IGY" in: Results of Investigations During the IGY Program (Seismol.
    Issled., No. 5), Izd. Akad. Nauk SSSR.
Kosminskaya, I. P., Zverev, S. M., Veitsman, P. S., Tulina, Yu. V., and Krakshina,
    R. M. (1963), "Basic outline of the structure of the earth's crust in the Sea of
    Okhotsk and the Kurile-Kamchatka zone of the Pacific Ocean according to
    data from deep seismic sounding," Izv. Akad. Nauk SSSR, ser. geofiz., No. 1.
Kosminskaya, I. P., Zverev, S. M., Veitsman, P. S., and Tulina, Yu. V. (1964), "Gen-
    eral outline of the structure of the earth's crust in the transition zone," in:
    Structure of the Earth's Crust in the Transition Region from the Asiatic Con-
    tinent to the Pacific Ocean [in Russian], Izd. Nauka, Moscow.
Kosminskaya, I. P., Krakshina, R. M., and Pavlova, I. N. (1964), "The northern and
    central parts of the Sea of Okhotsk," in: Structure of the Earth's Crust in the
    Transition Region from the Asiatic Continent to the Pacific Ocean [in Russian],
    Izd. Nauka, Moscow.
Kovylin, V. M., and Neprochnov, Yu. P. (1965), "Structure of the earth's crust and of
    the sedimentary sequence in the central part of the Sea of Japan according to
    seismic data," Izv. Akad. Nauk SSSR, ser. geol., No. 4.
Kozyrevskii, I. P. (1730), St. Petersburg Newspapers for 1730, No. 25.
Krasheninnikov, S. P. (1755), Description of the Land of Kamchatka [in Russian], Vols.
    1-2, St. Petersburg.
Kropotkin, P. N. (1953), "Recent geophysical data on the structure of the earth and the
    problem of the origin of basaltic and granitic magmas," Izv. Akad. Nauk SSSR,
    ser. geol., No. 1.
Kruzenshtern, I. F. (1809-1812), Journey Around the World in 1803-1806 by the Order
    of His Imperial Majesty Alexander I on the Ships "Nadezhda" and "Neva" [in
    Russian], Vols. 1-3, St. Petersburg.
Kuenen, P. H. (1952), "Indonesian abyssal depressions," in: Island Arcs [in Russian],
    Moscow.
Kuno, H. (1936), "On the crystallization of pyroxenes from rock magmas, with special
    reference to the formation of pigeonite," Japan. J. Geol. Geogr., Vol. 13, No.
    1-2.
Kuno, H. (1950), "Petrology of Hakone Volcano and the adjacent areas, Japan," Bull.
    Geol. Soc. Am., Vol. 61, No. 9.

Kuno, H. (1959), "Origin of Cenozoic petrographic provinces of Japan and surrounding areas," Bull. Volcanol., Ser. 2, Vol. 20.

Kuno, H. (1960), "High-alumina basalt," J. Petrol., Vol. 1, No. 2.

Kuno, H. (1964), "Igneous rock series," in: Chemistry of the Earth's Crust [in Russian], Vol. 2, Izd. Nauka, Moscow.

Kuno, H., Yamasaki, K., Iida, C., and Nagashima, K. (1957), "Differentiation of Hawaiian magmas," Japan. J. Geol. Geogr., Vol. 28, No. 4.

Lacroix, A. (1927), "The lithologic constitution of the volcanic islands of southern Polynesia," Memoir Acad. Sci. Paris, ser. 2, Vol. 59.

Lacroix, A. (1928), "New observations on the lavas of the Marquesas Islands and of Tubuai," C. R. Acad. Sci. Paris, Vol. 187, No. 7.

La Pérouse (1797), The Voyage of La Pérouse around the World [in French], Paris.

Larsen, E. S. (1941), "Igneous rocks of the Highwood Mountains, Montana," Bull. Geol. Soc. Am., Vol. 52, No. 11.

Lee, W. H. K., and Macdonald, J. G. F. (1963), "The global variation of terrestrial heat flow," J. Geophys. Research, Vol. 68, No. 24.

LePichon, X., Houtz, R. E., Drake, L. C., and Nafe, E. J. (1965), "Crustal structure of the mid-ocean ridges, 1, Seismic refraction measurements," J. Geophys. Res., Vol. 70, No. 2.

Lewis, R. Q., Nelson, W. H., and Powers, H. A. (1960), "Geology of Rat Island, Aleutian Islands, Alaska," U.S. Geol. Surv. Bull., No. 1028-Q.

Livshits, M. Kh. (1965), "On the problem of the physical make-up of the deep-seated material of the earth's crust and upper mantle in the Kurile zone of the Pacific ring," Geol. i Geofiz., No. 1.

Lyubimova, E.A. (1959), "On the temperature gradient in the upper layers of the earth and a possible explanation of the low-velocity layer," Izv. Akad. Nauk SSSR, ser. geofiz., No. 12.

Lyustikh, E. N. (1956), "On the role of volcanoes and hot springs in the energetics of the earth's crust," Izv. Akad. Nauk SSSR, ser. geofiz., No. 1.

Lyustikh, E. N. (1960), "Energy of formation of the earth's crust," Izv. Akad. Nauk SSSR, ser. geofiz., No. 3.

Lyustikh, E. N. (1965), "Convection in the mantle and spherical functions," Fizika Zemli, No. 8.

Macdonald, G. A. (1944a), "The 1840 eruption and crystal differentiation in the Kilauean magma column," Am. J. Sci., Vol. 242, No. 4.

Macdonald, G. A. (1944b), "Petrography of the Samoan Islands," Bull. Geol. Soc. Am., Vol. 55, No. 11.

Macdonald, G. A. (1960), "Dissimilarity of continental and oceanic rock types," J. Petrol., Vol. 1, No. 2.

Macdonald, G. A. (1961), "Volcanology," Science, Vol. 133, No. 3454.

Macdonald, G. A., and Katsura, T. (1962), "Relationship of petrographic suites in Hawaii," in: The Crust of the Pacific Basin (Amer. Geophys. Union Monogr., No. 6).

Macdonald, G. A., and Katsura, T. (1964), "Chemical composition of Hawaiian lavas," J. Petrol., Vol. 5.

Magnitskii, V. A. (1961), "The mantle and crust of the earth," Sov. Geologiya, No. 5.

Magnitskii, V. A. (1964), "Zone melting as a mechanism for the formation of the
    earth's crust," Izv. Akad. Nauk SSSR, ser. geol., No. 11.
Magnitskii, V. A. (1965), Internal Structure and Physics of the Earth [in Russian], Izd.
    Nedra, Moscow.
Magnitskii, V. A., and Kalashnikova, I. V. (1962), "On the general direction of de-
    velopment of the earth's crust," Izv. Akad. Nauk SSSR, ser. geofiz., No. 8.
Malahoff, A., and Woollard, G. P. (1964), Magnetic Surveys over the Hawaiian Ridge,
    Hawaii Inst. of Geophys., Univ. of Hawaii.
Markhinin, E. K. (1959), "Volcanoes of the island of Kunashir," Trudy Lab. Vulkanol.,
    No. 17.
Markhinin, E. K. (1961), "Volcanism of the Kurile Islands," Izv. Akad. Nauk SSSR,
    ser. geol., No. 6.
Markhinin, E. K., and Stratula, D. S. (1965), "Some new data on the volcanoes of the
    Kurile Islands," in: Quaternary Volcanism of Some Regions of the USSR [in
    Russian], Izd. Nauka, Moscow.
Marshall, P. (1911), "Oceania," Handb. der Regionalen Geol., Vol. 7, Part 1, No. 5.
Matuzawa, T., Matumoto, T., and Asano, S. (1960), "The crustal structure as derived
    from observations of the second Hokoda explosion," J. Seismol. Soc. Japan,
    Vol. 13, No. 2.
Memoirs of Siberian History in the Eighteenth Century (1882) [in Russian], Vols. 1-2.
Menard, H. W. (1954), "The East Pacific Rise," Sci. Am., No. 205, 52-56.
Mikumo, T., Otsuka, M., Utsu, T., Terashima, T., and Okada, A. (1961), "Crustal
    structure in central Japan as derived from Moboro explosion-seismic observa-
    tions," Bull. Earthq. Res. Inst., Vol. 39, No. 2.
Miller, G. F. (1758), Description of Ocean Voyages to the Arctic and the Pacific Oceans
    Made from the Russian Shore [in Russian].
Milne, J. (1879), "A cruise among the volcanoes of the Kurile Islands," Geol. Mag.,
    new ser., Vol. 6, No. 8.
Milne, J. (1880), "The Kurile Islands," Geol. Mag., new ser., Vol. 7, No. 4.
Milne, J. (1886), "The volcanoes of Japan," Trans. Seismol. Soc. Japan, Vol. 9, No. 2.
Miyatake, K. (1934), "On the eruption of a volcano on Harumukotan (Central Kurile
    Islands) on January 8, 1933," Bull. Volcanol. Soc. Japan, Vol. 2, No. 1.
Miyake, Y., and Sugiura, Y. (1953), "On the chemical compositions of the volcanic
    eruptives in New Britain Islands, Pacific Ocean," Proc. 7th Pacific Sci. Cong.,
    Vol. 2.
Monakhov, F. I., and Tarakanov, R. Z. (1955), "Characteristics of the Kurile-Kam-
    chatka earthquakes based on observations of nearby stations in 1952-1954,"
    Izv. Akad. Nauk SSSR, ser. geofiz., No. 5.
Mooser, F., Meyer-Abich, H., and McBirney, A.R. (1958), "Central America," Cata-
    logue of the Active Volcanoes of the World, Part 6, Napoli.
Müller (1774), "Geography and status of Kamchatka based on various written and verbal
    reports collected in Yakutsk in 1737, " in: Steller (1774).
Naboko, S. I. (1960), "Quaternary and Recent volcanism of Kamchatka and distinctive
    petrochemical features of the lavas," in: Petrographic Provinces, Igneous and
    Metamorphic Rocks [in Russian], Izd. Akad. Nauk SSSR.
Nelson, W. N. (1959), "Geology of Segula, Davidof, and Khvostof Islands, Alaska,"
    U.S. Geol. Surv. Bull., No. 1028-K.

Nemoto, T. (1934), "On the products of the eruption of Harumukotan Volcano (Central Kurile Islands)," Bull. Volcanol. Soc. Japan, Vol. 2, No. 1.

Nemoto, T. (1935), "On the igneous rocks of Uruppu (Central Kurile Islands)," Geol. J. [of Japan], Vol. 42.

Nemoto, T. (1936), "On the intrusive rocks of the Kurile Islands, in particular the granodiorites of Uruppu," Geol. J. [of Japan], Vol. 43, No. 508.

Nemoto, T. (1937a), "Geologic and petrologic study of the Central Kurile Islands, VI — Jigoku Volcano, Uruppu," Bull. Volcanol. Soc. Japan, Vol. 3, No. 2.

Nemoto, T. (1937b), "Geology of the island of Shimushiru (Central Kurile Islands)," Geol. J. [of Japan], Vol. 44, No. 525.

Nemoto, T. (1938), "The volcano on Ushishiru (Central Kurile Islands)," Geol. J. [of Japan], Vol. 45, No. 537.

Nemoto, T. (1958), "Volcanic activity in the Kurile Islands," in: Jubilee Volume for the Sixtieth Anniversary of Prof. J. Suzuki, Tokyo.

Nemoto, T., and Ishikawa, T. (1955), "Map of the volcanoes of the Kurile Islands," News of Volcanophysics, No. 2.

Neumann. and Padang, M. (1951), "Indonesia," Catalogue of the Active Volcanoes of the World, Part 1, Napoli.

Neverov, Yu. L., and Khvedchenya, O. A. (1962), "New data on the geology and ore mineralization of Urup," Trudy SakhKNII, No. 12.

Neverov, Yu. L., Sergeev, K. F., and Sergeeva, V. B. (1963), "Magmatic deposits of the main chain of the Kurile Islands," Trudy SakhKNII, No. 15.

Neverov, Yu. L., Sergeev, K. F., and Sergeeva, V. B. (1964), "On the 'exotic' rocks of the Greater Kurile chain," Geol. i Geofiz., No. 5.

Nikol'skii, V. M. (1956), "On the problem of survey prospecting in Kamchatka and the Kurile Islands," in: Materials for the Congress of Geologists of Eastern Siberia and the Far East on Methods of Geological Surveying and Prospecting [in Russian], Chita.

Officer, C.B. (1955), "South-west Pacific crustal structure," Trans. Am. Geophys. Union, Vol. 36, No. 3.

Ogloblin, N. N. (1891), "Two 'skaski' of Vladavets Atlasov on the discovery of Kamchatka," in: Readings in Imperial History and Russian Antiquities at Moscow University [in Russian], Book III, sec. 1.

Ogryzko, I. I. (1953), "Discovery of the Kurile Islands," Uchebnye Zapiski (Fak. Narodov Sev.), Vol. 2, No. 157.

Omori (1918), "Notes on eruptions of volcanoes in Japan," Rept. Comm. Earthq. Inv., No. 86.

Osborn, E. F. (1964), "Experimental investigations of oxygen pressure, water content, and the sequence of crystallization of basalts and andesites," in: Chemistry of the Earth's Crust [in Russian], Vol. 2.

Pakiser, L. C. (1963), "Structure of the crust and upper mantle in the western United States," J. Geophys. Res., Vol. 68, No. 20.

Pallas, P. S. (1781), "Brief reports and extracts of letters," Neue Nordische Beiträge, Vol. 1, No. 1.

Pavlova, I. N. (1964), "Southern part of the Sea of Okhotsk," in: Structure of the Earth's Crust in the Transition Region from the Asiatic Continent to the Pacific Ocean [in Russian], Moscow.

Perrey, A. (1864), "Documents on the earthquakes and volcanic phenomena in the Kurile archipelago and on Kamchatka," Ann. Sci. Phys. et Nat. d'Agric. et d'Industrie de Lyon, ser. 3, Vol. 8.

Perrey, A. (1865), "Note on the earthquakes of 1863, with supplements for the earlier years, from 1843 to 1862," Mem. Acad. de Belgique, Vol. 17.

Petelin, V. P. (1964), "Solid rocks of the deep-water trenches of the southwestern Pacific Ocean," in: Geology of the Floors of Oceans and Seas (Papers by Soviet Geologists at the 22nd Session, IGC) [in Russian], Izd. Nauka, Moscow.

Petrov, O. M. (1963), "Stratigraphy of the Quaternary deposits of the southern and central Chukotsk Peninsula," Byull. Kom. po Izuch. Chetvert. Perioda Akad. Nauk SSSR, No. 28.

Petrushevskii, V. A. (1964), Problems in the Geologic History and Tectonics of East Asia [in Russian], Izd. Nauka, Moscow.

Polonskii, A. (1871), "The Kuriles," Zap. Imp. Russk. Geogr. Obshchestva po Otdel. Étnografii, Vol. 4.

Popkova, M. I., Kaidalova, E. F., Petrovskaya, N. F., Klimovskaya, G. V., and Savrasov, N. P. (1961), Collection of Chemical Analyses of the Igneous Rocks from the Southern Part of the Far East [in Russian], Khabarovsk.

Pospelova, G. A. (1960), "Causes of the reversed magnetism of the volcanic rocks of Armenia and the Kurile Islands," Izv. Akad. Nauk SSSR, ser. geofiz., No. 1.

Powers, H. A. (1955), "Composition and origin of basaltic magma of the Hawaiian Islands," Geoch. Cosmoch. Acta, Vol. 7.

Powers, H. A., Coats, R. R., and Nelson, W. N. (1959), "Geology and submarine physiography of Amchitka Island, Alaska," U.S. Geol. Surv. Bull., No. 1028-P.

Pozdneev, D. (1909), Data on the History of Northern Japan and Its Relation to the Asiatic Continent and to Russia, Yokohama.

Pryalukhina, A. F. (1961a), "Information on the stratigraphy of the South Kurile Islands," Trudy SakhKNII, No. 10.

Pryalukhina, A. F. (1961b), "Stratigraphy of the South Kurile Islands," in: Materials of the Conference on the Development of a Unified Stratigraphic Scheme for Sakhalin, Kamchatka, and the Kurile and Komandorskie Islands (held in Okha, May 25-June 2, 1959) [in Russian], Gostoptekhizdat, Moscow.

Raitt, R. W. (1956), "Seismic-refraction studies of the Pacific Ocean basin," Bull. Geol. Soc. Am., Vol. 67, No. 12.

Raitt, R. W., Fisher, R. L., and Mason, R. G.(1955), "Tonga Trench," in: The Crust of the Earth (Geol. Soc. Am. Sp. Paper, Vol. 62), 237-254.

Richard, J. J. (1962), "Kermadec, Tonga, and Samoa," Catalogue of the Active Volcanoes of the World, Part 13, Napoli.

Richards, A. F. (1962), "Archipelago de Colon, Isla San Felix, and Islas Juan Fernandez," Catalogue of the Active Volcanoes of the World, Part 14, Rome.

Richardson, C. (1933), "Petrology of the Galapagos Islands," Bernice P. Bishop Mus. Bull., 110.

Rittmann, A. (1953), "Magmatic character and tectonic position of the Indonesian volcanoes," Bull. Volcanol., ser. 2, Vol. 14.

Rodionova, R. I., Fedorchenko, V. I., and Shilov, V. N. (1963), "Distinctive petrochemical features of the lavas of Ebeko Volcano on Paramushir (Kurile Islands)," in: Distinctive Petrochemical Features of Young Volcanism [in Russian], Izd. Akad. Nauk SSSR.

Rodionova, R. I., Fedorchenko, V. I., and Shilov, V. N. (1964), "The volcanic plateaus of Vernadskii Ridge (Paramushir, North Kurile Islands)," in: Plateau Basalts (Papers by Soviet Geologists at the 22nd Session, IGC) [in Russian], Izd. Nauka, Moscow.

Romankevich, E. A., Baranov, V. I., and Krishtianova, L. A. (1964), "Stratigraphy and absolute age of the Quaternary sediments of the western Pacific Ocean," in: Geology of the Floors of Oceans and Seas (Papers by Soviet Geologists at the 22nd Session, IGC) [in Russian], Izd. Nauka, Moscow.

Rudich, K. N. (1964a), "Trachyliparites of the Mayak extrusion dome (northeastern Asia)," in: Paleovolcanic Reconstruction of the Lavas and Ores of Ancient Volcanoes [in Russian], Alma-Ata.

Rudich, K. N. (1964b), "The late Quaternary volcano of Balgan-Tas (northeastern Asia)," in: Modern Volcanism of Northeastern Asia [in Russian], Izd. Nauka, Moscow.

Saltykovskii, A. Ya. (1963), "On some distinctive petrochemical features of the alkali olivine basalts of the continents and oceans," Sov. Geologiya, No. 10.

Santo, T. (1963), "Division of the Pacific area into seven regions in each of which Rayleigh waves have the same group velocities," Bull. Earthq. Res. Inst., Vol. 41, No. 4.

Sarychev, G. A. (1802), Journey of Naval Captain Sarychev in Northeastern Siberia, the Arctic Sea, and the Eastern Ocean in the Course of Eight Years During the Geographical and Astronomical Expedition Commanded by Naval Captain Billings from 1785 to 1793 [in Russian], St. Petersburg.

Sarychev, G. A. (1826), Atlas of the Northern Part of the Eastern Ocean [in Russian], St. Petersburg.

Sasa, Y. (1932), "On the geologic structure of Shikotan (Lesser Kurile chain)," Geol. J. [of Japan], Vol. 39, No. 465.

Sasa, Y. (1933), "Geological reconnaissance in the Northern Tishima Islands (North Kurile Islands)," Proc. 5th Pacific Sci. Cong., Vol. 3.

Savateev, D. V. (1958), "Volcanic sulfur deposits of the Kurile Islands," in: Information for the Investigation of Chemical Raw Materials in the Far East [in Russian], Vladivostok.

Sergeev, K. F. (1962), "Basic outline of the geologic structure of the islands of Paramushir and Shumshu," Trudy SakhKNII, No. 12.

Sergeev, K. F. (1963a), "On the problem of genesis of rocks of the spilite-keratophyre formation on Paramushir (Kurile Islands)," Dokl. Akad. Nauk SSSR, Vol. 152, No. 2.

Sergeev, K. F.(1963b), "Basic outline of the stratigraphy of the Tertiary deposits of the Greater Kurile chain," Dokl. Akad. Nauk SSSR, Vol. 153, No. 5.

Sergeev, K. F., and Sergeeva, V. B. (1963), "On the intrusive rocks of Vernadskii Ridge on Paramushir (Kurile Islands)," Dokl. Akad. Nauk SSSR, Vol. 153, No. 4.

Sgibnev, A. (1869a), "Data on the history of Kamchatka. The Shestakov Expedition," Morskoi Sbornik, Vol. 100, No. 2.

Sgibnev, A. (1869b), "Historical outline of the main events in Kamchatka (from material in the Siberian archives)," Morskoi Sbornik, Vol. 101, No. 4.

Shatskii, N. S. (1946), "The Wegener hypothesis and geosynclines," Izv. Akad. Nauk SSSR, ser. geol., No. 4.

Shebalin, N. V. (1961), "Intensity, magnitude, and focal depth of earthquakes," in: Earthquakes in the USSR [in Russian], Izd. Akad. Nauk SSSR.

Sheinman, Yu. M. (1946), "On the relation between magma types and tectonics," Sov. Geologiya, No. 2.

Sheinman, Yu. M. (1964), "Possible relations of magmas to the structure of the upper mantle," in: Chemistry of the Earth's Crust [in Russian], Vol. 2, Izd. Nauka, Moscow.

Shelekhov, G. (1812), "Route of March of G. Shelekhov from 1783 to 1790," [in Russian], St. Petersburg.

Shilov, V. N. (1962), "Eruption of Sarychev Peak Volcano in 1960," Trudy SakhKNII, No. 12.

Shilov, V. N., and Voronova, L. G. (1962), "Status of the active volcanoes of the North Kurile Islands in 1959 and some information on the eruption of Chikurachki Volcano in May 1958," Trudy SakhKNII, No. 12.

Shor, G. G., Jr. (1960), "Crustal structure of the Hawaiian Ridge near Gardner Pinnacles," Bull. Seismol. Soc. Am., Vol. 50, No. 4.

Shor, G. G., Jr. (1962), "Seismic refraction studies off the coast of Alaska: 1956-1957," Bull. Seismol. Soc. Am., Vol. 52, No. 1.

Shor, G. G., Jr., and Fisher, R. L. (1961), "Middle America Trench seismic-refraction studies," Bull. Geol. Soc. Am., Vol. 72, No. 5.

Simons, F. S., and Mathewson, D. E. (1955), "Geology of Great Sitkin Island, Alaska," U.S. Geol. Surv. Bull., No. 1028-B.

Smit Siblinga, G. L. (1943), "On the petrological and structural character of the Pacific," Verh. Geol.-Mijnbouw. Genootschap Nederland en Kolonien, geol. ser., Vol. 13.

Smith, W. C., and Chubb, L. J. (1927), "The petrography of the Austral or Tubuai Islands (southern Pacific)," Quart. J. Geol. Soc. London, Vol. 83.

Snow, H. J. (1897), Notes on the Kurile Islands, London.

Snow, H. J. (1902), "The Kurile chain," Zap. Obshchestva. Izuch. Amurskogo Kraya, Vol. 8, No. 1.

Snyder, G. L. (1959), "Geology of Little Sitkin Island, Alaska," U.S. Geol. Surv. Bull., No. 1028-H.

Sobolev, V. S. (1936), "Petrology of the traps of the Siberian Platform," Trudy Arkt. Inst., Vol. 43.

Sokolov, A. (1851), "The northern expedition of 1733-1743," Zap. Gidrograf. Depart., No. 9.

Solov'ev, O. N. (1961), "Aeromagnetic survey in the region of the Kurile-Kamchatka arc," Prikladnaya Geofiz., No. 29.

Solov'ev, O. N., and Gainanov, A. G. (1963), "Characteristics of the deep geologic structure of the transition zone from the Asiatic continent to the Pacific Ocean in the region of the Kurile-Kamchatka island arc," Sov. Geologiya, No. 3.

Solov'ev, S. L., and Shein, V. B. (1959), "Intensity of earthquakes from the data of the Far Eastern and continental stations of the USSR," Izv. Akad. Nauk SSSR, ser. geofiz., No. 9.

Steller, G. W. (1774), Description of the Land of Kamchatka [in German], Frankfurt-Leipzig.

Strahlenberg, P. J. von (1730), The Northeastern Part of Europe and Asia [in German], Stockholm.

Structure of the Earth's Crust in the Transition Region from the Asiatic Continent to
the Pacific Ocean (1964) [in Russian], Nauka Press, Moscow.

Sugimura, A. (1958), "The active zone: Izu-Shichito — northeastern Japan — Kurile
Islands," Tikyu Kagaku, No. 37.

Sugimura, A. (1960), "Zonal arrangement of some geophysical and petrological features in Japan and its environs," J. Fac. Sci. Univ. Tokyo, sec. II, Vol. 12, No. 2.

Suzuki, J., and Nemoto, T. (1935), "The chemical composition of the granitic rocks
of Japan," J. Fac. Sci. Hokkaido Imp. Univ., ser. IV, Vol. 3, No. 1.

Suzuki, J., and Sasa, Y. (1933), "Volcanic rocks of the northern Tishima Islands (North
Kurile Islands)," Proc. 5th Pacific Sci. Cong., Vol. 3.

Sykes, L. R. (1963), "Seismicity of the South Pacific Ocean," J. Geophys. Res., Vol.
68, No. 21.

Talwani, M., LePichon, X., and Ewing, M. (1965), "Crustal structure of the mid-ocean
ridges, 2. Computed model for gravity and seismic refraction data," J. Geophys.
Res., Vol. 70, No. 2.

Tanakadate, H. (1925), "The volcanic activity in Japan during 1914-1924," Bull. Volcanol., Nos. 3-4.

Tanakadate, H. (1931), "Volcanic activity in Japan and vicinity during the period between 1924 and 1931," Japan. J. Astron. Geophys., Vol. 9, No. 1.

Tanakadate, H. (1934a), "Morphological development of the volcanic islet Taketomi
in the Kuriles," Proc. Imp. Acad. Tokyo, Vol. 10, No. 8.

Tanakadate, H. (1934b), "Volcanic activity in Japan during the period between June
1931 and June 1934," Japan. J. Astron. Geophys., Vol. 12, No. 1.

Tanakadate, H. (1936), "Volcanic activity in Japan during the period between July 1934
and October 1935," Japan. J. Astron. Geophys., Vol. 13, No. 2.

Tanakadate, H. (1939), "Volcanic activity in Japan during the period between November
1935 and December 1938," Japan. J. Astron. Geophys., Vol. 16, No. 3.

Tanakadate, H. (1940), "Volcanoes in the Marianas Islands in the Japanese Mandated
South Seas," Bull. Volcanol., ser. 2, Vol. 6.

Tanakadate, H., and Kuno, H. (1935), "The volcanological and petrographical note of
the Taketomi islet in the Kuriles," Proc. Imp. Acad. Tokyo, Vol. 11, No. 4.

Taneda, S. (1962), "Frequency distribution and average chemical composition of the
volcanic rocks of Japan," Mem. Fac. Sci. Kyushu Univ., ser. I, Vol. 12, No. 3.

Tarakanov, R. Z. (1961), "Some results of the study of earthquakes of the Far East,"
Trudy SakhKNII, No. 10.

Tarakanov, R. Z. (1965), "Travel-time curves for P and S-P waves and a velocity cross-
section of the upper mantle from observations on Kurile-Japan earthquakes,"
Fizika Zemli, No.7.

Tatarinov, M. (1783), "New description of the Kurile Islands," Neue Nordische Beiträge,
Vol. 4.

Tatarinov, M. (1785), "Description of the Kurile Islands," in: Historical and Geographical Calendar for 1785 [in Russian], St. Petersburg.

Taylor, G. A. (1958), "The 1951 eruption of Mount Lamington, Papua," Australia Bur.
Miner. Resources Bull., No. 38.

Tazieff, H., Marinelli, G., and Gorshkov, G. S. (1966), Indonesia Volcanological Mission (November 1964 to January 1965), UNESCO, Paris.

Teben'kov, Capt. (1852), Hydrographic Notes on the Atlas of the Northwest Coast of America, the Aleutian Islands, and Some Other Localities in the North Pacific Ocean [in Russian], St. Petersburg.

Teleki, B. G. (1909), Atlas of the History of Cartography of the Japanese Islands [in German], Budapest.

Thayer, T. P. (1937), "Petrology of later Tertiary and Quaternary rocks of the North-Central Cascade Mountains in Oregon, with notes on similar rocks in western Nevada," Bull. Geol. Soc. Am., Vol. 48, No. 11.

Thompson, B. N., Kermode, L. O., and Ewart, A. (1965), New Zealand Volcanology — Central Volcanic Region.

Tikhmenev, P. (1861), Historical Survey of the Development of the Russian-American Company and Its Activities Down to the Present Time [in Russian], Vol. 1, St. Petersburg.

Tikhomirov, V. V. (1958), "On the problem of the development of the earth's crust and of the nature of granite," Izv. Akad. Nauk SSSR, ser. geol., No. 8.

Tokarev, P. I. (1958), "On the depth of the focal surface and the relation of earthquakes to topography in the Kurile-Kamchatka zone," Byull. Vulkanol. Stantsii, No. 27.

Tokarev, P. I. (1959), "On the relation of volcanic and seismic activity in the Kurile-Kamchatka zone," Trudy. Lab. Vulkanol., No. 17.

Tokuda, S. (1928), "On the echelon structure of the Japanese archipelago," Proc. 3rd Pacific Sci. Cong., Vol. 1.

Tomita, T. (1935), " On chemical composition of the Cenozoic alkaline suite of the Circum-Japan Sea region," J. Shanghai Sci. Inst., sec. 2, Vol. 1.

Tsuboi, C. (1954), "Gravity survey along the lines of precise levels throughout Japan, part IV. Map of Bouguer anomaly," Bull. Earthq. Res. Inst., Suppl., No. 4.

Tsuboi, K. (1932), "Petrographical investigations of some volcanic rocks from the South Sea Islands, Palau, Yap, and Saipan, Japan," J. Geol. Geogr., Vol. 9, No. 3/4.

Tulina, Yu. V. (1965), "Correlation of magnetic anomalies with seismic properties of the Mohorovicic discontinuity," Fizika Zemli, No. 3.

Tulina, Yu. V., and Mironova, V. I. (1964), "Southern and central portions of the Kurile zone of the Pacific Ocean," in: Structure of the Earth's Crust in the Transition Region from the Asiatic Continent to the Pacific Ocean [in Russian], Izd. Nauka, Moscow.

Udias, A., and Stauder, W. (1964), "Application of numerical method for S-wave focal method determination to earthquakes of Kamchatka-Kurile Islands region," Bull. Seismol. Soc. Am., Vol. 54, No. 6.

Udintsev, G. B. (1954), "New data on the topography of the Kurile-Kamchatka basin," Dokl. Akad. Nauk SSSR, Vol. 94, No. 2.

Udintsev, G. B. (1955a), "Topography of the Kurile-Kamchatka basin," Trudy Inst. Okeanol. Akad. Nauk SSSR, Vol. 12.

Udintsev, G. B. (1955b), "Origin of the topography of the floor of the Sea of Okhotsk," Trudy Inst. Okeanol. Akad. Nauk SSSR, Vol. 13.

Udintsev, G. B. (1957), "Topography of the floor of the Sea of Okhotsk," Trudy Inst. Okeanol. Akad. Nauk SSSR, Vol. 22.

Udintsev, G. B.(1960), "Topography of the floor and tectonics of the western part of the Pacific Ocean," in: Marine Geology (Papers by Soviet Geologists at the 22nd Session, IGC) [in Russian], Izd. Akad. Nauk SSSR.

Udintsev, G. B., et al., (1963), "New bathymetric map of the Pacific Ocean," Okeanol. Issled., No. 9.

Ueda, S. (1961), "An interpretation of the transient geomagnetic variations accompanying the volcanic activities at Volcano Mihara, Oshima Island, Japan," Bull. Earthq. Res. Inst., Vol. 39, No. 4.

Ueda, S., and Horai, K. (1964), "Terrestrial heat flow in Japan," J. Geophys. Res., Vol. 69, No. 10.

Umbgrove, J. (1962), "Island arcs," in: Island Arcs [Russian translation], Moscow.

Usami, T., Mikumo, T., Shima, E., Tamaki, J., Asano, S., Asada, T., and Matuzawa, T. (1958), "Crustal structure in northern district by explosion-seismic observation," Bull. Earthq. Res. Inst., Vol. 36.

Ustiev, E. K. (1958), "Late Quaternary volcanism in the South Anyui Range and the East Asiatic volcanic province," Trudy Lab. Vulkanol., No. 13.

Vasil'ev, V. G., Veitsman, P. S., Gal'perin, E. I. Gladun, V. A., Goryachev, A. V., Zverev, S. M., Kosminskaya, I. P., Krakshina, R. M., Panteleev, V. A., Solov'ev, O. N., Starshinova, E. A., and Fedotov, S. A. (1960), "Investigation of the earth's crust in the transition region from the Asiatic continent to the Pacific Ocean," in: Results of the IGY. Seismology [in Russian], No. 4, Izd. Akad. Nauk SSSR, Moscow.

Vebman, N. A. (1950), "Paths of differentiation in the Deccan traps," in: Geology and Petrography of the trap formations [in Russian], Moscow.

Veitsman, P. S. (1964), "The northeastern part of the Kurile-Kamchatka zone of the Pacific Ocean," in: Structure of the Earth's Crust in the Transition Region from the Asiatic Continent to the Pacific Ocean [in Russian], Izd. Nauka, Moscow.

Veitsman, P. S. (1965), "Peculiarities of the deep structure of the Kurile-Kamchatka zone," Fizika Zemli, No. 9.

Veitsman, P. S., Gal'perin, E. I., Zverev, S. M., Kosminskaya, I. P., Krakshina, R. M., Mikhota, G. G., and Tulina, Yu. V. (1961), "Some results of the study of the crustal structure in the region of the Kurile island arc and adjacent portions of the Pacific Ocean according to data from deep seismic sounding," Izv. Akad. Nauk SSSR, ser. geol., No. 1.

Venig-Meinesz, F. A. (1940), Gravimetric Observations at Sea [in Russian], Izd. GUGK.

Vergunov, G. P. (1961), "Outline of the geology and metallogenesis of the southern part of the Kurile archipelago," Trudy SakhKNII, No. 10.

Vergunov, G. P., and Pryalukhina, A. F. (1963), "Pliocene sediments in the Kurile Islands," Dokl. Akad. Nauk SSSR, Vol. 152, No. 6.

Vergunov, G. P., and Vlasov, G. M.(1964), "The Kurile Islands: magmatism and the phenomena of metamorphism," in: Geology of the USSR [in Russian], Vol. 31, Pt. 1.

Verhoogen, J. (1937), "Mount St. Helens, a recent Cascade volcano," Univ. Calif. Publ. Bull. Dept. Geol. Sci., Vol. 24, No. 9.

Veselov, K. E., Evdokimov, Yu. S., Zhilin, A. V., and Telenin, M. A. (1961), "On a gravimetric survey with a marine static gravimeter in the Sea of Okhotsk and the Pacific Ocean," Prikladnaya Geofizika, No. 29.

Vinogradov, A. P. (1962), "Zone melting as a method of study of some radial processes in the earth," Geokhim., No. 3.

Vlasov, G. M. (1958), "Volcanic sulfur deposits of Kamchatka and the Kurile Islands," in: Information from the Investigation of the Chemistry of Sulfur in the Far East [in Russian], Vladivostok.

Vlasov, G. M. (1958), "Quaternary glaciation in the North Kurile Islands," Geogr. Sbornik, Vol. 10.

Vlasov, G. M. (1959a), "High erosion surfaces in Kamchatka and the Kurile Islands," Materialy VSEGEI, nov. ser., No. 2.

Vlasov, G. M. (1959b), "Survey of the stratigraphy of the Tertiary formations of Sikhote Alin, Sakhalin, Kamchatka, and the Kurile Islands," in: Transactions of the Interdepartmental Conference on the Development of a Uniform Stratigraphic Scheme for the Northeastern USSR [in Russian], Magadan.

Vlasov, G. M. (1960), "Volcanic sulfur deposits and some problems of near-surface ore formation (in the case of Kamchatka and the Kurile Islands)," in: Information on the Natural Resources of Kamchatka and the Kurile Islands [in Russian], Magadan.

Wadati, K. (1935), "On the activity of deep-focus earthquakes in the Japan Islands and neighborhoods," Geogr. Mag.

Walker, F., and Poldervaart, A. (1950), "Karoo dolerites of the Union of South Africa," in: Geology and Petrography of the Trap Formations [in Russian], Izd. Inostrannoi Literatury, Moscow (translated from Bull. Geol. Soc. Am., Vol. 60).

Washington, H. S. (1929), "The rock suites of the Pacific and Atlantic basins," Proc. Nat. Acad. Sci., Vol. 15, No. 7.

Westerveld, J. (1952), "Quaternary volcanism on Sumatra," Bull. Geol. Soc. Am., Vol. 63, No. 6.

Weyl, R. (1955), "Contributions to the geology of El Salvador, VI. The lavas of the young volcanoes," N. Jahrb. Geol. Pal., Abh., Vol. 101, No. 1.

Willis, B., and Washington, H. S. (1924), "San Felix and San Ambrosio: their geology and petrology," Bull. Geol. Soc. Am., Vol. 35, No. 3.

Witsen, N. (1646), Koorte Beshriving van het Eylandt by de Japanders Eso Genaent Nevens de Manieren, Zeden, Ommegangh, Ende Gestalte des Selfs Inwoonderen; soo als hat Eerst in den Jare 1643 van't Schip Castricom Bezeylt Ende Ondervonden is, Amsterdam.

Woollard, G. P., and Strange, W. E. (1962), "Gravity anomalies and the crust of the earth in the Pacific Basin," in: The Crust of the Pacific Basin (Amer. Geophys. Union Monograph, No. 6).

Wyllie, P. J. (1963), "The nature of the Mohorovicic discontinuity, a compromise," J. Geophys. Res., Vol. 68, No. 15.

Yaffe, G. A. (1965), "Increase in activity of Pallas and Snow Volcanoes," Zap. Primorskogo Filiala Geogr. Obshchestva SSSR, Vol. 1, No. 24.

Yagi, K. (1960), "Petrochemistry of the alkalic rocks of the Ponape Island, Western Pacific Ocean," Internat. Geol. Cong., XXI Session, Part 13.

Zatonskii, L. K., Kanaev, V. F., and Udintsev, G. B. (1961), "Geomorphology of the submarine part of the Kurile-Kamchatka arc," Okeanol. Issled., No. 3.

Zavaritskii, A. N. (1939), "On the volcanoes in the vicinity of Mergen and their lavas," in: To Academician V. A. Obruchev on his Fiftieth Year of Scientific Activity [in Russian].

Zavaritskii, A. N. (1946a), "Volcanic zone of the Kurile Islands," Vestnik Akad. Nauk SSSR, No. 1.

Zavaritskii, A. N. (1946b), "Some facts that must be considered in tectonic syntheses," Izv. Akad. Nauk SSSR, ser. geol., No. 2.

Zavaritskii, A. N. (1950), Introduction to the Petrochemistry of the Igneous Rocks [in Russian], Moscow.

Zavaritskii, A. N. (1952), "One of the most important scientific problems of the earth," Vestnik Akad. Nauk SSSR, No. 6.

Zavaritskii, A. N., and Gorshkov, G. S. (1963), "The volcanic arc of the Kurile Islands," in: Zavaritskii, A. N., Selected Works [in Russian], Vol. 4, Moscow.

Zhelubovskii, Yu. S. (1964a), "Tectonics [of the Kurile Islands]," in: Geology of the USSR [in Russian], Vol. 31, Pt. 1.

Zhelubovskii, Yu. S. (1964b), "Stratigraphy [of the Kurile Islands]," in: Geology of the USSR [in Russian], Vol. 31, Pt. 1.

Zverev, S. M. (1964), "Results of the study of the sedimentary sequence in the Sea of Okhotsk and the Kurile-Kamchatka zone of the Pacific Ocean," in: Structure of the Earth's Crust in the Transition Region from the Asiatic Continent to the Pacific Ocean [in Russian], Izd. Nauka, Moscow.

# Index to Geographic Terms

# General Index